THE WORLD'S GREATEST
ALIEN ABDUCTION
MYSTERIES

D0308548

CHANCELLOR
PRESS

This 2001 edition published by
Chancellor Press, an imprint of Bounty Books
a division of Octopus Publishing Group Ltd,
2–4 Heron Quays, London E14 4JP

The material in this book has previously appeared in:
The World's Greatest Alien Abductions
(Bounty, Octopus Publishing Group Ltd, 1999)
The World's Greatest UFO Mysteries
(Hamlyn, Octopus Publishing Group Ltd, 1983)
The World's Greatest Ghosts
(Hamlyn, Octopus Publishing Group Ltd, 1994)

ISBN 0 7537 0425 0
ISBN 13 9780753704257

Printed in Great Britain by Mackays of Chatham

Reprinted 2001, 2002 (twice), 2004

Front cover picture acknowledgments:
Fortean Picture Library

Contents

The First
Abductions

An Interrupted Journey

Barney and Betty Hill were the archetypal Mr and Mrs America. They lived in the small town of Portsmouth, New Hampshire; Barney worked for the US Post Office, while Betty was a social worker employed by the state. Like many Americans, she believed in UFOs – her sister, Janet, had seen one four years before our story begins, in 1957, but Barney remained sceptical.

They had been up to Niagara Falls and had then driven on into Canada for a short holiday. But on the night of 19 September 1961 they heard that a hurricane was storming up the eastern seaboard; it was due to hit New Hampshire the following day. Rather than face the journey back through the chaos that the hurricane would leave its wake, they abandoned their plans to stay overnight in Montreal and turned for home.

They crossed the US border a little before 10pm, near Colebrook, New Hampshire, where they stopped at a restaurant for a bite to eat. They were tempted to stay in Colebrook overnight, but once fortified they decided that they would press on down Highway 3. They pulled out of the car park at 10.05pm. It was another 170 miles (274km) to Portsmouth and Barney reckoned that they should be home by 2.30am – or 3am at the latest.

Thirty-five miles down the highway, just south of Lancaster, Betty noticed a bright light in the sky. It was to the left of, and slightly below, the full moon. Then she noticed another light, which appeared to be becoming brighter and brighter. It seemed to be moving, but Betty could not be sure that that was not due to the movement of the car. Barney slowed down to take a look, but dismissed the light as a satellite that had gone off course.

The light stayed with them for several miles. It occasionally disappeared behind trees and hills, only to reappear again. The Hills had their dachshund, Delsey, with them, in the back of the car. The dog grew restless (animals are frequently spooked by UFO activity, the Hills learned later). They decided to pull over and let the dog out for a run. By stopping they could also observe the light while they were stationary and thus ascertain whether it was moving or not.

Always a 'worrier', as Betty put it, Barney was a little nervous about stopping: he said that he was afraid of the bears that inhabited the hills. Betty laughed at him and they then found a place to stop which had good all-round vision. On getting out of the car they saw that the light was moving all right, but they could still not make out what it was. It was certainly not a star, as at one point it moved in front of the moon. They got the dog back into the car and drove on.

Barney suggested that the lights belonged to the aircraft - a helicopter, perhaps. Whatever the airborne vehicle was, it now appeared to be following them. They stopped again a couple of times to try to identify it. It soon became clear that it was neither a satellite nor an aircraft as it seemed to hover without motion before setting off with a burst of incredible speed. At one point it appeared to be circling them.

Betty was convinced that it was UFO because it was festooned with multi-coloured lights that flashed out beams. Looking through a pair of binoculars that they kept in the car she could see what she took to be an alien spacecraft. It had a fuselage, but no wings, and whatever powered it made no noise at all.

They were now on a remote section of the road. Not even the official rangers patrolled there and people had been known to freeze to death if their car broke down, so private car-recovery firms covered the area. They passed a motel and for a moment considered stopping, but the film *Psycho* was in everyone's minds back then and no car had passed them for miles.

As they drove past a resort area called Indian Head, curiosity finally get the better of Barney and he suddenly stopped in the middle of the road. Leaving Betty in the car, he took the binoculars and got out to take a better look at the craft, which had dropped down level with the height of the trees. Barney could see a double row of windows and approached within 50 feet (15m) of the aircraft. Through the windows he saw around half a dozen beings looking at him. He thought that they were wearing shiny, silver uniforms, but what he remembered most clearly was their curious, hypnotic eyes. Betty screamed for him to return to the car, but Barney stared on, oblivious to her.

What seemed to be landing gear then appeared beneath the craft. This broke the spell and Barney turned and ran back to the car. He jumped in and set off at high speed, convinced that the aliens intended to capture them. Betty looked around, but the craft had disappeared. It was then that they heard a strange beeping sound as a tingling feeling of drowsiness came over them. The drowsiness left them and they found themselves 35 miles (56km) beyond Indian Head, with no recollection of having travelled there. 'Now do you believe in UFOs?' Betty asked casually. Barney replied that he did not.

A signpost indicated that they were now 17 miles (27km) from Concord, New Hampshire, and their sense of disorientation left them. By the time they reached home the sun had almost risen. They checked their watches, only to find that they had stopped, but the clock in the kitchen said that it was five o'clock. Two hours had gone 'missing'.

They had a light snack, went to bed and slept until three in the afternoon. By the time that they awoke Barney had put the whole thing out of his mind, but Betty phoned her sister and told her that she had seen a UFO. Janet had a neighbour who was a physicist. He suggested that Betty check the bodywork of her car with a compass in case it had been subjected to any strong, electromagnetic field. Betty did just that and found a dozen magnetic hot spots on the boot which coincided with shiny patches on the paintwork. Barney, however, was not impressed.

Janet had another friend, who was a former chief of police. He suggested that Betty call the United States Air Force and report the sighting, so Betty called Pease Air Force Base. The officer to whom she spoke seemed interested in her description of the craft, but Betty did not describe the occupants; only Barney had seen them. Nevertheless, a report was forwarded to the staff of Project Blue Book, the official UFO study group at Wright-Patterson Air Force Base in Ohio.

Although Barney pushed the incident to the back of his mind, Betty could not. She went to the library and began reading up on UFOs. One of the books that she read was *The UFO Conspiracy* by Donald Keyhoe, a former major in US Marine Corps. On 26 September 1961, just one week after her own sighting, she wrote to Keyhoe, who was now the director of the National Investigations Committee on Aerial Phenomena (NICAP), the leading UFO-research group of its day. Three days after writing the letter, Betty began to have a series of disturbing dreams in which she was taken aboard an alien spacecraft and subjected to a medical examination.

The NICAP took her letter seriously and sent scientific adviser Walter Webb to interview the Hills. He was impressed by them. They were pillars of the community and stalwarts of the local church. A mixed-race couple – unusual in New England during the 1960s – they were also active in the National Association for the Advancement of Colored People. Barney sat on the New Hampshire Civil Rights Commission, too. Webb was furthermore struck by the way in which the couple constantly played down what they had seen.

Webb's report prompted two more NICAP investigators, Robert Hohman and C D Jackson, to visit the Hills. James MacDonald, a friend of the Hills who had recently retired from the US Air Force Intelligence, was also present at the meeting. It was MacDonald who floated the idea of using hypnosis to discover what had happened during the Hills' missing two hours. Captain Ben Swett from Pease Air Force Base also thought that this might be a good idea.

In the summer of 1962 Barney became ill. His symptoms included exhaustion, high blood pressure and stomach ulcers. His doctor thought that the cause of his problems might be psychological and sent him to see Dr Duncan Stephens, a psychiatrist. Barney found it difficult to open up to Dr Stephens on the subject of his UFO sighting, which was one possible cause of the anxiety from which he was plainly suffering. When he finally did, however, he brought up the idea of using hypnosis to find out what had really happened. Dr Stephens agreed and sent him to see Dr Benjamin Simon, a hypnotherapist in Boston.

Barney's first consultation with Dr Simon was in December 1962. Although the appointment was for Barney alone, Betty came along, too, because in view of her dreams she thought that both of them might need treatment. The hypnotic sessions began in January 1963 and continued for six months. Dr Simon used regressional techniques with which to take the Hills back to the missing two hours. He treated them separately, so that each of their accounts would not be tainted by the other's experience or perception of it. And it was only towards the end of the six months that he played any of the tapes of the session s back to the Hills to help them to come to terms with what had happened.

The story that emerged under hypnosis was as follows. After Barney had sped away from the craft he grew calmer. Then he had inexplicably turned off the main road down a track, seemingly heading towards a light that he thought was the setting moon. But the track was blocked by a gaggle of figures. Betty thought that they were men whose car had broken down, while Barney feared that they were about to be robbed.

One of the figures held a strange device in its hand, which it pointed it at the Hills. At that moment the engine of the Hills' car cut out. As the figures approached, Betty, realising that something was wrong, got out of the car and tried to run, but they grabbed her. It was then that she saw that they were not men at all, but small, hairless, ashen-coloured beings, with large heads, high foreheads and huge eyes. Barney just remembered that they were humanoid, but he saw little beyond the strange, slanting eyes. They had a hypnotic effect on him and slowly he closed his own.

The Hills were taken on board the alien's spacecraft, which had landed nearby. They were led to separate rooms, where they were subjected to medical examinations. Samples of blood and skin were taken from them and the aliens were particularly interested in Barney's false teeth. A suction device was also placed over his groin, which caused him considerable discomfort. (It presumably took a sperm sample). The shape of the suction cup matched a circular wart that subsequently developed on Barney's groin.

The aliens unzipped Betty's dress and took off her clothes. They then made her undergo a gynaecological examination. A long needle was next inserted into her navel, in what they told her was a pregnancy test. They also showed her what has since come to be known as the 'Star Map', a chart delineating their trade routes and the details of their expeditions. Betty asked where their home planet was. The larger being, whom Betty took to be their leader asked her where the Earth was on the map, but Betty could not show them. The leader then said that there was little point in telling her where they had come from because she did not know where her own planet was on the map. Betty reproduced the Star Map under hypnosis and amateur astronomer Marjorie Fish believes that she has found a match of it in space. Her interpretation of Betty's Star Map leads her to believe that the aliens were from Zeta Reticuli, a binary-star system some 37 light years from Earth.

The redoubtable Betty was eager to take something from the craft that would prove that she had had an encounter with aliens, and therefore persuaded them to give her a book. However, the leader took it back before she left the ship, having presumably decided that the Hills would not be allowed to remember the encounter. Betty in turn became upset, and swore that she would remember, whatever happened. The Hills were then dressed and taken back to their car. They watched the alien spacecraft depart before taking to the road again. Something implanted by the aliens in their subconscious minds then wiped all recollection of the experience from their memories.

Dr Simon had never heard anything like this tale and had no idea of what to do next. He believed that the Hills were telling the truth, but was terrified that if the story got out his own reputation would be ruined and hypnosis would be discredited as a psychiatric tool. His fears were not unfounded in 1965 a series of unauthorised articles published in a Boston newspaper leaked the story, which made the Hills sound like kooks.

The Hills were also concerned that the implausible story of their abduction would tarnish their reputations if it were made public and confided their fears to Dr Simon. Together they subsequently approached the well-known local writer John G Fuller and asked him to put their side of the story in writing. He agreed, and they consequently collaborated on the book *The Interrupted Journey: Two Lost Hours Aboard a Flying Saucer*. When *Look* magazine picked up the story in 1966 the two issues that carried the Hills' tale broke all records at the news-stands. The story was syndicated world-wide and offers for the film rights came flooding in – the lowest bid was $300,000. Their tale was eventually recreated in a made-for-TV movie called *The UFO Incident*, starring James Earl Jones as Barney and Estelle Parsons as Betty.

Betty took a lie-detector test on national television, which indicated that she was telling the truth. Detailed scrutiny of the Hills' lives revealed them to be people of the highest integrity. Their story seems unshakeable, and although Barney died in 1969, 30 years later Betty was still telling the same tale. So, is it true?

The UFO-sceptic Robert Sheaffer investigated the details of the Hills' abduction. Trained as he was in astronomy, he pointed out that the first light that they had seen, close to the disc of the full moon, was, in all likelihood, Saturn and that the second, brighter light was probably Jupiter. (Indeed, around 30 per cent of all supposed UFO sightings are actually of celestial bodies, which viewers frequently report appear to move, wiggle and dart about. During World War II, for example, crews of the US Air Force (USAF) B-29 bombers flying from the Mariana Islands on bombing missions over Japan, reported that they were being followed by an enemy aircraft that was fitted with bright searchlights. They believed that it was picking them out for attack by Japanese fighter planes. For several weeks the B-29 gunners tried to shoot it down, only giving up when they discovered that they had been blasting away at the planet Venus. USAF planes on UFO-intercept missions have also since found themselves chasing celestial bodies).

Another UFO-sceptic, the author Philip J Klass, interviewed Dr Simon just weeks after the Hills' story appeared in *Look* magazine. Simon said openly that he did not believe what the Hills had told him, even though he stood to make money from the sales of the book, as well as from

any film deal. He said that he had also told the senior editors of *Look* magazine that he did not accept that the Hills
had been abducted by aliens, although he had a financial
stake in the articles that they were about to publish, too.

Dr Simon believed that it was a simple case of *folie à deux*
(joint madness): both Barney and Betty Hill were absolutely
convinced that the abduction had happened, but it was simply a shared delusion. Regressive hypnosis, he said, is not
a magical getaway to the truth, but rather an aid to human
beings' all-too-fallible memories. (Indeed, dreams, fantasies
and material from books and films can easily become mixed
up with memories of real events; however, statistics indicate
that some 50 per cent of alien-abduction cases are 'discovered' under regressive hypnosis.)

Dr Simon furthermore said that Betty had not started
dreaming about being abducted by aliens until she began
reading up about UFOs in the local library. Although
Barney was not interested in Betty's nightmares, she told
her friends about them – often when he was within
earshot, reading the paper or watching TV; he also heard
her relating them to the UFO investigators who visited them.
In this way he had absorbed Betty's delusion, as if by osmosis.

Dr Simon said that he had been surprised when Betty
turned up at his office with her husband when Barney
came for his first therapy session. The root of Barney's
psychological problems, Dr Simon believed, was the guilt
about leaving his first wife and their children – who were
black – for Betty, a white woman. And it was Betty who
had forced the subject of UFOs onto the agenda.

Reviewing the tapes of the Hills' sessions under hypnosis, Dr Simon pointed out that most of the details of their
trip back from Montreal tallied with each other's version,
which is what would be expected in a shared experience.
But this mutual agreement was not true of the abduction:
for example, Betty said that the aliens had spoken to her in
lightly accented English – like the aliens in Hollywood sci-

fi movies – while Barney said that they had had no mouths and had communicated by means of telepathy. Even their reactions to the experience were different. Barney had been truly terrified; under hypnosis, he had screamed out for help and had had to be reassured by the psychiatrist. But Betty had been calm throughout: although she had recounted her feat that they were about to be dissected like laboratory mice her voice had betrayed no emotion.

The magnetic 'hot spots' on the Hills' car do not bear close examination either. A compass always reacts erratically around a large body of metal, such as a car. And Betty was, by her own admission, a believer in UFOs.

But what of the missing two hours? According to their account the Hills had turned off the road, possibly because they believed that a UFO was chasing them. It was late at night, they were tired and it probably took them some time to find their way back to the right road. Two hours can easily pass quickly during a rambling detour.

Nevertheless, the Hills' story became the world's most famous alleged alien-abduction case, and it opened the floodgates, leading to a deluge of other such claims.

Flights of Fantasy

It seems that Betty and Barney Hill were not the first earthlings to have been abducted and that such instances may even have been occurring for centuries. The celebrated US astronomer Carl Sagan unearthed a folk tale dating from 1645. Sagan reported:

A Cornish teenager, Anne Jeffries, was found groggy and crumpled on the floor. Much later she recalled being attacked by little men, carried paralysed to a castle in the air, seduced and returned home. She called the little men fairies. The following year she was arrested for witchcraft.

There are those of the UFO lobby who even claim that the 'angels' who appeared to Joan of Arc in 1425 and told here that she would lead her countrymen against the English were aliens. (Had she not been abducted, as they claim, England might still have a Continental foothold today.) And there are passages in the Old Testament Book of Ezekiel, as well as in the Hindu Ramayana and Mahabharata, that could be interpreted as telling of alien abductions.

On 25 October 1595, a Filipino soldier, Gil Perez, was found in the main square of Mexico City. He was dressed in a uniform bearing Filipino military insignia. The night before, the terrified man said, he had been standing guard in Manila when a thick fog had enveloped him. He lost consciousness and later awoke in Mexico City – 1,000 miles (1,609km) away. He also mentioned that the governor of the Philippines had been assassinated. Perez was promptly arrested as a deserter, but months later the news arrived that the governor of the Philippines had indeed been killed. It was then confirmed that Perez had been on sentry duty in Manila on the night in question and that he had mysteriously disappeared. He was consequently released.

In the early seventeenth century an 18-year-old nun at a convent in Agreda, Spain, called Sister Mary, claimed that she had been flown to the New World, where she had converted an Indian tribe named the Jumano to Christianity. She also said that she had seen the Earth spinning below her during her trip. She wrote details of her journey in a diary, which her superiors burnt. She was warned that she was risking being charged with heresy with such talk, but she still refused to retract her story.

Then, in 1622, Pope Urban VIII and King Philip of IV of Spain received letters from Father Alonzo de Benavides, a missionary who had been sent to New Mexico. He complained that his journey had been a waste of time because the Indians had already been converted to Christianity.

They said that a lady in blue had shown them rosaries and crosses and had taught them how to celebrate Mass.

When Father de Benavides returned to Europe in 1630 he heard about Sister Mary's claims and visited her in Agreda. Questioning her, he quickly discovered that the places that she said she had visited, and the people whom she had met, corresponded exactly with his own experiences in the region. She even knew details of the local customs and folklore that few outsiders were party to. But the piece of evidence that clinched the truth of her tale, according to Father de Benavides, was a chalice that he had brought back from New Mexico, which the nuns recognised as one that had been missing from the Agreda convent.

Sister Mary was lucky: others who told similar stories suffered the punishment for heresy. In 1655, for example, a businessman who had mysteriously turned up in Portugal was hauled in front of the Roman Catholic Inquisition. He claimed to have been standing outside his office in Goa, then a Portuguese territory in India, when he was suddenly whisked into the air; seconds later, he found himself in Portugal. The Inquisition damned his tale as being evidence of witchcraft and burnt him at the stake.

A Very Victorian Abduction

The modern versions of UFO reports began to be made in 1896 and 1897, when people across the US Midwest saw huge, flying craft, the descriptions of which revealed a marked similarity to the airships that were then being developed in Europe. An attempted alien abduction was also reported at that time. According to the *Stockton Evening Mail* in 1897, Colonel H G Shaw had been travelling by buggy through the Californian countryside with his companion, Camille Spooner. They were outside Lodi, early in the evening of 25 November 1896, when the 'horse stopped suddenly and gave a snort of terror'. Shaw then saw three pale, thin, 7-foot-tall (2.1m) alien beings, who emitted a soft, warbling sound.

'They were without any sort of clothing', he said. 'But they were covered with a natural growth as soft as silk to the touch and their skin was like velvet. Their faces and heads were without hair, the ears were very small, and the nose had the appearance of polished ivory, while the eyes were large and lustrous. The mouth, however, was small and it seemed they were without teeth.'

When he touched them, Shaw found that they were almost weightless. He said that he thought that they weighed less than 1oz (28g). Using their small, nail-less hands and long, narrow feet, they tried to lift him, 'probably with the intention of carrying me away' to their ship, a '150-foot-long [46m] cigar-shaped craft with a single large rudder'. But they abandoned the attempted abduction because Shaw was too heavy. Then they flashed lights at their ship and made off towards it in a loping motion, touching the ground only every 15 feet (4.6m) or so. 'With a little spring, they rose to the machine, opened a door in the side and disappeared', he said.

Shaw maintained that the aliens were from Mars and that they had been 'sent to earth for the purpose of securing one of its inhabitants'. (Presumably they later found someone lighter.) There was certainly an alien abduction two days afterwards, if the abductee is to believed. John A Horen claimed to have met an alien who took him onto an 'airship'. They toured southern California and then flew to Hawaii. But Horen was a well-known practical joker and his wife claimed that he had been sleeping beside her throughout the night in question.

The Missing Regiment

A second wave of UFO sightings occurred in Europe between 1909 and 1913. This time there were no reports of aliens trying to abduct anyone; however, something strange did happen in Gallipoli, Turkey, in August 1915, during World War I. A British regiment was climbing a hill near Suvia Bay when members of a New Zealand regiment who were watching them saw six or seven clouds hovering near the hill. One of them slowly descended and the British regiment climbed into it. When the last man had disappeared the cloud lifted to reveal no sign of the British soldiers. After World War I was over, the British authorities asked the Turks what had happened to their soldiers. The Turks said that they had never captured an entire British regiment – and had never even fought the regiment concerned.

Early Abduction Attempts

In 1933 'ghost rockets' were reported to have appeared in northern Sweden. And during World War II 'foo fighters' – luminous balls of light – were seen following aircraft. But no one was 'spacenapped'.

The modern era of ufology began in June 1947, when pilot Kenneth Arnold was flying over the Cascade Mountains in USA's Washington state searching for the wreckage of a C-46 Marine transport plane which had crashed. Between Mount Rainer and Mount Adams, 5,000 feet (1,524m) above him, he saw a formation of flying objects travelling at between 1,300 and 1,700 miles (2,092 and 2,736km) an hour – faster than any plane of the day was capable of. They were about 23 miles (37km) away and their formation was spread over 5 miles (8km). Another search plane was also visible – a DC-4 – and Arnold estimated that the unidentified craft were about two-thirds of its size. They were shaped like rounded boomerangs, but Arnold told the press that they travelled 'like a saucer would if you skipped it across the water'. From then on, unidentified flying objects became known as 'flying saucers'.

A month later, the so-called 'Roswell Incident' occurred. It was reported that a flying saucer had exploded and that the wreckage had come down on ranch land belonging to William 'Mac' Brazel. The remains were collected by the US Air Force, which took them to the Roswell Army Air Force Base. In the interests of national security, Brazel was asked not to talk about the incident. What the USAF hauled away from the crash site has been a source of controversy ever since, but according to some accounts the debris included the dead bodies of aliens who looked remarkably like the creatures that the Hills later described.

Undeterred by this setback, aliens attempted an abduction towards the end of that month. On 23 July 1947 Jose Higgins, one of band of men working near Bauru, Brazil, saw a large flying saucer land nearby. All of them, with the exception of the courageous Higgins, ran away. He stood his ground and was confronted by a group of seven-foot-tall (2.1m) aliens with round, bald heads and big eyes, whom he found strangely attractive. They tried to lure him aboard their ship, but he first asked them where they were from. In response, they drew a map of the solar system and indicated that they came from Uranus. Higgins summoned up all his willpower and managed to give them the slip. While he was hiding in the bushes he saw them cavorting about; when they tired of doing this they went back to their ship and flew off.

Another attempted alien abduction failed in France in

1950, when, on 20 May, a woman rushing home to make dinner at around 4pm was blinded by a brilliant light which also paralysed her. Two huge, black hands grabbed her from above and she thought that she was being carried off by a giant bird. The talon-like hands gave her an electric shock; they were cold, as if they were made of metal. The fingers around her neck stifled her screams and she was dragged off into a field. Suddenly her abductor dropped her and she heard him making off through the bushes. Stumbling back to the path, she felt a violent pain down her back, as if she had been burned, and noticed a strange, metallic taste in her mouth. In the distance she could see several houses, but before she could reach them she heard a great noise and was bent double by a 'violent windstorm'. Then she saw a bright light rising in the sky. After pushing on, she at last came upon the house of a lock-keeper. The lock-keeper and his family had observed the bright light, too, and the marks that they noticed on her face were consistent with her story.

The intended abductee had seen what she took to be a shooting star the previous night, but it had appeared to be rising in the sky rather than falling.

A Willing Victim

One of the first people who claimed to have been whisked off in a flying saucer said that he had gone quite willingly. His name was George Adamski. Born in Poland in 1891, he had emigrated to the United States with his parents at the age of two and the family had settled in New York. At the age of 22 Adamski joined the 13th Cavalry and served in the US Army until 1919. Then he moved to California, where he became interested in all things mystical, eventually styling himself 'professor of Oriental Mystical

Philosophy'. His 'Royal Order of Tibet' set up a farming commune at Valley Center in 1940.

In 1953 Adamski published the book *Flying Saucers Have Landed*. In it, he claimed to have made numerous sightings of flying saucers, both in flight and on the ground. His credibility was enhanced by the fact that he claimed to work at the Mount Palomar Observatory, which then boasted the largest telescope in the world. (This was true, up to a point: Adamski worked at the hamburger concession a the observatory).

Adamski said that he had seen flying saucers since 1946 – the year before Arnold had inadvertently coined the name. On 20 November 1952, he took two families – the Baileys and the Williamsons – UFO-spotting in the Californian desert. Leaving the others in their cars, Adamski trudged away from the road into the desert with his telescope. (It seems that aliens abhor crowds and traffic.) At around 12.30pm, when he was about a quarter of a mile (40m) from the road, he saw a figure whom he took to be a man who was not prospecting. The figure gestured to him and, thinking that he might need help, Adamski approached him. But the creature turned out to be no prospector and Adamski believed him to be an alien. He was about 5 feet 6 inches tall (1.65m), weighed around 135lbs (61kg) and Adamski judged him to be 28 years old. Furthermore, his eyes were grey-green, his hair was blond and flowing, his teeth were pearly white, his skin was tanned, his features were oriental and he was dressed in a one-piece, chocolate-brown jump suit.

They communicated using a mixture of telepathy and sine language, Adamski said. Adamski pointed at the sun and then counted off the numbers one, two and three on his fingers, indicating the three closest planets to the sun – Mercury, Venus and Earth. The alien gestured two, which Adamski too to mean that he came from Venus. (It has subsequently been proved, however, that Venus is entirely unsuited to supporting life. Even toughened space probes cannot withstand the bone-crushing atmospheric pressure and the constant downpour of concentrated sulphuric acid on the planet.)

The alien took Adamski to see his flying saucer, which was parked nearby. It was very small, but the alien explained that it was only a 'scout ship'. Adamski stumbled and nearly fell under the ship, but the alien grabbed him. Telepathically, he learnt that the electromagnetic power that supported the ship would have caused him serious injury; as it was, his arm was flung beneath the craft and he lost all feeling in it for several years afterwards.

The alien also conveyed a message to Adamski. (It was the same message that the silver-suited alien played by Michael Rennie delivered in the 1951 film *The Day the Earth Stood Still*.) The people of Earth, he said, were endangering the other inhabitants of the solar system with their nuclear-weapons' tests and they should therefore cease them.

Adamski told the creature that he was puzzled by the fact that aliens never landed in built-up areas. They were afraid that they would be torn apart by a mob, the alien explained. However, he also revealed that 'space brothers' were walking among humans unnoticed, so it was only the landing that was dangerous.

Adamski's contact with aliens continued on a regular basis. He frequently met 'space brothers' in the cafes and bars of southern California. They had problems in communicating though because the language that they spoke sounded to Adamski like some ancient dialect of Chinese, which Adamski believed was the original human language. In his second book, *Inside the Spaceships*, Adamski revealed that he had become so friendly with the aliens that they took him on jaunts around the solar system in their spacecraft. Interplanetary travel was conducted in huge 'mother ships' that were several miles long; the small scout ships were usually stored on board the mother ships and were used by the aliens to hop around a planet once they had arrived on it. Adamski even produced an 8mm colour movie of one such mother ship that he claimed to have filmed at Silver Spring, Maryland. The film could not be authenticated, however.

During the course of his travels, Adamski said, he had been introduced to Martians, Saturians and Jovians, who were all remarkably humanoid in shape. They did not explain how they lived on Saturn and Jupiter. (These giant planets have no solid surface, being constituted of gas, and the pressure is such that it would crush a human like a juggernaut squashes a hedgehog.) Despite these omissions, Adamski lectured around the world on what he had seen.

One of the most interesting places that Adamski claimed to have visited was the far side of the moon. It was covered with forests, mountains and lakes and people lived there, he said. But in 1959 the Soviet probe Lunik III sent back pictures of the far side of the moon which showed it to be as sterile and cratered as the side that we can see. Adamski alleged that this revelation was a communist plot and that the Soviet Union had doctored the pictures in order to deceive the West. However, because the data subsequently collected by the US National Aeronautics and Space Administration's (NASA) space probes revealed that none of the places that Adamski claimed to have visited bore the slightest resemblance to his descriptions, many ufologists began to wonder whether Adamski had ever been abducted at all.

The Sexy Spacenapper

Adamski was actually pipped to the post in the alien-abduction stakes. The year before Adamski's *Inside the Spaceships* appeared, Truman Bethurum published *Aboard a Flying Saucer*. Bethurum said that he had been carrying out maintenance work on a highway in the Nevada desert when he met five aliens. They took him back to their massive spacecraft, which was 300 feet (91m) wide.

On board he met the captain, a female alien called Aura Rhans, whom Bethurum described as a 'queen of women' and 'tops in shapeliness and beauty'. These aliens were much more advanced than the ones that Adamski had met. They spoke perfect English and Bethurum stayed up all night swapping interplanetary stories with Aura Rhanes. He was not as lucky as many people who have met attractive aliens, in that Aura Rhanes did not have sex with him.

Bethurum's aliens looked sufficiently human to be able to mingle with earthlings unnoticed; some time later, Bethurum saw Aura Rhanes drinking a glass of orange juice in a restaurant, but she ignored him. The aliens, Bethurum said, were from a planet called Clarion, which was free from disease, war and other human scourges. It was quite close to Earth, but lay on the other side of the moon, which meant that it was permanently hidden from human gaze. (However, human spacecraft have travelled around the other side of the moon since then and have reported no sign of Clarion.)

A New York Return Trip

In 1954 Daniel Fry published *The White Sands Incident*. In it, he claimed that he had taken a trip on an alien space-craft. The incident had taken place in 1950, although he later said that it was in 1949. At the time he had been at the White Sands Proving Grounds in New Mexico, where the US military was trying out the rockets – and rocket scientists – that they had captured from the Germans in 1945.

One day Fry saw a large UFO land in an isolated area near White Sands. As he approached it he heard the cautionary words 'Better not touch the hull, pal: it's still hot'. These words emanated from an alien entity who was circling the Earth in a massive mother ship; the UFO that Fry could see was merely a freighter that was used for carrying cargo. He was invited aboard and the alien gave him a guided tour of the craft's propulsion system. His name was A-Lan, although he said that Fry could call him Alan if he liked. They got on famously and A-Lan offered to take Fry on a trip. They did not go anywhere exotic, like Mars, Jupiter, the far side of the moon or Clarion – instead Fry was treated to a trip to New York and back, which took just half an hour.

A-Lan explained that he was not an alien at all, but an earthling: he was descended from an ancient race which had abandoned Earth to roam the skies after a disastrous war between the inhabitants of the lost continents of Lemuria and Atlantis. Then A-Lan delivered his Michael Rennie speech and asked Fry to relate it to the world, which he accordingly did in *The White Sands Incident*. He then gave up his job in order to lecture on his experiences and to run a foundation, based on the wisdom of A-Lan, called Understanding.

Fry was often asked why it was that if creatures from outer space really wanted humans to mend their ways they did not land on Earth openly and then pursue their agenda with the appropriate authorities through the usual channels. Fry explained that change could only come about through the work of a number of enlightened individuals, not by means of aliens descending to Earth to solve humankind's problems.

Invasion of the Abductors

Aircraft mechanic Orfeo Angelucci saw his first flying saucer in 1955, as it landed in a field near Los Angeles. As seems typical, the aliens welcomed him on board and explained the workings of their electromagnetic system to him. They also gave him the Michael Rennie lecture.

Angelucci subsequently began to see aliens all over California, hanging out in cafés and bus stations. These highly developed aliens were apparently quite happy living among inferior humans.

They later took him to Neptune, Angelucci claimed, and revealed that Jesus Christ had been an alien. The Angelucci discovered that he himself had been alien in a previous life. He shared this information in his 1955 book *Secret of the Saucers*, which also contained a chilling message for humankind: mend your ways or face oblivion in 1986 (this date coincided with the return of Halley's Comet).

Fortunately, the world must have heeded Angelucci's extraterrestrial message and have mended its ways, because 1986 passed without calamity: Halley's Comet came and went and the Earth survived unscathed.

A Trip to the Moon

Howard Menger had been plagued by visits from beautiful, blonde, female aliens since his childhood. In 1932 the first one turned up near his home in New Jersey and told him that aliens would always be with him. Despite his long acquaintanceship with these extraterrestrial nymphets, he had no sexual contact with them (which was, perhaps, just as well, since some of them claimed to be as much as 500 years old). Menger was, however, concerned about the effect that such a level of ageing would have on their figures and therefore bought them bras, only to be informed that they did not wear such things. (Indeed, when you think about it, if you lived in the weightless conditions of extraplanetary space, a bra would be entirely unnecessary.)

The male aliens whom Menger saw were tall, blond, good-looking, clad in one-piece jump suits and extremely old. Menger would cut their hair for them so that they could mingle with earthlings without attracting attention (this was in the pre-hippie days). These aliens had visited Earth many times before and had had a hand in many of the major developments made by ancient civilisations. Disturbingly, Menger said that they were responsible for the rise of the blood-thirsty Aztecs.

They took Menger on a trip to the moon, where the air, Menger said, was similar to Earth's. To prove that he had made the trip, he brought back some 'lunar potatoes', which were unfortunately confiscated by the federal government because it is illegal to import agricultural products into the continental United States.

In common with other aliens, Menger's friends told him how their propulsion system operated. Menger made an Earthly prototype of it, but it did not work. (It is strange that no matter how many times they explain it to humans no earthling has yet managed to reproduce the extraterrestrials' mode of transport. You would think that they could abduct a scientist or an engineer who might stand a chance of understanding it.)

Unlike many other abductees, Menger made money both from lecturing and his 1959 book *From Outer Space to You*. But he was luckier than some because his aliens taught him how to make extraterrestrial music, and Menger's *Music From Another Planet* is still available by mail order.

The Missionaries

As well as trying to alert us to the possibilities of advanced forms of propulsion and the dangers of nuclear weapons, it seems that aliens are also concerned about our spiritual well-being. In 1954 Marion Keech reported that she had heard from beings on the planets Cerus and Clarion (from which latter planet Truman Bethurum's spacemen came). Apparently, the Son of God, formerly known as Jesus, was now living on Clarion under the alias Sanada. He had spotted a fault in the Earth's crust, so he contacted Keech and warned her that a great flood was going to wipe out Salt Lake City, Utah, on 21 December 1954. Keech and her followers gathered to await a great spaceship that was supposed to take them to safety. It failed to show up, but fortunately there was no flood either.

Several cults have been formed around the idea that aliens will come to Earth to save humankind. Among them was Heaven's Gate, whose 39 members committed suicide in March 1997 believing that an alien spacecraft travelling behind the comet Hale-Bopp would then rescue their liberated spirits and take them to a heavenly haven.

The London taxi driver George King said that he was

first contacted by aliens when he was alone in his west London flat. He was told that he had been chosen to be the voice of the interplanetary Parliament. In order to prepare him for his role he was taken on several trips to Mars and Venus. On one of these extraterrestrial sojourns, he said, he had saved the world by diverting a menacing meteor. Unfortunately, some of the people to whom he revealed his mission in the back of his cab were sceptical about his claims. He therefore decided to go somewhere where they were more open-minded about such things and moved to California, where he set up the Aetheris Society. He and his followers found that by using the power of prayer alone they could charge up batteries that would relieve suffering all over the world.

While he was living in Los Angeles King revealed that he was visited regularly by enormous spaceships. Fortunately, they were fitted with a cloaking device which rendered them invisible, otherwise the sight of these huge ships hovering above the city would certainly have engendered panic. Even in their invisible state they apparently presented no hazard to aircraft.

Is God an Astronaut?

George King's transfer of the centre of his operations to California was a good move: aliens seem to flock to the Golden State, especially if they have some quasi-religious message to impart.

On 30 January 1965 a UFO landed near the Californian home of Sid Padrick (maybe the aliens were looking for St Patrick, but Sid was as close as they got). He heard aliens telling him to 'fear not' – the same message that the heavenly host had vouchsafed to the shepherds when it appeared to them to announce that Jesus had been born.

In Padrick's case, the aliens invited him on board their flying saucer and gave him a guided tour. These beings were not as attractive as some that other abudctees had met: they were essentially human in appearance, but had pointed noses and chins. Like other aliens, they said that they came from a planet in the solar system that was hidden from human view. It again had no crime, disease or any of the other drawbacks of life on Earth; everyone there 'lived as one'. But Padrick's extraterrestrial experience was unique in one respect: in an area of the ship known as the 'consulting room' he was asked whether he would like to 'pay his respects to the Supreme Deity'. Were they kidding? Padrick asked. Was God really on board? Apparently he was for after Padrick had knelt and prayed that night he truly felt the presence of God.

Padrick related his experience to officials of the United States Air Force at Hamilton Air Force Base. They took copious notes, which were then passed on to the Project Blue Book at Wright-Patterson Air Force Base. One question that the USAF was particularly interested in was how a huge spacecraft could escape radar detection. Padrick told them that the aliens' craft had a hull which absorbed energy. Energy-absorbing paint is now used on the USAF's Stealth bombers, so maybe Padrick's contact with aliens has been of some benefit to humankind. However, although some abductees do still find themselves meeting godly aliens, by 1965, the year in which Padrick met his maker aboard a flying saucer, some aliens had begun to turn decidedly kinky.

Sexual Encounters
of the
Alien Kind

Looking for a Latin Lover

In 1957, four years before the Hills' 'close encounter of the
fourth kind', a Brazilian man named Antônio Villas-Boas
had what can only be described as a 'close encounter of the
fifth kind.'

Close encounters of the first kind are simple UFO sightings. Close encounters of the second kind are sightings of
UFOs accompanied by the physical evidence of an alien
craft – such as the grass having been pressed down where
it landed, broken twigs or scorch marks having appeared
on the ground or injuries having been inflicted on animals
or humans. Close encounters of the third kind are those in
which the human observer has contact with intelligent
beings from spacecraft, while close encounters of the
fourth kind are alien abductions. And close encounters of
the fifth kind involve contact of a more intimate character
– at least that is what Villas-Boas said had happened to
him.

The 23-year-old law student had experienced close
encounters of the first kind twice before. The first was on
5 October 1957, when Villas-Boas and his brother saw a
bright light shining down outside the bedroom window of
their home near São Francisco de Sales in central Brazil.
Villas-Boas worked on the family farm in order to pay his
way through law school. The second sighting took place
nine days after the first, when he was out ploughing on the
night of 14 October. He again saw a bright light, which
appeared to be hovering over the field 150 feet (46m)
above him. This time he chased the light in his tractor.
The light darted around so much, however, that Villas-
Boas eventually gave up his pursuit. The light then broke
up into multiple beams before disappearing completely.
Again, his brother had also witnessed the phenomenon.

On the next night Villas-Boas was out ploughing again when he saw a red light coming straight towards him. This time he was alone. The light came so close that he could see that it was attached to an egg-shaped craft which had three 30-foot-long (9m) legs. When it landed Villas-Boas was seized with fear. He tried to drive away, but the tractor stalled and its lights and motor died. Four aliens then clambered out of the spacecraft. They grabbed him, dragged him to the spacecraft's ladder and forced him to climb aboard. The aliens were about 5 feet (1.5m) tall, with disproportionately large heads. (As the epidemic of alien abductions increased, this became the typical description of the extraterrestrial perpetrators of encounters of the fourth and fifth kinds.)

On board the alien spacecraft, Villas-Boas was forcibly stripped. His body was anointed in a strange, colourless, viscous liquid. The aliens wore grey overalls, gloves with five fingers and helmets and talked in a strange language which he could not understand. In another room, two other aliens cut his skin and took a sample of his blood. Then they left him alone and naked for about half an hour in a small, round room. Gas was pumped into it; it smelt foul and Villas-Boas vomited. He recovered, however, when a naked female alien came in. She was humanoid, around 4 feet 6 inches (1.35m) tall, with fair skin and high cheekbones; her eyes, though blue, were oriental in shape. Her face was wide, with a pointed chin. Her hair was blonde – evidently bleached blonde, because the hair in her armpits, as well as her pubic hair, was bright red. Despite her strange appearance Villas-Boas became sexually aroused. He said:

> She had the most beautiful body that I have ever seen on a woman: high, shapely breasts and a narrow waist. She had broad hips, long thighs, small feet, thin hands and normal fingernails. She was much smaller than I am, and the top of her head

only reached my shoulder. Alone with this
woman, who made it clear to me what she wanted,
I got very excited ... I forgot everything, grabbed
her and responded to her caresses. It was a normal
act and she behaved like any other woman.

They had sexual intercourse twice. Satisfactory though
the experience was, however, Villas-Boas said that he
would have preferred a human women because he would
rather have made love to a woman to whom he could talk.
She did not know how to kiss either, but he imagined that
her playful bites on his chin amounted to the same thing
(maybe she was just avoiding his vomit-tainted breath).

Despite having had sex, she then took another sperm
sample from Villas-Boas and stored it in a test tube. Before
she left, she rubbed her stomach and pointed upwards.
Villas-Boas interpreted this to mean that she was going to
have his baby, which would be raised somewhere in the
skies. The disgruntled Villas-Boas later whimpered:

That was what they wanted of me, a good stud to
improve their stock. In the final analysis, that was
all it was. I was angry, but then I decided not to
worry about it. After all, I had enjoyed some pleasurable moments, even though some of the grunts
that I heard coming from the alien woman's mouth
at certain moments nearly spoilt everything, giving
me the disagreeable impression that I was having
sex with an animal.

Once his moment of post-coital contentment had
passed, Villas-Boas realised that no one would believe his
story without some sort of proof. He therefore tried to
steal a small device from the craft, but because he had
nowhere in which to conceal it he was apprehended.
Nevertheless, once he was dressed again, the aliens gave
him a quick tour of the craft. Like other extraterrestrials,
they proudly showed off their propulsion system, which
was housed in a central dome and gave off a strange, green

light. Then they took him to the ladder and made it plain that he was free to go. After Villas-Boas' feet were placed safely back on the ground the spacecraft then rose slowly into the air before taking off at a dizzying speed. The whole abduction had lasted for little more than four hours.

During the days following his abduction Villas-Boas suffered from nausea and headaches. He felt an unpleasant, burning sensation around his eyes and could not sleep at night. Strange wounds developed on his arms and legs, which healed to leave scarring. The scars were the only tangible proof of his story, but he showed them to no one. Villas-Boas was also reluctant to tell anyone about his abduction, confiding only in his mother.

In 1958 he read an article about UFO sightings in a publication called *Cruzeiro*. He consequently contacted a journalist, João Martins, who took him to Rio de Janeiro to see Dr Olavo Fontes, the Brazilian representative of the Aerial Phenomena Research Organization, a world-wide UFO group that had been formed in 1952. Dr Fontes was impressed by the symptoms of trauma that Villas-Boas exhibited. Fontes diagnosed the physical problem that he had suffered as 'radiation poisoning or exposure to radiation'. He was also impressed by Villas-Boas' willingness to undergo all sorts of tests – along with the financial loss that he thereby suffered by taking time off work for them – in the hope of finding some sort of explanation for his experiences.

In 1961 the story was picked up by Dr Walter Buhler, a Brazilian UFO expert. He interviewed Villas-Boas and submitted a detailed report of his findings to the English-language journal *Flying Saucer Review*. Despite the editors' worries that they might be dealing with the sexual fantasies of a deranged young man they eventually published the story in 1964. Years later, they appeared in print in Brazil, but it was not until 1978 that Villas-Boas made any public statement about his abduction, confirming the details of the encounter on Brazilian televisions. He did not give lectures on the significance of his experience, however, nor did he found a cult. He had no message or warning to convey from the aliens whom he had met, nor did he write a book about his encounter or attempt to profit from it in any way. After the investigation of his abduction was completed he qualified as a lawyer and spent the rest of his life working in the legal profession, living near Brasilia with his wife and four children.

Although he was unwilling to talk about his experience again, Villas-Boas had seemingly cleared up one question about aliens: at least we know why some of them are here. He died in 1990, saying that 30 years earlier the US government had invited him to examine the wreckage of a crashed UFO; it is assumed by many that the wreckage came from Roswell.

Alien Adultery

One night in 1981, in Birstall, West Yorkshire, Jane Murphy went to bed exhausted. A few hours later she awoke. She did not open her eyes, but she knew that something was wrong: she could not hear her husband's snoring and the feeling in the room was not right.

When she eventually opened her eyes she found herself in a field near to her mother's house; she did not know how she had got there. There was a huge, metallic UFO floating above the field and then she saw a group of humanoid figures approaching her. One of them held a cloth in its hand which it put over her nose and mouth. She feigned unconsciousness, but the aliens gave her an injection that knocked her out anyway.

She came round in a strange room, surrounded by aliens. One of them was seven feet (2.1m) tall; he looked like a male human, but his eyes were totally black. A young alien woman told her that she must bathe, so she slipped off her nightdress and got into a strange tub that seemed to match the contours of her body exactly.

There was a table in the centre of the room. All the aliens had now gone, except for the tall one, who brought his face close to hers. 'I just stared into his big, black eyes', she said, 'and I knew what was going to happen.' She is unsure of whether he raped or seduced her. She was completely mesmerised by his eyes and did not resist. In fact, he lay back on the table and she climbed on top of him. Although they embraced sexually, there was very little movement and she was not even conscious of his penis inside her. She did not think about the physical act, but nevertheless felt all the sensations of normal, human sex. 'Inside me it was all happening', she said. 'At the time, I felt it was the best sex I had ever had. It seemed so strange, lying on top of this stranger, not moving yet having sex and enjoying it.'

The only thing that repelled her was his strange, inhuman smell. She looked deep into his eyes and said 'Why me?' The answer came telepathically: 'Because we love you'. She was disappointed that his profession of love was so cold: 'There was no emotion in it at all', she said. Even so, she climaxed and said that it was the best sex that she had ever had.

Afterwards, the other aliens, including several females, examined her while her sexual partner got off the table and left the room. Then she was given a gynaecological examination with a long instrument, but felt no pain. She was later taken on a tour of the ship and saw other humans on board. She was also invited to try out some pills and was given a drink. Murphy had the impression that the aliens communicated by means of telepathy.

Suddenly she found herself back in her own bed. It was 6.30am and the alarm clock had just gone off. She would have dismissed the whole incident as having been a bad dream had she not had a strong desire to take a bath and wash the strange smell of her alien lover off her body: it was not pleasant, in fact, not human at all. When she had a bath she discovered puncture marks on her body, where, she said, the aliens had injected her. Then she felt a distinctive sensation in the pit of her stomach which she remembered experiencing when she had been pregnant in the past. When her period was late she grew frightened.

She went to the doctor, who ascertained that she was not pregnant, but had a strange, vaginal infection, which was eventually cleared up with a course of powerful antibiotics. That was by no means the end of the story, however, for within a month the aliens were back, swarming all over the house and bombarding her with questions about human reproduction. She would wake at night to find them standing by her bed. Although she had no recollection of it, she was also certain that they had carried her off again.

Soon she became plagued by a dream in which she gave birth to a blond, black-eyed, alien baby. She thought that she was going mad – indeed, she was certainly on the verge of a nervous breakdown. Her GP suspected that she had been taking LSD, and although her husband said that he did not believe that she had been abducted by aliens, her stories of having great sex with a man from outer space were putting a strain on their marriage.

In desperation, Murphy contacted the British UFO Research Association's hot line, whose members debriefed her and consequently discovered that it seemed that she had been abducted numerous times, the first occasion being when she was just 16. She may well, they believed, have given birth an alien child which was itself then abducted.

A Case of Rape

According to the British UFO Research Association's investigator, Barry King, a married woman from Taunton, Somerset, was driving her car near her home one night when the engine mysteriously cut out. She got out to look under the bonnet and was then grabbed from behind.

She awoke to find herself naked, tied to table and covered with a blue blanket. Three aliens in blue tunics were examining her; they were around 5 foot 6 inches tall (1.65m), with fair skins and round, emotionless eyes. Two of the aliens left. The third gave her an injection in the thigh, which made her numb; then he raped her, coldly, without emotion and she passed out. When she awoke, she was beside her car again, which now worked perfectly. King found her story perfectly credible; she had even told her husband.

Visiting Voyeurs

Yet it seems that aliens don't just want to have sex with humans – they like to watch them making love, as well. Indeed, they seem fascinated with human reproduction. David M Jacobs, of Philadelphia's Temple University, reported that a young woman who was abducted in 1988 was forced to have sex with a man who appeared to be unconscious. Without the aid of hypnosis, she later recalled that she was made to climb upon him. She straddled the comatose man and one of the aliens who was watching her told her – telepathically – to kiss him. She did not want to do so, so she put her hand on his chest and pretended to the aliens that that was a human kiss. They seemed satisfied with her explanation and then told her to touch his penis. 'Like I was supposed to scoot down below, you know', she said. 'Down low on the table'.

She said that she didn't want to have oral sex with the man and tried to persuade the aliens that earthlings did not do that. They did not force her, but their will was hard to resist. Although the man was completely immobile otherwise, he nevertheless got an erection. The aliens next wanted her to have intercourse with him. She said that she did not want to because it was immoral, but found herself doing so anyway. 'It's totally mechanical', she recounted. 'It's really bad. I don't think I have feelings or anything'.

After the man had ejaculated, the woman's mind went blank; Jacobs thinks that she was being mind-scanned at that moment. Only after that did she find herself becoming aroused (fortunately for her, the man's penis had remained erect).

'It's so weird because it's not like any normal pattern that I've ever been through', she said. All the time she had been worried that she could become pregnant by the unconscious man.

Sex with an Older Man

A 15-year-old girl reported having been abducted by aliens in 1959 and having been forced to have sex with an older man. She had been totally powerless. An alien had started off the assault by staring into her eyes. 'He's completely penetrating me.' She said, 'every bit of me is in my eyes. He's in my eyes…I can't do anything…He's spreading into my brain…totally invading me.' The invasion went on for a horribly long time. Worse still for the girl, 'He's making me feel feelings, sexual feelings…It must be that he's making me feel things because I don't feel them.'

Other aliens were standing around watching, and there

was a human man at the end of the table who was erect. He was a big man with a bit of a paunch and was middle-aged, with receding hair. She said that the man was 'absolutely out of it'; his mouth was hanging open and his eyes were glazed over. The brainwashing alien moved aside so that she could see the older man, then he moved back again. By this time she said that her body was responding sexually, but 'I mean, I didn't know what it was at the time, what a sexual response was'. 'I knew it was very strange. It had pleasurable parts to it, but it wasn't a pleasurable situation, obviously', she said.

The man was amused too, and the 'inevitable followed'. The aliens seem to have had the idea that she had just ovulated, and the man climbed on top of her. The alien who had initiated matters, she said, then 'zapped' her. 'All of a sudden I'm really sexually excited', she recalled. 'Overwhelmingly sexually excited.'

When the man reached the point of ejaculation the aliens pulled him off her and put a metal instrument up her vagina. She described how the aliens had then taken an egg from her before leading the man away. She claimed that this had happened to her several times.

An Alien in Essex

On the night of 28 September 1985 Ted Johnson was driving down a deserted road in deepest Essex when he saw a strange light in the distance. After he had stopped and had got out of the car to see what it was he heard a low, whistling sound. The light approached so close to him that it hurt his eyes. Then there was a green flash, which knocked him to the ground and partially blinded him.

When he got to his feet he was greeted by a strangely dressed female being. Her eyes were large and her nose small; she had a slit instead of a mouth and she was glowing. In the darkness behind her Johnson spotted other luminous humanoid forms. The female alien led him to her spacecraft, where she stripped off. 'She was naked and we had sex', Johnson claimed. 'It was not different from human sex.'

So alien-human sexual encounters do not just happen in Brazil.

Spacenapped in Somerset

The 33-year-old Mrs Milan had moved from Turin, Italy, with her husband and had settled in Somerset. On 16 October 1973 the daughter of a friend asked her to drive her to her mother's house in Wellington because her mother was ill. Mrs Milan was making dinner at the time, but promised to take her later that evening.

A number of people then dropped by, so Mrs Milan finished dinner later than she had expected. Her friend's daughter had already set off to visit her mother, but Mrs Milan decided to go and see her friend anyway and left at around 10.45pm.

At around 11.15pm she found herself driving on a lonely stretch of road near Langford Budville. A yellow light then appeared in the field to her right and next her headlights dimmed and her engine cut out. After the car had coasted to a halt Mrs Milan got out and took a look at the engine. She suddenly felt a blow to her shoulder and turned around to see a metallic man, along with flashing lights. The next thing that she knew she was standing in a field looking up at a huge flying saucer, with the yellow light that she had seen from the road issuing from its windows. It was 20 feet (6m) high and 40 feet (12m) across. Grey and metallic, it stood on thick legs and emitted a humming sound. Then she blacked out again.

She awoke inside a room. She was naked and her arms and legs were strapped to a table. The metallic man stood immobile in the corner while three creatures wearing surgeons' masks and gowns performed a detailed examination of her body. They seemed to be recording their findings on a series of illuminated colour cubes along the edge of the bed. A pen-shaped probe was used to prod her. Nail parings and blood samples were taken from her body and a fair-skinned, round-eyed man passed a small scanning device over her. A suction device was fitted over her sexual organs (this was the only part of the proceedings that caused her discomfort). Throughout, she felt that she was being treated like a laboratory animal.

The creatures then left, but she was unable to free herself. Next, one of them came back and stuck a needle into her thigh, which paralysed her. He then proceeded to rape her. Because she had been immobilised she could not struggle, but she still felt the cold, clammy alien's every move. Afterwards, he pulled the needle out and the other creatures helped her down form the table. Then she collapsed.

When she awoke again she was standing by her car, fully dressed. In a state of turmoil, she got into the car and found that it worked perfectly well. Somehow she managed to drive home. Arriving at about 2.30am, she told her husband the whole story in floods of tears. They agreed to tell no one. However, four years later, in 1977, she met UFO investigator Barry King and decided to put her experiences on record.

Task Accomplished

On the night of 3 March 1978, the 19-year-old Jose Alvaro was on his way over to his father's house in Fragata Pelotas, Brazil. His father was away and it was Alvaro's task to check that the house was secure.

Before he got there he was struck by a beam of blue light. Some time later, a passer-by saw him lying on the ground, as if he was in a drunken stupor. Alvaro awoke in a daze; images of war were racing through his brain, along with the phrase 'The task is accomplished'. Other witnesses had seen strange lights hovering in the area that night.

Under hypnosis, Alvaro recalled having been led into a room by an alien female with large eyes and silver hair. She forced him to have sex with her. There was another task that he would have to perform later, she said.

Strangely enough, before she knew of this Alvaro's mother had reported having had a strange dream, in which she was told that her son who was unmarried was about to become a father but that the child would be born in another world.

Aliens in Scotland

On 17 August 1992, the painter and decorator Colin Wright and the mechanic Garry Wood were delivering a satellite dish to friend in a small village in West Lothian. At around 11.30pm they were rounding a bend in their vehicle when they saw a UFO hovering around 20 feet (6m) above the road, surrounded by thin mist. As they drove under it a tube of light descended from the craft and suddenly everything went dark. The next thing that they knew they were skidding down the wrong side of the road at 70 miles

(113km) an hour.

Both men felt high, and when they reached their friend's house they discovered that it was 1am: they had lost over an hour. Wright discovered that his safety belt was unbuckled, although he remembered doing it up at the start of the journey. The next day Wood began to suffer a series of headaches and Wright found that his eyes were sore.

Under hypnosis, Wood recalled seeing a floating examination table and a pool of gel, in which the aliens appeared to be living. He remembered one alien – a small 'grey' one, with large eyes-emerging from the goo. The creature appeared to be translucent and he could identify its skeleton and internal organs. He also saw a naked woman, who seemed cold and upset and certainly not as obliging as those who had descended to Earth in Latin America.

Love Bites?

Many abductees seem to end up with strange marks on their bodies. In the early hours of 29 November 1982 in Botucatu, Brazil, 39-year-old Joao Valeiro da Silva got out of bed for a drink of water. He was in the kitchen, he said, when he was hit by a beam of intense light, in which he floated up into a UFO.

On board, he found himself surrounded by strange creatures. He was approached by a naked woman who touched his cheek, and it was at this point that he passed out. His family discovered him the next morning lying naked on the floor, with his clothes piled up next to him. His body was covered in strange marks and there were lesions on his penis, which, he said, he had never noticed before. Clearly an understanding soul, his wife put him to bed and nursed him better.

Making Babies

The Birmingham Baby

It is interesting that the alien abduction of Antônio Villas-Boas in 1957 occurred within days of the first artificial satellite, *Sputnik* I, orbiting the Earth (perhaps because humans invaded their word the aliens decided to get their own back). And it seems that aliens are not just entering the Earth's atmosphere because they want to have sex with earthlings: they want to impregnate human women, too. This sort of encounter appears to have happened first to an ordinary woman living in suburban Birmingham.

On 16 November 1957 the 27-year-old mother of two Cynthia Appleton suffered a 'missing-time' experience in her home at Fentham Road, Aston. Later she learnt that this was an alien attempt at contact that had failed. The aliens tried again, however. On 18 November, at 9pm, her front room was filled a rosy glow and the air was suffused with static electricity. It was then that an alien materialised, reeking of ozone. (A blackened newspaper indicated that there had been some sort of electrical discharge).

The alien was tall, blond and perfectly proportioned – of the 'Nordic' type that is commonly seen in Britain. He wore the usual alien attire – a one-piece, silver jump suit – and communicated by means of telepathy. 'Do not be afraid', he said. They were soon locked in telepathic conversation. He came from the Gharnasvarn, he said, which Mrs Appleton took to mean Venus. The Gharnasvarnians apparently wanted to be friends with humans, but this was proving impossible because of humankind's aggressive nature and nuclear weapons. The alien illustrated what he was saying using three-dimensional images that he created in the air. (Holography was then in its infancy and it is unlikely that a Mrs Appleton would have known about it). Although the alien showed Mrs Appleton a holographic

47

image of his ship she never actually saw the craft.

The second time that he appeared he materialised inside the house again. After that, however, he arrived by more conventional means, turning up at the front door saying that he had driven there. Furthermore, as his exotic suit would have made him conspicuous, he took to wearing an ordinary cloth suit and a Homburg hat (which must have looked a little old-fashioned, even during the 1950s).

The alien visited her eight times in all. He seemed to have come in order to relate the secrets of the universe to her, because she was instructed on the structure of the atom, how to cure cancer and how to make a laser. (Although Albert Einstein had theoretically predicted the possibility of making lasers in 1917 the first working model was not built until 1960.) However since she was not a professor of physics Mrs Appleton was not well equipped to absorb exactly what the alien was telling her. She asked why he had chosen to convey this information to the world via her and was told that it was because she had a brain that aliens could 'tune into'. For her part, Mrs Appleton found her selection something of a burden.

But worse was to come: in September 1958 the alien told her that she was going to have a baby which would 'belong to the race on Gharnasvam'. He did not go into the details of how this would conceived, but when she went to the doctor she discovered that she was indeed pregnant. The child was furthermore born on the day that the alien had predicted; he had forecast its weight correctly, too, within an ounce. And the child also had fair hair.

A psychologist from Manchester interviewed Mrs Appleton and wrote up the case. She was also questioned by a priest, who found her to be a 'very trustworthy woman'. Her daughter said that she had seen the alien, too (but she was only four at the time). Apparently it also emerged that the alien had once burnt his hand and had given Mrs Appleton some salve to put on it; when she washed the wound a piece of skin had come off. Scientists at Birmingham and Manchester universities studied the skin but could not identify it, although they concluded that it was more like an animal's skin than a human's.

The Lost Love Child

Elizabeth Klarer claimed that she had had numerous encounters with perfectly charming aliens between 1954 and 1963. Born in 1910, she had had a varied career, as a piano teacher, a meteorologist, a pilot and an intelligence officer in the South African Air Force. One of the extraterrestrials whom she had met was a handsome male humanoid, she said. She fell in love with him, but when she became pregnant she was whisked off to his home planet to have the child. She was then returned to Earth, but the aliens decreed that her baby had to stay behind. The planet may have been Clarion, from which the aliens that Truman Bethurum saw had supposedly come, because Klarer said that – like Clarion – it was free from war, poverty and disease and was therefore obviously the best place in which to bring up a child.

The Mystery Miscarriage

On the evening on 19 August 1979 36-year-old Lynda Jones was walking along the banks of the river Mersey, near Manchester, with her two children, aged 5 and 15. Suddenly the youngest shouted 'Mum, the moon is coming towards us'. Lynda looked up to see a huge, grey, Frisbee-shaped craft hurtling across the sky towards them. It came into view from behind some trees and she noticed some golfers playing on the greens as approached. 'It was not an optical illusion', she said. 'The object had a spinning effect and it seemed to be coming at us at an angle.'

For a moment she thought that it was a plane that was trying to make an emergency crash-landing at the nearby Manchester Airport. She pushed the children to the ground, telling them to lie flat and brace themselves. But there was no explosion, only silence. She saw the UFO pass slowly overhead. Then it dropped silently down behind the embankment that ran along the side of the river as a flood defence. The silence struck her more than anything else: 'It was more than just silence', she said, ' it was complete stillness.' She realised that not only had there been no noise from the object, but there had also been no other sound either – neither bird song nor any traffic noise form the neighbouring motorway, which was always busy, day and night.

The UFO was now hidden behind the embankment. She wondered whether it had crashed and had then caught fire, so she ran up to the top of the embankment to take a look. Jones knew nothing about UFOs and what she saw amazed her: it was a crescent-shaped craft, 60 feet (18m) across and made from a fine network of metal. 'It looked so old-fashioned', she said. 'It really did look like something out of the Bible.'

The UFO hovered about 2 feet (60cm) above the ground and seemed to be constantly disappearing and reappearing. A light was somehow suspended from the underside of the craft. She was drawn towards it, and as she approached it became brighter. Then an orange ball of light appeared from the far side of the object. It rotated slowly and began to move menacingly in her direction. But she still continued to walk towards the spacecraft. 'Now it gives me the shivers', she recalled.

Suddenly her daughter screamed out 'Mum, come back. Come back.' The child's voice caused her to snap out of her trance and she simultaneously had a terrible premonition that it was Judgement Day.

Jones grabbed her children by the hand, turned on her heels and ran. Her five-year-old son was too small to keep up, so she swept him up into her arms and continued to flee. The spacecraft followed them and her terrified daughter looked back at it. 'Don't look back, just run', shouted Jones. The ground seemed to undulate beneath them. It was the strangest sight that she had ever seen, but they kept on running until they got home.

When they burst through the front door Jones' husband, Trevor, was home. He asked what the matter was with her eyes. When she looked in the mirror the skin beneath them was red and scaly. The fact that Trevor was home at all was even more peculiar. He usually worked until 10pm and did not get home until half-past. Jones had seen the UFO at around 9pm and the journey home would have taken only 10 minutes at walking pace. She had therefore lost nearly an hour. Jones said that she had got the impression that the UFO had made time slow down.

Afterwards, she found that her menstrual cycle had been disrupted and that there were marks on her body. Eighteen months later Jones underwent regressive hypnotism. Just before she had turned to run she recalled seeing an alien emerge from the UFO. She then had the feeling of floating before finding herself in a room with six small creatures of oriental appearance. They had slanting eyes, dark hair and yellow or olive skins. She had the feeling that she had seen one of them before. They laid her on a table and examined her. Bright lights were shone into her eyes, but the exact details of the examination were wiped from her mind.

Some weeks after the incident Jones noticed a waxy discharge form her vagina and her doctor told her that she had had a miscarriage, although she protested that she had not been pregnant. She was then sent to hospital to see a specialist, who found scar tissue on her Fallopian tubes of the type caused by an ectopic pregnancy – that is, when a fertilised egg develops in the Fallopian tube rather than in the womb. At first she could not account for this, but then she remembered another strange incident that had occurred one day in 1972 when she was out with Trevor. She had not seen a UFO on that occasion, but when they pieced the events of the day together they seemed to have lost six hours in 'missing time.'

But that was not the end of the story. Five years after her original sighting she was awoken during the night by the same humming sound. She then saw a tall, blonde female, with blue eyes, in her bedroom, who effortlessly floated her out of the bedroom window,

She had little further memory of the incident and did not undergo regressive hypnosis. However, she did not know that she had been examined and also had the sneaking feeling that something 'very important' had happened.

Three years later, she was again awoken during the night by what she believed was the hand of a three-year-old child holding her own. She cried out and her husband switched on the light. As he did so, a ball of light flew out of the bedroom.

Holding the Baby

In 1988 Barbara Archer was a 21-year-old university student who was studying to become a journalist when she became plagued by feelings of fear and anxiety. She began to experience strange flashbacks of bizarre events in her life of which she had no memory, so she therefore sought professional advice.

Under hypnosis, she recalled an experience that had happened to her when she was 16. She had been getting ready for bed one night when she noticed a light shining in through the window. She closed the curtains, but the light seemed to illuminate the whole room. She peeked out of the window but could see no source for the light. She looked out of the other window, too, and then felt the strange sensation that there was someone in the room with her.

She saw a small creature by the wardrobe, which she took to be male. Although she was puzzled by the light, she was not shocked to see him there. He touched her on the wrist, which reassured her. Then she began to float out of the window; she said that it was like being in a lift with no walls. She could clearly see first the driveway, and then her house, then the rest of street below her. She was scared of heights and her levitation made her feel nauseous; she hoped that she would not be sick.

She floated upwards until she was underneath some sort of flying saucer. It was dark grey and metallic. Then she noticed that the light that she had seen was coming out of it. The alien whom she had noticed in her bedroom was still with her. More were waiting inside.

She also recalled having been abducted when she was 12. That time she had found herself in a room containing 40 or 50 tables. After she had been subjected to a physical examination a tall alien had come over to her and had looked deep into her mind. This made her feel happy and she lay back and relaxed; she did not feel sick, just a little cold. Although she was scared of the smaller aliens, she liked the taller one and thought that he liked her, too. There was a very sexual element to this feeling and although she was only 12 Archer suddenly felt very womanly, very grown up. She got the feeling that the alien could read her mind and really understood her.

She remembered that she had also been abducted at the age of 16, when she was suffering from anorexia, which had annoyed the aliens because she had stopped menstruating.

When she was 21, Archer went on holiday to Ireland, from where she was yet again abducted. This time the aliens had become cross with her because she had not taken off her clothes quickly enough. Later she had been taken to a nursery on board the spacecraft, where there were about 20 babies in cribs. Some were in nappies; others were slightly older and were dressed in simple smocks. They all looked rather odd because they did not have much hair and because their skin was an unnatural, grey colour.

The aliens had told her that she could hold one, and had picked out a baby girl for her. The child had had big eyes, which were shaped like an alien's but were not as ugly. Archer remembered feeling very protective and maternal. The alien nurse had then told her to feed it and she had obediently put the baby to her breast. Afterwards they had taken the baby away from her. When Archer was told that she had to leave she had felt bad about leaving the baby behind. She had asked that aliens whether she could see it again, but they did not give her an answer.

An Non-maternal Abductee

In 1957 the nine-year-old Jill Pinzarro was returning home from the library on her bike when she decided to take a short cut through the park. It was years later, when she was a minister, that she revealed what happened next.

She had stopped by a bench, parked her bike and sat down to do some reading. There was still an hour to go before dinner. Then she wandered into the nearby wood, but did not know why. Suddenly she felt the presence of someone beside her before seeing a strange glow. She walked towards it and found herself standing under something. She noticed a ladder and climbed up it into an alien spacecraft, where she was stripped and examined.

When she returned to the park bench it was already dark. Scared, she put her books in the basket into her bike and set off for home. It was late and her parents were frantic; they had already called the police, who were out looking for her. But when she arrived at her house her parents were so relieved that they were not even angry with her.

She was 11 when she was next abducted, the aliens being particularly interested in the scab on her knee – the result of falling off her bike. A tall alien stared into her eyes, which she found reassuring: the alien, she thought, really cared about her and would not let her come to any harm. This feeling was not initially accompanied by any sexual sensations and she thought that the alien was a female. However, when the alien touched her on the forehead, she felt calm and was willing to surrender sexually to the creature.

When she was abducted in 1980, at the age of 32, she was given a baby to hold. It was about two and a half months old, she reckoned, and had little hair and light skin. The alien nursemaid said that the child needed nurturing, but that they were not very good at it and therefore needed her to do it. The baby was quiet and seemed to enjoy Pinzarro holding it. When she had to give it back she felt an acute sense of loss. She had bonded with it but thought that this was strange because she did not consider herself to be a very maternal person. She had only ever wanted one child, which, indeed, she had – back on Earth.

Abducting Embryos

Lynn Miller, a Mennonite woman who worked as a wait-ress, had been the victim of numerous inexplicable events for all of her life. But things began to make sense in 1986, when she was 31. She was driving to Cape May, New Jersey, with her son, when she saw a huge UFO hovering above the road near Tuckahoe. She stopped the car and was promptly abducted. Separated from her son, she was forced to strip and climb onto an examination table. A tall alien then looked into her eyes and placed his hands on her head. She felt an overwhelming surge of love for him; there was a strong sexual element to this, a feeling of pleasure. Then the alien conducted a gynaecological examination on her. The alien next poked a long needle into her womb through her navel. It hurt, but the aliens put their hands on her head, which took the pain away. The aliens told her that they were implanting something in her, but did not say what.

Some time later, she was abducted again, and this time and a long, black instrument that looked like a speculum was pushed into her to dilate her vagina. It had suction cup on it and was attached to some sort of machine. She felt a tearing sensation inside her and complained to the tall alien who was performing the operation that they were hurting her, but they continued anyway. The tall alien then pulled something out of her and put it into a fluid-filled container that was handed to him by a smaller creature. They showed it to her and she recognised it as being a foe-tus. The tall alien told her that it was her child and that they were going to raise it. She protested that they had no right to do so, but the alien took no notice and placed the foetus into an incubator before telling her to get dressed.

The aliens also took her into another room, where they showed her a chart and told her to memorise the names inscribed on it. Miller was puzzled by this, but an alien told her that it would help them. When she wondered why they needed help, the alien explained that there was a war on and she needed to know these names. Although she looked at the chart she made no active effort to remember the information on it. However, the alien moved close to her and she got the impression that he was telepathically storing the names in her memory.

The next thing that she remembered was that she was again on Earth, miles from where she had stopped, in a part of New Jersey that she did not recognise. She got back into her car and drove home. Two hours were missing form her memory of events.

She additionally recalled having been abducted when she was six. Being a Mennonite, she was not allowed to have vaccinations and consequently came down with diphtheria. For religious reasons, too, her mother would not permit her to be taken to hospital; the doctor visited her every day, but there was little that he could do. Miller's condition deteriorated steadily for two weeks until it reached the point at which the doctor did not expect her to survive the night. But the next morning her mother discovered her playing happily on the floor. The doctor was called and found that her temperature had returned to normal and that the symptoms of diphtheria had vanished.

During the night, Miller remembered, she had been abducted by aliens. They had told her that they had come to cure her. They wound a tube around her body and made her stand in a strange machine. The aliens then sat around her in chairs and watched as a curtain of blue light descended from the top of the machine. Afterwards she was told that she had been 'cleansed'. The disease was certainly gone and although the doctor ordered her back to bed for another week it was a struggle to keep her there: all she could think about was getting up and playing.

Siring Aliens

James Austino was a student at Temple University, Philadelphia, when he came to the conclusion that the strange events that had plagued him throughout his life might be due to alien abduction. Fortunately for Austino, the distinguished ufologist David M Jacobs was also working at Temple University, where he was conducting research into the alien-abduction phenomenon, so Austino underwent regressive hypnosis with Jacobs.

He remembered that in 1980, when he was 14, he had been with his girlfriend, Monique, when the aliens came to get him. Monique was 'switched off' and Austino alone was taken. He next remembered lying on a table, which had a bright light above it. An alien pulled it down, over his groin area, before starting to fumble around. Austino found this a bit embarrassing, but something in his mind kept telling him that it was alright. The alien seemed to being showing his genitals to a colleague.

Austino asked what was happening and the alien then put his face close to his and implanted in his mind an image of the naked Monique. He got an erection, whereupon they attached his penis to a machine and took a sample of his semen.

In this case there was no witness to the procedure, but some female abductees claim to have seen semen samples being taken from their boyfriends if they were abducted along with them. And one poor mother witnessed aliens taking a semen sample from her teenage son. Luckier abductees report having been bonded to female aliens and getting the impression that they were making love to a human female.

Next, Austino was shown what his semen would be used for. The aliens took him to a room that was full of incubators, which looked like fish tanks with tubes coming out of them and were illuminated by a strange blue light. There were 60 or 70 of them. He was allowed to inspect them more closely and saw that there were foetuses inside. They looked like little bald hamsters with black eyes, he said. The foetuses floated in a liquid that gurgled and bubbled, and they had wires attached to them. As far as Austino could tell, they were human foetuses, although their eyes were very different.

He was abducted again when he was 21, at which time he was given a mind scan. Then, when he was 23, he was abducted for a third time and saw a young woman wearing a white smock who was helping the aliens. She took a dip in a large, oval pool containing a luminous-green liquid and urged him to join her. So he jumped in; the pool was about 5 feet (1.5m) deep. The aliens who were watching looked pleased.

The woman encouraged Austino to lie back and relax. On doing so, he sank to the bottom, but found that he could still breathe. The woman then got out, whereupon he began to feel numb and almost blacked out. He therefore pushed himself up and two aliens helped him to get out of the pool. He observed that the liquid did not run off his body like water: it was more viscous and had to be slopped off by hand.

Paternal Instinct

In 1989 the 34-year-old alcohol-rehabilitation counsellor Andrew Garcia was abducted. After having his mind scanned he was introduced to a five-year-old girl who looked like his niece only with a big, black eyes.

She looked deep into his own eyes, which evoked all sorts of intense emotion in him: he wanted to hold the child and love her. She touched him, but then pulled back, which upset him. He did not want her to leave. The aliens who were watching did not understand this and studied his emotional reaction intensely. A tall alien then stared into his eyes, whereupon Garcia calmed down. 'This is how it has to be', he was told. The alien said that he could see the child again another time.

Abductions
in Mind

I Met a Man Who Wasn't There

Maureen Puddy, a 27-year-old mother from the Australian town of Rye, had an invalid husband and two young children when, in June 1972, her seven-year-old son was involved in an accident. His leg was injured and he was taken by air ambulance to a hospital in Heidelberg, near Melbourne.

On the night of 5 July 1972 she as driving down the Mooraduc Road on her way to visit the boy in hospital. Between the towns of Frankston and Dromana she spotted a blue light in the sky that seemed to be following her. She initially thought that it could have been an air ambulance, so she stopped to take a look, but when she got out of the car she saw that it was no helicopter. It was a giant UFO, shaped like two saucers, one above the other, around 100 feet (30m) in diameter. It made a low, humming sound, glowed all over with a blue light and hovered just 40 feet (12m) above the ground. Mrs Puddy got back into her car and drove off. The UFO followed her and chased her for 30 miles (48km) down the deserted road before suddenly peeling away and shooting off like a steak of light into the distance. She reported the matter to the local police.

On 25 July 1972, driving down the same stretch of road at 9.15 at night, Mrs Puddy saw the UFO again. But this time the car's engine died on her when she tried to escape. Out of her control, the car was dragged onto the grass verge and the UFO now appeared to hover directly over it. Mrs Puddy was bathed in blue light and felt as if she was in the centre of a vacuum.

It was then that she felt a telepathic communication being beamed down to her. The voice that she heard in her head, she said, sounded like it was translating what it had to say from a foreign tongue. The voice said 'All your tests will be negative'. Mrs Puddy had no idea what this was supposed to mean. Then it said 'Tell the media, do not panic, we mean no harm'. And finally: 'you now have control.'

The car now roared back to life. The eerie, blue glow and the strange, vacuum effect disappeared and Mrs Puddy sped off. She went straight to the police again, who this time put her in touch with the Royal Australian Air Force. It checked its records but could identify no unusual aircraft movements in the area on that night, and then told her to say nothing to anyone. However, Mrs Puddy did contact Judith Magee, of the Victoria UFO Research Society.

Magee revealed that there had been some other peculiar sightings in the sky in the area on that night: 21-year-old engineering manager Maris Ezergailis had reported seeing a blue streak of light in the sky from nearby Mount Waverley. 'It was like a meteor trial, but an unusually broad one, travelling horizontally', he said.

When Mrs Puddy heard of this she became excited. She had seen a broad, blue streak of light just like that when the UFO had shot off after her first encounter with it. The problem was that Ezergailis had observed the streak at 10pm, so if he had seen the UFO streaking off Mrs Puddy had lost 45 minutes during her encounter.

Mrs Puddy was worried by the aliens' instruction to 'tell the media'. She feared that the aliens would come back again if she did not do what they said. They did. On 22 February 1973 Mrs Puddy was doing the housework when she heard a voice. It said 'Maureen, come to the meeting place'. At first she thought that there was someone at the door, but throughout the day she heard the voice over and over again. Eventually she came to the conclusion that the aliens were telepathically planting the voice into her head. It was thus plain that the aliens wanted to talk to her again.

She called Magee, who agreed to meet her, along with her fellow UFO investigator Paul Norman, out on the

Mooraduc Road at 8.30pm. When Mrs Puddy arrived she was flustered, and claimed nearly to have been run off the road. Magee calmed her down and asked her to drive them back to the place where the incident had occurred. When Magee got into Mrs Puddy's car she felt a tingling sensation, 'like a mild electric shock'.

On the way, Mrs Puddy explained to the two investigators that a figure wearing a gold suit had materialised inside the car, between the two front seats, before disappearing again. When they reached the spot where she said that this had happened she claimed that she could see it again, this time standing outside the car. Magee and Norman could see nothing. Norman even got out of the car and stood on the exact spot where Mrs Puddy claimed that she could see the figure. Mrs Puddy said the figure had stepped backwards to let Norman by.

Mrs Puddy then said that the figure had gestured to her that she should follow him. Magee volunteered to go, too, but Mrs Puddy gripped the steering wheel hard and replied that she would go nowhere. It was then that Mrs Puddy announced that she was being abducted. While the startled UFO investigators looked on, Mrs Puddy began describing the inside of the UFO, whose crew, she asserted, had kidnapped her. 'I can't get out', she yelled. 'There are no doors or windows.' She saw a strange 'mushroom' in the room. Inside it it looked like jelly was moving about. 'He wants me to close my eyes', she said. Then she fell into a hypnotic trance and continued describing what she saw inside the ship. Fortunately, the trance-like state did not last very long. 'He's gone', said Mrs Puddy, coming round. 'I can tell. It feels different.'

Magee wrote up the case for *Flying Saucer Review* only a few weeks after the event. Although Mrs Puddy had not physically gone anywhere during the abduction, there was one that Judith Magee was convinced of: she had not been making it up and had been genuinely upset by her experience.

A Meeting of Minds

Vaunda Hoscik was first abducted by aliens when she was 14. Since then it had happened to her hundreds more times. Although she believed that they were conducting regular tests on her, she thought that she was benefiting from her contact with them, too.

Hoscik was born in 1975, in south London. Of eastern European extraction, she believed herself to be psychic. In 1989 she went to bed one night at 10pm; after over four hours of peaceful sleep she woke suddenly at 2.45am. She sat up in bed, but then found herself overcome by a creeping paralysis. Only her eyes were unaffected. She was suddenly pushed back onto the bed with considerable force. There were four grey, foetus-like figures in the room. They had dark, slanted eyes and were about 4 feet (1.2m) tall, with long, thin arms. She tried to scream, but found that she could not do so.

The next thing that she knew she was in a brightly lit room surrounded by computer screens. The room also contained a number of tables on which other humans lying. She was scared and confused, but one of the aliens then telepathically reassured her, telling her not to worry because they wanted to help her. She was subjected to a number of psychological tests involving numbers and shapes. Then she found herself in her bedroom. She had been gone for three-quarters of an hour.

On the following night the aliens came for her again and conducted a series of tests on her. Their telepathic instructions were so clear that she got the impression that they were talking to her rather than implanting their thoughts directly into her mind. However, she sometimes found that they were going too fast and therefore begged them to slow down.

The aliens told her that they came from Zeta Reticuli (which is where Betty Hill's Star Map indicated that her aliens were from). They said that they abducted a lot of humans for testing, but usually chose those with psychic powers. To the aliens, humans were primitive creatures. They themselves had evolved far beyond humankind, but had lost their emotions due to cloning. They would be happy to exchange some of their advanced technology in order to get them back.

Hoscik was a secretive child and told no one of her multiple abductions. Her father was ill and her mother worked long hours to support the family. They did not notice that their daughter was disappearing in the middle of the night, even though she claimed that at one time she was abducted every night for 18 months. She explained that she did not tell anyone about her abductions because she was afraid that she would be called a liar. However, she had devised a plan with which to prove that she had been abducted by aliens. On one occasion she had taken book with her to show the aliens, hoping thereby to discover whether she could carry things to the spacecraft and back. When she found that she could she bought herself a camera and one night took some surreptitious photographs of the aliens at work. When she took the film to be developed, however, the pictures did not come out properly. Realising that she was now out of pocket, the aliens kindly left the exact amount that she had spent on the camera film and developing under her pillow.

Hoscik believed that the aliens benefited her as much as she helped them: while she taught them about feelings they boosted her intellect. Soon she was over-achieving at school, which caused her problems because she began to be bullied. In order to stop this, she consciously began under-achieving. This strategy had the added advantage of discouraging her nocturnal visitors, who did not appear when she was not being intellectually stimulated.

After leaving school she drifted from job to job. Like many alien abductees, she was unsettled and aimless. Eventually the abductions ceased altogether. Then, in 1995, she met Chris Martin. She knew that he was someone special because six years before the pair met the aliens had shown her a picture of him. Martin was a UFO enthusiast and a science-fiction buff. He asked her whether anything strange had ever happened to her. She said that it had and, for the first time in her life, admitted to being an abductee.

Martin wanted her to resume contact with the aliens, and she knew how to. Together they went back to the house in which she had been brought up. They took some intelligence tests with them. Martin was amazed at how quickly she got through them. This intense intellectual stimulation in turn attracted the aliens. But this time they did not abduct her, simply establishing remote telepathic communication with her and using her to channel alien information to Earth.

A number of aliens whom Hoscik called 'grey twos' channelled through her. She named them Antholas, Minnie and Crispin because she could neither remember nor pronounce their real names. Antholas told her that there were 16 different types of alien besieging the Earth. Although some of them were experimenting on humans, he said, they were also afraid that humans would conduct experiments on them, which was why they revealed themselves in such a covert fashion. The aliens reckoned that one in ten humans believed in them; when everyone was convinced of their existence then they might learn to trust humankind enough to reveal themselves fully. They abducted people so that they could get to know them on a personal level, after which those individuals would spread the word of their existence. At the same time, Antholas said, they were trying to extend the capacity of the human brain as much as possible, in order to enable humankind to approach their own level of learning.

Minnie was a younger alien who had been entrusted to Hoscik for the sake of her emotional education. Older aliens told Hoscik that Minnie was her child, who had been fathered by an alien. At first, Hoscik refused to accept this, but then came to believe that she did indeed have a hybrid child which had been taken from her. She consequently severed all contact with the aliens, destroyed all records of her communications with them and also broke up with Martin.

A Family Tormented

British UFO investigator Tony Dodd was approached by Darren and Tracey Jones, a young couple with four children in 1997. Tracey was psychic. From childhood she had had the sensation that she was being watched and she also had a phobia about being grabbed by unseen hands in the dark.

The family had been bothered by odd, abduction-related events for years. In 1993 they were living in a cottage in Yorkshire. One evening, at around 9pm, they had heard a strange, humming noise coming from outside. Suddenly the cottage was bathed in light. The experience seemed to last only for a few seconds, although the clock said that a quarter of an hour had passed. The following morning they found that the padlocks on the stable doors had been forced and that the wrought-iron gate had been buckled and twisted.

Tracey had had a chest operation. The wound had become badly infected and she wore a bandage that had to be changed daily. One morning she awoke to find that the bandage was gone; they could not find it anywhere. But the pain from the wound had also vanished – as had the infection. On another night the underwear that she had been wearing underneath her pyjamas similarly went missing and never turned up.

They began to think that the house was haunted. Darren saw strange figures in the night. Other inexplicable things also happened and the electrical appliances and plumbing system in the house seemed to have a mind of their own. No one else who had lived in the house previously had reported experiencing such occurrences, however.

When the Jones family moved to another old house, overlooking the Yorkshire moors, they were plagued by UFO sightings and bizarre, 'missing-time' experiences. Two-year-old Marcus began having nightmares. Tracey saw strange men in monks' habits standing at the end of her bed and awoke during the night to find herself paralysed. In the morning, though, she was perfectly fit, except for the strange marks that she and the rest of the family developed on their bodies: they were bruises that looked like finger marks.

In 1997 Darren got a job in Dubai and initially went out there alone. It was a peaceful time for them, with no odd occurrences. But when Tracey and the rest of the family arrived in Dubai everything became as crazy as before. The electrical appliances again went haywire and the family had more 'missing-time' experiences. Marcus had the idea that he could fly; he said that a man lived in his bedroom light and took him and his sister, Georgina, up into the light to play. One night Georgina saw a small man surrounded by light standing at the foot of her bed. The following day she had finger marks on her body. When she drew the man whom she had seen, Marcus recognised him as being the man who lived in the light. By now the man had become malevolent, however, and after Marcus complained that the man was hurting his tummy he had to sleep in his parents' bedroom.

After they returned to Yorkshire the man in the light

continued to bother Marcus. The youngest child then began having nightmares, which became associated with the marks that were now being found on his body, too. Strange things happened around the house and the whole family heard unexplained bangs during the night, as well as unfamiliar strange voices; Tracey thought that she recognised the voice of a stillborn child that she had lost years before.

Like many women abductees, Tracey suffered from nosebleeds. Under hypnosis, she recalled that the two men that she had seen wearing monks' habits had taken her to a round room, where she lay on a bed, unable to move even though nothing seemed to be restraining her. A strange apparatus was being used to operate on her chest.

Dodd believed that Tracey – and probably also the rest of the family – had been abducted many times, but put his investigation on hold until the children were old enough to undergo regressive hypnosis.

Psychic Connections

Luli Oswald was the stage name of the Brazilian concert pianist Margarida Henriquieta Marchesini. The mother of seven and a charity worker, she was also a well-known psychic.

On the evening of 15 October 1979 she was driving from Rio de Janeiro to Saquarema with a 25-year-old male student. It had been raining earlier, but by then it was about 9.30pm and the skies had cleared. As they drove along the coastal road the topic of UFOs came up in conversation and Oswald asked the young man what he would do if he saw one. At that moment they noticed a dome-shaped object, with three lights on it, out to sea. It could not be a

UFO, Oswald thought – that would be just too coinciden-
tal. But the lights were keeping space with them, so, in
order to be on the safe side, they turned off the coastal road
with the intention of following an inland route. However,
once in the hinterland they lost their way.

The student, who was driving, took a wrong turning
and they found themselves on the coastal road again.
Once on it, they had no sensible option but to continue.

By now it was about 11.30pm and they were in a hurry
to get home. Suddenly they saw the lights again: this time
they were rising out of the sea, pulling a huge column of
water with them. At that point the car's engine cut out.
They were terrified, but worse was to come. Above them,
on the cliff tops, was another UFO – this one was long and
pencil shaped, with orange windows along its side. It sent
three balls of dazzling, white light hurtling down the cliffs
towards them. They panicked and were unable to decide
whether they would be safer in the car or underneath it. It
made no difference, as they both blacked out.

The next thing that they knew the danger had passed.
They were still in the car, but it was now in a side road
leading to a farm a little further down the main highway.
They set off again, both of them feeling the need for something with which to settle their nerves. Oswald knew of a
petrol station that sold coffee; when they got there, however, it was shut. It was already 2am.

Oswald experienced some bad after-effects following
that night. Her eyes burnt for a couple of days. She was
unable to pass water, and when she finally did she felt a
burning sensation. There was something wrong with her
watch, too: it gained minutes every day, rendering it virtually useless. Unsettled, she went to her local priest, who
put her in touch with the Brazilian UFO investigator Irene
Granchi. A US ufologist was also visiting Brazil at the
time, and he checked out the Fiat 147 that they had been
driving. It was heavily magnetised down one side.

Oswald then agreed to undergo regressive hypnosis.

Under hypnosis, she remembered that they had come face to face with the UFO. It had bathed the car in a beam of light, which had lifted the vehicle off the ground and had pulled it into the spaceship. She next recalled being in the room, surrounded by ugly creatures who were 4 feet 6 inches (1.35m) tall, with grey skins, pointed features and long, thin arms. They removed her clothes and examined her, using strange lights whose beams hurt her eyes. They then took samples of her hair and gave her a complete gynaecological examination. However, they telepathically told her that she was of no physical use to them (perhaps because she was then past child-bearing age); she had been chosen, they said, because she was psychic. For the purposes of their research they were interested only in the young man. She saw him lying unconscious on a table; the aliens were examining his sexual organs and taking samples of his semen.

The aliens then volunteered a curious piece of information: they told her that they were from a 'small galaxy near Neptune'. Oswald was an educated woman and knew that this made no sense – to a human mind, at least.

Tall Ships

On 15 August 1992 the transatlantic businessman Harry, along with Janet a computer-banking executive, went to see the tall-ships' parade at the Albert Dock in Liverpool. After that they went back to Janet's flat, where they spent the night together.

At around 3am Janet awoke in a state of absolute terror. She was convinced that she had not had a nightmare, but had actually been somewhere in her sleep; the moment before she woke up she had had the sensation that she had been placed back into bed. Janet woke Harry, who thought that he saw something in the bedroom.

Janet gradually began to recover some memories of what had happened. She remembered lying down in a brilliantly lit room. There were a number of creatures standing around her. (She could not picture their faces and believed that her mind was somehow blocking out the details). The beings gave her something to calm her down, which paralysed her. After that they started to so something to her – she did not know exactly what – which she did not like.

The next incident that she recalled was being in a room with a strange man who had compelling eyes. He seemed very knowledgeable and began telling her things. She could not remember precisely what, but had the strange sensation that he had talked to her before.

Then she found herself in a corridor with a man whom she had known when she was 12. He had been very bright in those days, but now appeared demented. She was taken into the room that he had just left and was terrified that she might come out of it in the same state as him. She was reassured, however, to see a human being in the room who seemed to be a kindly doctor. He explained that they were going to have to operate on her in order to put something right. She put up a fight, but the next thing that she remembered was being back in bed.

Janet was convinced that she had conceived a baby that night. She had been trying to become pregnant for some time, but without success until then. She also came down with all sorts of strange ailments including headaches, lethargy, sore eyes and blocked sinuses. For his part, Harry had a strange mark on his leg, which quickly faded.

One night both Janet and Harry dreamt that when they had been watching the tall ships on the quay they had been beamed up into a flying saucer, which then flew off. In the morning they ran to the bathroom and vomited.

Janet's ill-health persisted until, on her doctor's advice,

she had the pregnancy terminated, whereupon she quickly got better. Then, on 15 November 1992, she again awoke at 3am. Three of the creatures whom she had seen during the original incident were in the bedroom. She tried to scream, but could make no sound. She believed that the aliens were angry with her because she had aborted the baby.

Under regressive hypnosis, Janet recalled having seen small, hairless 'playmates' in her room when she was a child. They had scared her. She also remembered having flown up to a special 'playroom', which had taken her to the places that she wanted to visit. She had furthermore been shown pictures of the Earth and had been told that it was in danger. Other pieces of wisdom had been imparted to her either from boxes or from a tall man, with whom she had been left alone and who had implanted something into her, she believed. Both she and her mother had suffered from terrible nosebleeds, which had persisted during Janet's life. She also told of 'missing-time' experiences, which also seemed to have affected her brother and his girlfriend.

Harry had seen a UFO once, he said. He had also had a vivid dream about being strapped down onto a table and being examined; he had felt his head being cut open. He later saw a girl whom he recognised from his class at school being operated on.

Both Janet and Harry said that they had been followed by 'men in black' (who often appear to witnesses or victims of alien abductions).

The Swedish Experience

On 23 March 1974 a man named Anders left a political celebration in Vallentuna, Sweden, and set off to walk home to Lindholmen, a distance of about 3 miles (4.8km). He had had a few drinks, but claimed that he was still sober.

It was a moonlit night and he decided to take a short cut over a hill. As he neared the peak he saw a light coming towards him; thinking that it was a fast-moving car, he leapt off the road. He slipped down the embankment and landed on the grass at the bottom. Then he looked up, he realised that what he had seen was not a car at all. The light was hovering directly above him.

The next thing that he remembered, he was outside his own front door. His wife answered his frantic ringing of the doorbell and found him bleeding from a wound on his forehead. His cheeks were also burnt.

The next day, he called the Swedish ministry of defence, which put him in touch with a UFO investigator. Under hypnosis, Anders recalled that when he had flung himself from the road he had not actually hit the ground – some unseen force had lifted him into the air and had transported him upwards, into a spaceship. The craft had been manned by four 'semi-transparent' aliens. They had the features of Indians, he said, although they possessed no noses or ears. In fact, they could have been wearing hoods, he thought. Anyway, they were surrounded by an eerie glow. They had approached him with an instrument that he did not like the look of. He had tried to fight them off, but they had overpowered him and had placed the instrument against his forehead, whereupon he had experienced a burning sensation. Then they had dropped him off outside his home.

In this case there was no missing time, but a woman

cyclist had seen a strange light at the time that Anders said that he had been abducted, as had a courting couple who were out for a drive. They thought that it was a light shining from the window of a water tower, but there was no water tower in the vicinity. On the following night more strange lights appeared in the sky. People watching television reported interference; there was also something wrong with the phones.

In common with other abductees, his experience subsequently gave Anders the sensation of 'oneness with the world'. He felt a pricking feeling inside his head, too – at the spot where the aliens had placed the instrument – and suspected that they had implanted something inside his brain. Anders furthermore began to exhibit paranormal talent. He was sensitive to magnets located 1 foot (30cm) away, while rock crystal placed at a distance of 6 or 7 feet (1.8 or 2.1m) from him elicited a sucking feeling in his head.

A year after his abduction Anders had a dream, in which he was told to 'search in yttrium'. Anders thought that yttrium was a place; it is, however, a rare metal that is used as a component in colour television screens. UFO investigators discovered that Anders was sensitive to yttrium, especially along certain notional lines within his body that were similar to those delineated in acupuncture charts.

One of Anders' new psychical abilities was the ability to dowse. He accordingly dowsed the abduction site and discovered that a number of ley lines converged on it. When a chart recording Anders' biorhythms was analysed it was found that he had been abducted at the exact moment when his three fundamental energies – emotional, intellectual and physical – had peaked simultaneously. This triple maximum is believed to occur only once every 46 years.

A Change For the Better

John Day laughed at UFO stories ... until the night he drove into a strange green mist and lost three hours of his life. It happened in December 1978 as Day, a 33-year-old father of three, and his wife Sue, a 29-year-old nursery nurse, were returning to their Essex home after visiting her parents in Harold Hill.

Normally the journey took 30 minutes. They had set out at 9.20 pm. But when they arrived, the clock on their mantelpiece showed 12.45 am. In the days that followed, both had recurring dreams of being on treatment tables and undergoing examinations by strange beings. The nightmares became so vivid that they were afraid to go to bed.

Finally Day contacted a UFO group, who introduced him to dentist and hypnotist Leonard Wilder. Under hypnosis, Day's subconscious revealed an amazing tale.

A white light had followed the car, landed in a field beside their route, and beamed them and the car aboard what seemed to be a spaceship. Day said he found himself in a giant room, standing beside three aliens, all 7 feet tall and wearing silver-grey one-piece outfits which looked like body stockings. Balaclava-type hoods covered the bottom half of their faces, and they stared at him with bright pink eyes that had no eyelids.

'I found I knew what they wanted me to do,' Day said. 'I think they communicated with me by telepathy. I knew they wanted me to walk across the room, so I did. There was a doorway leading to another room. We all went through it.

'This was obviously an examining room, and they asked me to lie on what looked like an operating table. A metal arm swung over me, scanning my body. Then three other beings, squat and ugly like dwarfs, appeared. One started

78

to prod me with a pen-shaped object.

'After a while, the examination seemed to be over. I asked if I could look round the ship, and they agreed. All the furniture was moulded to the wall. On one table, I saw a pile of cubes with magnets on them. They looked like some sort of game.

'At the end of the tour, the beings left me alone in another room. Suddenly an incredibly beautiful woman walked in ... she had golden hair and was surrounded by a sort of grey mist. She walked towards me, but when I took a step towards her, she vanished. Next thing I knew I was back in the car, driving along the road.

'I had never really believed in UFOs before this happened. Now I'm convinced aliens are here and only show themselves when they want to. The ones I met kept telling me they were friendly. I enjoyed meeting them immensely.'

Hypnotist Wilder said: 'I have no doubt that Mr Day is telling the truth – when I first hypnotised him, I conditioned him to tell only what really happened.'

Wife Sue declined to be hypnotised – she said she did not want to relive the experience. But later, discussing her husband's statements, she recalled something of what had happened to her.

She told John Clare, a reporter for the *News Of The World* newspaper: 'When I lay on the operating table they painted me with a mauve liquid. Then they washed it off. They prodded me all over with a pen-like object and didn't spare my blushes. Then I screamed.

'One of the tall beings came over and put his hand on my forehead. I went out like a light. Later they took me on a tour round the ship. They showed me a screen and said, "This is Earth." They pointed out England on it. Then we seemed to zoom in and they showed me where I lived.

'I told the beings I didn't want to go back. I asked if I could stay on the craft and they agreed. I saw John climb into the car and it started to vanish. As it disappeared, I said I had changed my mind, and wanted to go back. Then I found myself sitting in the car.'

British UFO investigator Barry King said: 'We have made exhaustive inquiries and are convinced that these two did indeed have a close encounter of the third kind. Some of their descriptions are similar to those given in other cases of abduction. We can find no reason to doubt the authenticity of their story.'

Abductions
USA

A Fishing Trip

The 18-year-old Calvin Parker and the 45-year-old Charles Hickson were fishing from a disused pier on the Pascagoula river, in Mississippi, on 11 October 1973. First they heard a strange, buzzing sound and then they saw an egg-shaped spacecraft 10 feet (3m) wide and 8 feet (2.4m) high that was emitting an eerie, bluish-white glow and was hovering above the river some 40 feet (12m) from the bank.

The craft landed and three aliens floated out (they had legs, but did not use them). The creatures were about 5 feet (1.5m) tall, with odd, conical protrusions instead of ears and noses and slits where their mouths would have been. They had claws for hands and horrible, grey, wrinkled skin (although the two men said that this may have been a uniform).

Two of them grabbed Hickson and the third seized Parker, who fainted. The two humans were then taken aboard the spacecraft, where Hickson was given some sort of medical examination, which involved being scanned by what seemed to be a large eye that floated in mid-air. Hickson was consequently paralysed, except for his eyes. Parker, he assumed, was being examined elsewhere. After about 20 minutes Hickson was dumped unceremoniously back on the pier, where he found Parker crying and praying nearby. The craft rose up vertically and shot out of sight.

At first Hickson and Parker kept quiet about their abduction, being afraid that they would attract ridicule. Then they decided that the government ought to know about it and accordingly called Kessler Air Force Base in Biloxi; a sergeant there told them to speak to their local sheriff. After going to the local newspaper's office, which they found closed, they went directly to the sheriff's office to report the abduction. They both gave a detailed description of the incident to the sheriff, Fred Diamond, without being hypnotised. Diamond then left them alone in his office for some time and eavesdropped on them via a hidden microphone, thinking that if they were hoaxers they would use the opportunity to go over their story together once again, or else to have a laugh at the sheriff's expense. Instead, however, he heard Parker praying and then telling Hickson that he wanted to see a doctor. Hickson himself was evidently awe-struck by the power of the craft that he had seen. Neither gave any hint that they did not believe everything that they had reported. 'If they were lying they should be in Hollywood', said Diamond later.

The two were subsequently subjected to lie-detector tests conducted by the Pendleton Detective Agency in New Orleans. The sceptical machine operator, Scott Glasgow, was determined to prove that they were lying. Instead, however, they easily passed. 'This son of a bitch is telling the truth', Glasgow exclaimed afterwards.

They were also interviewed at Kessler Air Force Base and their case was investigated by Dr J Allen Hynek, of the USAF's Project Blue Book, too. Hickson submitted himself to regressive hypnosis under Professor James A Harder, of the University of California, but the session had to be stopped when Hickson became too distressed to continue. He nevertheless handled the aftermath of the abduction well; by contrast, Parker had a nervous breakdown and required psychiatric care. 'I had to learn to accept what had happened', said Hickson. 'I saw what happened to a man who could not accept it. This thing almost destroyed his life.'

Unlike other abductees, Hickson did not write a book about the incident, nor did he try to exploit his experience in any way. 'I had a chance to make a million dollars', he explained. 'I was offered all kinds of money to let them do a movie. I declined. I am still declining. Making money is not what this experience is all about.'

Travis Walton

On 5 November 1975 the twenty-two-year-old Travis Walton was one of a group of seven men who were felling trees on Mogollon Rim, in the Sitgreaves-Apache National Forest near Snowflake, Arizona. At around 6pm they finished for the day and were driving home in a lorry belonging to the foreman, Mike Rogers, when they noticed a strange light in the trees.

As they rounded a corner they saw a glowing, disc-shaped UFO hovering about 15 feet (94.6m) above the trees. The lorry screeched to a halt and Walton leapt out to investigate while the other men stayed inside. As the intrepid Walton walked towards the UFO it began spinning and emitting an electronic, beeping sound. A beam of light suddenly flashed from it, hitting Walton in the chest and flinging him 10 feet (3m) backwards. His co-workers then sped off in the lorry, leaving him lying unconscious on the ground; it was not until they had seen the UFO lift off and fly away that they plucked up the courage to return. When they did, Walton was nowhere to be seen.

Ninety minutes after the abduction they reported Walton's disappearance to Deputy Sheriff Chuck Ellison. He organised the men into a search party, but three of them refused to go into the woods again that night. 'One of the men was weeping', said Ellison. 'If they were lying they were damned good actors.'

For the next two days they made a detailed search of the forest, but found no sign of Walton. Rumours began to circulate that the loggers – who included Walton's brother,

Duane – had murdered the missing man, buried his body in the forest and had then invented the UFO-abduction story to cover their tracks. In order to clear their name the loggers insisted on taking lie-detector tests, which the Navajo County sheriff, Marvin Gillespie, organised on 10 November. The tests were conducted by Cy Gilson, of the Arizona State Office of Public Safety. Although one man was too worked up to enable the test to be declared valid the other five passed. 'I think they did see something they believed was a UFO', said Gilson.

On 11 November Walton's sister received a call from a pay phone. It was her missing brother, who asked her to send someone to fetch him from a nearby petrol station. He was subsequently found lying on the floor of the phone booth in a highly confused state.

When he had calmed down sufficiently he was interviewed with the aid of hypnosis. The first part of his story matched those of his co-workers. He had felt no fear, he said: when the lorry had drawn to a halt he had been excited and had jumped out to run towards the glow. 'Then something hit me', he said. 'It was like an electric blow to my jaw and everything went black ... When I woke up I thought I was in hospital. I was on a table on my back and as I focused I saw three figures ... They weren't human.' Walton said that the aliens looked like foetuses. They were about 5 feet (1.5m) tall, with large, hairless heads, huge, oval eyes, small ears and noses, slits instead of mouths and skin as white as mushrooms. They wore tight-fitting, tan-brown robes.

The terrified Walton had grabbed a transparent-plastic tube and had lashed out, but the aliens had scampered away. Then a smiling man – whom Walton took to be a human being – had appeared; he was wearing a helmet shaped like a fishbowl. He had led Walton into a hangar, where a number of smaller, disc-shaped craft were parked. Three other human-looking creatures had met him there and had taken him to another examination table. A mask had been put over his face and he had then blacked out.

'When I woke again I was shaky. I was on the highway', he said. 'It was black, but the trees were all lit up because just a few feet away was the flying saucer.' The UFO climbed up into the sky and this time Walton ran away from it. 'I recognised [that] I was in a village a few miles from my home in Heber', he recounted. 'When I found a phone booth I called my sister.'

Walton thought that he had been gone for only a few hours and could not explain where he had been for the missing five days. He was examined by psychiatrist Dr gene Rosenbaum, of Durango, Colorado, who concluded 'This young man is not lying. There is no collusion involved. He really believes these things'. Walton also took and passed a lie-detector test which was administered by George J Pfeifer. Leading UFO experts endorsed his claims, while the *National Enquirer* awarded Walton and his co-workers $5,000 for the most impressive UFO story of 1975 (the paper's star prize was $100,000 for concrete evidence of at least one extraterrestrial visitor).

However, the UFO sceptic Philip J Klass was not convinced. He conducted an investigation of his own and discovered that Walton, Duane and their mother, Mary Kellett, were avid UFO buffs who frequently reported seeing UFOs. Walton's abduction had also occurred just two weeks after a dramatisation of the abduction of Betty and Barney Hill was aired on US TV.

Klass noted that Mike Rogers and Duane Walton had been interviewed by a UFO investigator while Walton was still missing, during which they had shown no concern for his well-being, even though they claimed to have seen him being zapped by a massive electric-shock. They were both convinced that he would return unharmed. Duane even expressed a 'little regret because I have not been able to experience the same thing'.

Duane and his brother, Klass discovered, had regularly discussed making contact with aliens – they had seen enough UFOs to believe that this was a real possibility. They had agreed that one should avoid moving directly underneath a UFO, but decided that the opportunity to climb aboard one would be too great to pass up. They had even made a pact that if one of them was abducted by aliens he would try to persuade them to come back and pick up his brother. Alien abduction was such an ever-present possibility for Travis Walton that he had told his mother shortly before he went missing that if he should ever be abducted she was not to worry: he would return safe and sound. And he was right: although he was groggy when he was discovered, he had no burns from the lightning-bolts with which the aliens had zapped him.

Klass also discovered that before Walton had been polygraph-tested by Pfeifer he had undergone another test with the highly respected Jack McCarthy. He had flunked the first one and McCarthy had accused him of 'gross deception, as well as of resorting to such tricks as holding his breath in order to influence the outcome. McCarthy also suspected that the UFO investigator who had interviewed Walton while under hypnosis had planted some suggestions into his mind. The *National Enquirer*, which had paid for the polygraph test, had persuaded McCarthy to sign a hastily prepared secrecy agreement, but it had been dated wrongly, rendering it legally invalid.

Klass furthermore uncovered the fact that Walton had pleaded guilty to charges of burglary and forgery five years before his abduction. His co-defendant was Charles Rogers, the brother of the loggers' foreman. The Walton family also had a reputation as being practical jokers and hoaxers and Klass even unearthed a possible motivation for such a hoax: the logging crew was behind a schedule and Mike Rogers faced paying out on a penalty clause in his contact unless an 'act of God' – which a UFO abduction may have been interpreted as – prevented him and his men from completing the work on time. There was, of course also the prize money that the *National Enquirer* was offering.

In contrast to Klass, such respected UFO experts as the USAF's Dr Hynek tend to believe Walton. 'It fits a pattern', he told ABC-TV's 'Good Night America' programme in 1975. 'If this were the only case on record then I would have to say, well, I couldn't possibly believe it. But at the Center for UFO Studies now we have some two dozen similar abduction cases currently being studied. Something is going on.'

The story was made into the film *Fire in the Sky*, starring James Garner. 'If I hadn't believed Travis Walton's story implicitly after talking with him for many hours, I wouldn't have touched the project with a bargepole', Garner said.

'Dissected Like Frogs'

Shortly after the screening of '*The UFO Incident*' – the made-for-TV movie about the Hills' abduction, which was aired for the first time on 20 October 1975 – a young woman from North Dakota named Sandy Larson reported that she, her boyfriend and her young daughter had been abducted by aliens two months before. They had been 'stripped naked and all parts of our bodies examined'. But the aliens did not stop there: 'Even our heads were opened up and our brains looked at', she said. 'We were dissected like frogs.' Apparently their dissections had no ill after-effects and just a few hours later the three abductees were delivered home as right as rain. They did not even have any scars with which to substantiate the extraordinary tale.

Webbed Invaders

The 21-year-old David Stephens was also abducted by aliens just a week after the Hills' movie was aired and described the incident in the July 1976 issue of *Official UFO* magazine. He said that he had been driving in his car with an 18-year-old-male friend at the time. Only Stephens had been abducted, however, and his friend had waited for him in the car.

Under hypnosis, Stephens described the aliens as being about 4 feet 6 inches (1.35m) tall. Their heads were shaped like mushrooms, their noses were small and flat and they had no visible mouths or ears. Unusually, they wore flowing black robes and had white eyes. Stranger still, they possessed webbed hands with only three fingers and a thumb.

Late Invaders

Following the abduction of Travis Walton, Air Force Sergeant Charles Moody came forward. He claimed to have been abducted several months earlier, on 13 August 1975. The aliens had spoken perfect English with an American accent, he said, although they did not move their lips. One of them had assured him that 'within three years his people will make themselves known to mankind'. At the time of writing they are more than 20 years overdue – although it must be conceded that they may not necessary have meant Earth years.

Alien Aliens

Steve Harris and Helen White were also abducted from a small town in northern California during he mid-1970s. They recalled the incident while being hypnotised by Dr James A Harder, who was by then working at the Aerial Phenomena Research Organization. Harris said that the alien responsible spoke in English with a slight German, or Danish, accent. He also described the creature, who appeared to be 'human, except that his ears and mouth were slightly smaller and he was fluorescent-looking'. White said that he was blond-headed fellow, with wavy hair and wearing a long kind of thing that looked like a raincoat.'

An Alien Examination

On the night of 11 June 1976 the 28-year-old Hélène Giuliana, who worked for the mayor of Houston, Texas, as a maid, was coming home from the cinema when her car broke down on the bridge at Romans. She then saw a huge, orange glow in the sky. After it had vanished she found that she was able to drive on, but when she got home it was 4am – several hours were unaccounted for. Giuliana's missing time came to the attention of UFO investigators, and under hypnosis she recalled having been carried into a room by small aliens with big eyes. They had strapped her to a table and had examined her, paying particular attention to her lower abdomen.

The Influence of the UFO Incident

The stories related above represent just a handful of the alien-abduction reports that were made following the screening of the NBC-TV movie *The UFO Incident*. Indeed, over a hundred were reported during the two years after its airing. There had been only 50 similar reports filed during the previous 30 years; the aliens, some believed, were thus clearly intensifying their activities.

Kidnapping a Cop

On 3 December 1967 Patrolman Herbert Schirmer was patrolling the outskirts of Ashland, Nebraska, when an eerie feeling that something was wrong came over him. Seeing a bull charging at a gate, he stopped to make sure that the gate would hold. Later, near Highway 63 – the road that leads to Ithaca and Swedenburg – he noticed a light ahead of him. Believing it to be a truck, he flashed his headlights at it, whereupon the light shot up into the night sky. When he returned to the precinct at the end of his shift he reported seeing a UFO. After he got home he felt ill; a buzzing sound in his ears stopped him from sleeping and there was inexplicable red welt on his neck.

Ufologists are always interested in sightings by reliable witnesses, such as policemen, and Schirmer's case came to the attention of the USAF-sponsored UFO-investigation team at the University of Colorado. Reviewing his report, the team members observed that there were 20 minutes missing form his police log, and in order to find out what had happened during the missing time they suggested that Schirmer undergo regressive hypnosis.

While hypnotised, Schirmer recalled having seen an

alien spacecraft coming in to land and its landing gear tele-
scoping from the bottom of the craft. He had tried to drive
away but his police vehicle could not start – either that, or
something in his mind was preventing him from driving
away. He therefore sat there helplessly while a hatch
opened and the aliens got out of the craft. They were
between 4 feet 6 inches (1.35m) and 5 feet (1.5m) tall and
wore tight-fitting, silver suits. Their hands were covered
by gloves, their feet by little, silver boots, while on their
heads they wore thin, silver hoods, like an old-fashioned
pilot's helmet. Each had a small antenna over its ear. Only
their faces were left uncovered, and Schirmer noted that
their skin was light grey in colour, their noses were flat,
their mouths mere slits and their eyes opened and closed
like the shutter of a camera.

As a trained cop, Schirmer instinctively went for his
gun, but found that it would not budge from his holster.
The aliens then surrounded him. There was a flash, and he
passed out. On board the alien ship, the creatures communicated with him in broken English, both vocally and
by means of telepathy. They explained that they had been
studying human languages.

They told him that their ships were vulnerable to the
ionisation caused by radar. They came from a neighbouring galaxy, but had hidden bases on Earth; one was under
water, near Florida. They then showed him a film, which
incorporated a shot of three alien 'warships' flying against
a background of stars that included the Plough. They had
come to Earth to study human behaviour, they said.
Unfortunately, despite their advanced state, they survived
by stealing electricity from overhead power lines, but in
such small quantities that the power companies did not
notice.

Before he was released, the aliens' leader looked deep
into his eyes and implanted the suggestion into his mind
that he would not reveal what had happened on board the

ship to anyone. Schirmer was next told precisely what to say in his report of the incident. The alien leader also said that they would return and visit him twice more.

Schirmer submitted himself to a polygraph test, which he passed with flying colours. Other psychological tests were carried out and Schirmer was given a clean bill of health. He was regressed again by Dr Leo Sprinkle, a professor of psychology at the University of Wyoming, who discovered more hidden memories. But still the team at the University of Colorado would not accept Schirmer's case as having been a genuine abduction, because although he genuinely believed that he had been taken on board a UFO by aliens, he could present no further corroborating evidence.

The Nightly News

One night in the autumn of 1973 Mike Bershad was driving home to Baltimore down Route 40. He had been visiting his girlfriend in the nearby town of Frederick, Maryland. On the road, he had suddenly been overwhelmed by the feeling that he was being watched. He could not remember whether he had seen a light in the sky or whether he had pulled the car over, all he could remember was that he had wished that the car could have gone faster. When he had arrived home it had been much later than he had expected.

Bershad was so disturbed by whatever it was that had happened to him that he sought out the one-time abstract-expressionist artist and full-time uflologist Budd Hopkins, who arranged for Bershad to undergo regressive hypnosis. During the session it was discovered that Bershad had indeed pulled over: he had parked on the hard shoulder and had got out of the car. Some black-clad figures, with chalky, putty-like skin, had then emerged from behind a fence. The aliens had next placed a brass clamp over his shoulder that had held his head in place; it was connected to a flying saucer. Bershad had then been led up a ramp into the craft, which gave off a quiet, humming sound. He had walked around inside for some time before finding himself lying on a table in a round, white room. He was wearing some kind of nappy. A round, metal bulb, which was extended on an arm from the ceiling, seemed to soothe him. All sorts of blades and other instruments stuck out of it and one of them dug into his back painfully. Then the aliens had pushed a tube into his stomach. Stirrups grabbed his thighs, before lifting and spreading his legs, while an alien examined what was between them.

Hopkins wrote the story of Bershad's abduction and it made the NBC television channel's nightly news.

A Rash of Abductions

In 1974, after seeing a TV series about UFOs, a young woman (who is known in the literature of ufology as 'Sara Shaw') approached the UFO investigator Ann Druffel. She reported a strange sighting that had occurred over 20 years before, when she was 21 and staying in a cabin in Tujunga Canyon, California.

On the night of 22 March 1953 Shaw had been woken by a bright light, which she first took to be a motorcycle's headlight. But the light was in the sky, so she looked at the clock to check whether the time was approaching dawn. Shaw was with her 22-year-old friend, 'Jan Whitley'. Whitley got up and went to the wardrobe to get a dressing gown. Shaw looked at the clock again – 20 minutes had passed and Whitley was still standing by the wardrobe. Then Shaw realised that she had read the clock wrongly:

two hours and twenty minutes had actually passed. Shaw had initially been kneeling on her pillows while looking out of the window, but now she was sitting on her bed. She had a sense that something strange and terrible had happened.

Druffel arranged for Shaw to undergo regressive hypnosis. Under hypnosis, Shaw remembered looking out of the window and seeing a group of individuals walking through the yard. They entered the house and Whitley fought them while Shaw watched. She recalled being inside a dome-shaped chamber inside an alien craft. Whitley continued to struggle while the aliens tried to take off her pyjama top. The next thing that Shaw remembered was that she was floating above a table, undergoing a physical examination. The aliens used some sort of machine before examining her with their hands. They seemed particularly interested in an old, surgical scar. She also felt that in some mysterious way they had marked her with an 'invisible number four'. The two young women were then floated back to their bedroom. They were told that they would forget everything that had happened to them.

There was one obvious way in which Shaw's story could be corroborated, and UFO investigators accordingly tracked down Whitley. She had not forgotten the abduction: it had come back to her in 1956 in a series of vivid dreams. Curiously, Whitley had another friend, Emily Cronin, who reported an abduction experience that had happened in 1956 that had involved her and Whitley, but not Shaw. They had been travelling in a car together when they had pulled into a lay-by. They had seen a bright light, which then paralysed them. Cronin recalled having been watched by someone through the back window of the car. She had eventually managed to move her finger, which had broken the spell.

Whitley revealed that she had had another, related experience in 1967 or 1968, during which she had been unable to move and had seen the faces of the aliens whom she and Shaw had first encountered in 1953 floating above her. She also reported that she had seen her first UFO during the early 1940s, when she was still in her early teens.

Similar events took place in nearby Panorama City, California, during the 1970s. In 1970, for example, Lori Briggs awoke to find herself paralysed. She then realised that she was not alone in her flat. Furthermore, her nocturnal visitors were not human: they were short creatures, with long, thin hands. While lying on her back, unable to move, she found that she was looking directly into the eyes of an alien.

Five years later Briggs had been awoken in the same way; this time the creatures had told her to come with them. Under hypnotic regression, she recalled having woken to find the aliens standing next to her bed. They had large, hairless, egg-shaped heads – so large that their thin bodies could not have supported them under ordinary circumstances. They had small noses and mouths, but big eyes, which seemed to have bright light shining from them. When they had asked her to come with them she had not wanted to go, but had found that their will was stronger than hers. The creatures could hover in mid-air, and along with them she had floated through the walls to a dome-shaped craft. She had next found herself inside the alien ship, where there was an examination table which appeared to be made of pink stone. There was a very bright source of light under it, which she had believed to be some sort of X-ray apparatus. She had lain on the table while the aliens examined her. When that was finished, she had said that she wanted to go home and had pressed a button that opened the door. The aliens had courteously said that they would accompany her. Once back in bed, she had found herself paralysed again, but the next morning she had woken with only the memory of being incapacitated and of something staring at her.

However, on this second occasion Briggs had not been alone. She had had a girlfriend named Jo Maine staying with her. Under hypnosis, Maine remembered having seen a bright light and having had the sensation of floating. She was in a tube, she thought, surrounded by light. Then she was lying flat in the dark; some sort of examination was going on. When she was dressed again they had moved on somewhere else, before she had suddenly found herself back in bed, with Briggs lying beside her.

A Spaceman in the House

On 17 October 1973 Pat Roach, a divorcée who lived in Lehi, Utah, awoke. She knew that something strange had happened, but had no idea what it was until her two children – Bonnie, aged thirteen, and Debbie, aged six – told her that spacemen had been in the house. Roach did not believe them, of course, and was afraid that a burglar was on the prowl. She therefore called the police, who searched the house and garden and combed the area. They found nothing. However, Roach and her children were too scared to stay in the house alone, so they spent the rest of the night with friends.

Two years went by, but Roach and her children could not forget the incident, which was preying on their minds. So she wrote to the US magazine *Saga*, asking for help and saying that she just wanted to find out what had happened. Her letter mentioned spacemen, so she was put in touch with Dr James A Harder, the noted hypnotist who investigated alien abductions.

Under hypnosis, Roach recalled that two aliens had been standing near her bed when she awoke. They were small, with large heads, pasty, white faces and big eyes.

They wore silver uniforms, with a tight-fitting cap or helmet, gloves and a Sam Browne-style belt that carried a small pouch. They had touched her on the arms and had lifted her up. As they had carried her away she had seen her children trying to fight off other aliens in their room. One of them seemed to be a female wearing a maxi skirt, who had long hair which was held back in a headband.

The aliens seemed to have put no great effort into Roach's abduction, and she had apparently floated effortlessly into the waiting craft. She had not really seen the outside of the spaceship, but had merely glided up a stairway and through a hatch. Inside, she had been separated from her children before being stripped and placed on a floating table, where she was given a gynaecological examination. The aliens had showed off some of their technology to her. Then they had hypnotised her; they had appeared particularly interested in her life story and her emotions. Finally, they had given her clothes back to her and had told her to dress.

Under hypnosis, Bonnie, Roach's older daughter, said that she remembered her mother being examined by aliens. A balding human with greying hair was helping them. She had thought that he was a doctor and said that she had known that he was human because he had 'regular ears'.

Abducting the Righteous

Three women – all devout, born-again Christians – were abducted from their car near Stanford, Kentucky, on 6 January 1976. They had been returning from a birthday party to their home town of Liberty when they saw an intense, red glow in the sky to the east. The glow came closer, until it was hovering at tree-top level near the car. All three women saw a domed disc, which flashed three shafts of bluish-white light onto the highway.

The driver, 44-year-old Mary Louise Smith, stopped the car and tried to get out, but the 36-year-old Mona Stafford pulled her back. The lights on the spaceship then went out and the car raced down the road of its own accord. Smith could not slow it down – it was like it was being pulled by an invisible force. An instant later, they were deposited in Houstonville, 8 or 9 miles (12.8 or 14.5km) away. They then realised that they were missing some time and also noticed blistered paint on their automobile and burn marks on their bodies.

All three were regressed by Dr Leo Sprinkle. Smith had a memory block that he could not get past, while the 48-year-old Elaine Thomas only remembered having been observed by short creatures. They were 'like humans, except that they were only about four-and-a-half feet [1.35m] tall and they had fingers that looked like the edges of birds' wings – complete with feathers'. They had put a 'cocoon' around her neck that had choked her and had placed a bullet-shaped object above her left breast.

The regression of the youngest abductee, Stafford, yielded much more information. She remembered having been removed from the car and having been taken into a hot, dark room. Then she had found herself lying on a bed, with her arms clamped. Several small creatures surrounded her; they had grey skin and large, slanting eyes. The rest of their features were obscured by the surgical caps and gowns that they were wearing. A white light seemed to force her back as an 'eye-like' device examined her. Her eyes hurt; her feet were twisted and bent backwards; and she felt something 'blowing up her insides'. (This description would certainly be consistent with a gynaecological examination, but – perhaps because they were modest, Christian women – none of them mentioned this aspect of their abduction.)

All three women insisted on taking lie-detector tests, which were performed by Detective James Young, the senior polygraph operator for Lexington, Kentucky. He said 'It is my opinion that these women actually believe they did experience an encounter'.

Stafford subsequently recovered more memories. She recalled having been taken from one ship to another. She also remembered having been inside a cave or volcano. Later still, she reported another encounter with an alien: a creature dressed in a gold robe had issued telepathic commands to her, which she refused to obey at first. But she had then come to believe that she was under the control of the aliens.

Smith also reported a second encounter. She had woken one night seized with a strange compulsion to drive to Stanford, to the place from which they had been abducted. Once there, she had got out of her car and had stood there for a while, unable to leave. Suddenly, at about 3am, she had run back to her car and had driven off. She had then noticed that three rings were missing from her fingers and later remembered feeling a tugging at her hand while she had stood there waiting.

The Allagash Abductions

Jim Weiner, his brother, Jack, and two friends, Charlie Fotz and Chuck Rak, were on a canoeing trip to the Allagash Wilderness Waterway, Maine, on the night of 20 August 1976. Before they went out on to the lake, they built a large fire on the shore to act as a beacon so that they could find their way back to their camp. When they were on the water they saw a light in the sky that was coming towards them. Eventually a spacecraft hovered over them and the light that it beamed out played over the water around them. They observed it for a while before rowing frantically back to shore, where they stood and watched the alien craft fly off. By this time all that remained of the fire was embers; they had no idea how it could have burnt itself out so quickly. Just how long had they been on the water?

After the holiday all of the men had nightmares concerning strange creatures looking at them – they were particularly interested in the men's genital areas. Weiner, who was particularly disturbed by the dreams, met the ufologist and author Raymond Fowler at a symposium, who recognised in Weiner's story the telltale signs of an abduction.

Under hypnosis, Weiner recalled having been taken on board the ship and the aliens examining him. At first he could not remember their faces. (Fowler believed that aliens have special powers with which to induce amnesia.) During a second session of hypnosis Weiner recounted the bedroom visitations and other paranormal events which seemed to follow him around.

Fowler interviewed the other men who had been on the trip, too, who also remembered seeing the UFO. Under hypnosis, they recalled having been taken from the canoe. Once inside the ship, they had been stripped. Three of them had sat on a bench while samples of blood, skin and semen were taken from the fourth. The room reminded them of a hospital – cold and sterile. And they all talked about their bodies having been probed with medical instruments. After the examination was finished they had been allowed to dress and were then sent back to their camp. Next morning, they had had little recollection of the event. They did not talk about it – it was almost as if it had never happened.

Strangely enough, none of the men could initially remember the faces of the aliens, but after persistent questioning they finally came up with a description. The aliens were shorter than humans. They had large heads, which were out of proportion to their thin bodies. Their chins were pointed and their eyes large. Their legs and arms were spindly. And their hands had four fingers.

Abductions
UK

A Secret Weapon

Albert Lancashire claimed to have been the first British victim of an alien abduction. It happened in 1942 – during World War II – when he was on guard in a sentry box outside a top-secret radar base at Cresswell, just north of Newbiggin, in Northumbria. He was 27 years old at the time.

First he saw a light over the North Sea and then a strange cloud rolled in, he initially thought that he was being attacked by some new, German, secret weapon. As he stepped out of the sentry box to investigate he was hit by a beam of yellow light and felt the sensation of floating. The next thing that he knew was that he was waking up in a confused state and was lying on the ground.

He forgot about the incident until around 1963, when UFOs became all the rage, and he began to get the feeling that he had been in one. In October 1967 there was wave of UFO sightings in Britain and at around that time he also read of the Hills' abduction. This prompted a series of strange dreams in which he awoke in a strange room to find an oriental woman lying on a bed; he was given a pair of goggles and a man dressed in white was also there.

Swimming With the Dolphins

In January 1981 Linda Taylor, who was in her late twenties, was travelling home by car from Southport to Chorlton, in Manchester. At around 7.30pm she was driving around the East Lancs Road, which was usually quiet, when a huge light appeared in the sky. It appeared to be keeping pace with her car, which jerked about and then slowed right down.

An old car of a peculiar model that Taylor did not recognise then appeared in front of her. It was dangerously close, so she took her eyes off the light for a minute. When she looked back, the light had turned into a huge flying saucer that hovered over an area stretching roughly from Leigh to Worsley. The old car had mysteriously vanished from the road. Taylor pulled into a petrol station and pointed out the UFO to a man who was filling his car (the man was never traced). At that point the UFO tilted sideways and shot off skywards.

When Taylor reached home she felt unwell and furthermore discovered that her coat was missing; two hours of her time also seemed to be unaccounted for. She underwent hypnosis, but recovered no memories. However, she had a recurring dream following the incident. In it, she was in a room with a blond-haired, blue-eyed man wearing a white suit who took her to visit a dolphin, which he said was sick. Taylor was supposed to touch it and make it well again.

After her encounter Taylor became a UFO investigator.

A Rejected Abductee

Many are called, but few are chosen – or so the Bible says. When it comes to UFO abductions, however, it seems that many are called and many are chosen. However, one who was definitely not chosen was the then 77-year-old Alfred Burtoo, who, on 12 August 1983, was fishing on the banks of the Basingstoke Canal, near the army barracks in Aldershot. At around 1.15am Burtoo noticed a bright light land some way down the towpath. Two figures then approached him and Burtoo's dog growled warningly. The figures were aliens, about 4 feet (1.2m) tall and

dressed in pale-green overalls and helmets. They beckoned to him and he obediently followed them.

They led him down the towpath to a spacecraft, whose
shape was similar to a spinning top. He climbed a ladder
and entered it. Once inside, a voice told him to stand
under an amber light. Then he was asked his age to which
he responded truthfully. You are too old and infirm for our
purposes', the voice said. 'You may go.' So Burtoo clambered back down the steps and returned to his fishing.
When he looked back he saw the UFO start to glow; it
hummed like a generator and then sped off into the sky.

Burtoo died in 1985, but stuck to his story to the end.

The Photofit Abductor

On 1 December 1987 Philip Spencer, a former police officer,
was walking across Ilkley Moor, in Yorkshire, to visit his
father-in-law, who lived in a village on the other side of the
moor. It was a 5-mile (8-km) journey and he had left the
town of Ilkley at around 7.10am. A keen amateur photographer, he had his camera with him, intending to take
some pictures of the town from the top of the moor. He
also carried a compass in case he got lost.

Near an overgrown quarry he heard a low, humming
sound. At first he thought that it came from a low-flying
aircraft, but then a small, green creature approached him.
It immediately scuttled away while Spencer shouted at it.
As it stopped and turned he took photograph of it before
giving chase. On rounding an outcrop he saw a huge, silver UFO, which consisted of two flying saucers that were
connected together. The UFO was the source of the humming, which now increased in intensity. Suddenly it shot
skywards and disappeared into the clouds.

The dazed and confused Spencer retraced his steps to

Ilkley, which was, to his surprise, bustling. Although he thought that it could not be any later than 8.15am the town-hall clock said that it was 10am. Fearing that he was losing his mind, Spencer now realised that his camera contained proof of his sanity. He therefore travelled to nearby Keighley, where he knew that there was a one-hour photoprocessing service. The wait was agonising, and when he was finally given the pictures he immediately tore open the package that contained them. The quality of the photograph was poor, but the image was unmistakable: it showed a green creature with large ears, short legs and long, thin arms – in short, the alien that he had seen that morning.

Spencer was now in a quandary. He was trying to rejoin the police force and the last thing that he needed was for it to become generally known that he had seen an alien – the force would be sure to doubt his mental health or would possibly brand him a hoaxer. But because he was now convinced that he had indeed come into contact with an alien he wanted to know more. And he also wanted to discover what had happened during the 'missing time'.

He therefore contacted the Manchester UFO Research Association under a false name, using a post-office box as his address. When he eventually met the investigators he was happy to hand over his photograph, as well as the negative for testing. They quickly concluded that the negative had not been tampered with, that the picture was not a photograph of a photograph and that no photographic trickery been used. The photo was genuine.

Spencer also noted that his compass had been useless since the encounter, and the investigators found that its polarity had been reversed. In order for this to have been done, a pulsed, magnetic field must have been used; without the help of rare and expensive equipment the process of producing such a field is dangerous and possibly even lethal – assuming that you know how to do it in the first place.

Six weeks after the encounter Spencer was visited by men from the Ministry of Defence who wanted to take the picture away, but it was with the investigators at the time. He also began to have dreams in which he was lost amid a starry sky, and he started to wonder if the dreams had anything to do with his missing two hours. He consequently agreed to undergo hypnotic regression.

Under hypnosis, he revealed that he had been abducted. He recalled that when he had first seen the alien creature he had become paralysed. Then he had felt himself levitating and floating over the quarry. When he had seen the UFO a door had opened in its side. Then everything had gone black.

When he awoke, he was in a brilliantly lit room. He was told not to be afraid: they were not going to harm him. Then he was placed on an operating table and a beam of light was passed over him. He closed his eyes and then felt something uncomfortable in his nose. Next, he was given a guided tour of the ship. On looking through a porthole he could see the stars; below him lay the Earth, and he realised that he was in space. After that he was shown a couple of films. One was full of apocalyptic images, but he would not reveal what the other one was about.

Afterwards, he had woken up in the same spot from which he had been abducted. He had spotted the alien walking away from him and it was then that he had shouted out, the alien had turned and he had taken the picture of it. Although the picture was blurred, under hypnosis Spencer gave a clear description of the alien, which matched the image. He also said that the aliens were about 4 feet (1.2m) tall, with big eyes and pointed ears. They had two toes on their feet, in a funny, V-shaped arrangement.

Curiously enough, it was not the first time that little

green men had been seen in the area: it had happened less than 500 yards (457m) from where Spencer had encountered his alien, at the White Wells Spa. In 1815, when the caretaker was opening up the bathhouse there, he had put his door key into the lock; the key, he said, had then revolved of its own accord. Inside, he had been shocked to see that the bath was already occupied by 'a lot of little creatures dressed in green'; the creatures had then taken off their green garments in order to wash. On seeing him, they had all run off. The caretaker had been paralysed with fear and had not chased after them.

After the investigation Spencer was re-employed by the police force on the strength of character references from former colleagues, which said that he was honest and trustworthy. He had certainly had no reason to lie about what he had seen on Ilkley Moor – indeed, by telling the truth he had risked losing any chance of resuming his career with the police.

Lost in Dorset

In September 1990 James and Pamela Millen went camping in Dorset. One night they woke up at around 3am. James noticed that there were a number of orange balls of light dancing above the next field and called Pamela to point them out to her. The couple found them hypnotic, but James managed to tear himself away to get his camera. Neither of them is very clear about what happened next. When James returned he thought that he had been away for just a couple of minutes, but there was now a pile of cigarette butts at Pamela's feet. When they looked at their watches it was 5.40am: some two-and-a-half hours had gone missing.

A vague recollection of what had happened during that missing time returned later, however. James recalled lying on his back on something cold in a white, circular room and being attended to by a man in white robes. He did not know what the man was doing to him, but believed that it was something beneficial.

Another Near Miss

Elsie Oakensen, the head of a teachers' centre in Daventry, encountered a UFO on her way home on 22 November 1978. It had been a disturbing day: at lunchtime she had felt an odd, tightening sensation around her head. On setting off on the 6 mile (9.7 km) journey at 5.15pm she found that one of her sidelights was not working and she therefore had to drive with her headlights dipped.

When she reached the traffic lights at Weedon she turned right, on to the A5 towards Towcester. She noticed two bright navigation lights – one red, one green – ahead, in the sky above the road. At first she thought that a low-flying aircraft was going to crash on to her, but then she realised that the lights were stationary. As she drove underneath them she saw that they marked the ends of a huge, grey, dumb-bell-shaped craft. It was hovering soundlessly above the point where she would turn off the road to follow the lane to her home, in the village of Church Stowe. (Although this huge spacecraft was hovering above a busy main road at around 5.30pm no one else reported having seen it.)

Oakensen watched the craft in fascination as she climbed the hill into Church Stowe. Suddenly her car's engine cut out and the headlights failed. Bright lights flashed at her from above; the beams were so brilliant that she could see nothing in the darkness apart from them. When she turned the ignition and touched the accelerator pedal the car started again as if it had never stopped, but when she arrived home it was 5.45pm. The journey – which she covered twice a day and which usually took her 30 minutes – had taken 45 minutes. (On the following day she timed the trip just to make sure how long it took and discovered that some 15 minutes had indeed gone missing on the night in question.) When she reached home she discovered that her sidelights were working again; they did not malfunction after that. Later that evening Oakensen again felt a tightening sensation around her head.

Although no one else had seen the craft hovering above the A5 there had been another UFO sighting that night. Four young women who were driving in a car 4 miles (6.4km) from Church Stowe had seen two streaks of light flash across the sky. Then they had noticed a red and a green light some way to the south that seemed to be keeping apace with the car. The car's engine had begun to lose power, but the two lights had next vanished incredibly quickly. The engine had returned to normal and the car had functioned perfectly well after that.

Oakensen subsequently underwent hypnosis and while hypnotised remembered again feeling the tightening sensation around her head when her car's engine had cut out. The pain had been intense and she had become hotter and hotter. Then two figures had appeared; they looked grey against the glow. Oakensen came to believe that she had been visited by aliens, who had scanned her. She somehow felt that she had been promised a return visit, but had ultimately been rejected. (Like Alfred Burtoo, she may have been too old.)

Oakensen later became deeply involved in the UFO-related movement. She joined a witness support group whose members counselled each other; together they experienced a growing spiritual awareness. She also became interested in psychic phenomena, corn circles, ley lines, pendulums, crystals and natural healing. She wrote poetry and practised as a medium. She furthermore published a book, *One Step Beyond… A Personal UFO-abduction Experience,* all of which goes to show how much a chance encounter with an alien being can change a person's life.

Alien Assassins?

On 11 June 1980 a man was found dead in a coal yard in the West Yorkshire mill town of Todmorden. The body was that of 56-year-old Zigmund Adamski, one-time Polish prisoner of war who had managed to escape from captivity and had settled in England after the war. He had worked as a miner for 30 years and had recently not been a well man, but during the months before his death he had been on leave and the rest had done him the power of good.

Adamski's body turned up five days later – almost to the minute – twenty-five miles (40km) away. Nobody knew what had happened to him in the meantime. Nobody knew how he had come to be in the coal yard either. His body was discovered by Trevor Parker, the owner's son, at 3.45pm. It was lying in a hollow at the top of a pile of coal, near Todmorden railway station and within view of the line. Parker was adamant that it had not been there at 8.15am, when he had been at the same spot, and the coal yard had been locked during the intervening hours.

There were other mysterious aspects to the case, too. The time of death was estimated to be anywhere between 11.15am and 1.15pm on the day of the body's discovery, which meant that it had been lying in full view for at least two-and-a-half hours of daylight – that is, unless it had been put there after Adamski had died. There was no obvious cause of death; there were a few, minor, external wounds, but none of them would have been fatal. Parker had noticed a large, brown burn on Adamski's neck and the pathologist concluded that he had been burnt with a corrosive substance. Yet this would not have killed him either, and besides, the burn had occurred two days before he died, according to the pathologist. And where had he been for the five days since he had gone missing? He had certainly not been sleeping rough. Indeed, he had been well taken care of: he had only one day's growth of beard on his chin, although his stomach was empty, suggesting that he had not eaten on the day of his death. The police were baffled.

However, the publicity surrounding Adamski's death was enough to attract ufologists, although Zigmund Adamski was no relation of the first alleged modern alien-abductee, George Adamski (the name is quite common in their native Poland). But there had been a wave of UFO activity over the Pennines during the late 1970s and ufologists were on the alert. Although some suggested that Adamski had been abducted and then dropped onto the coal heap from above, no one could discover any evidence with which to take the case further forward.

Alan Godfrey, one of the policemen who had first been called to the coal yard, was not looking for an extraterrestrial explanation – he was a sceptic who took no interest in UFOs. A down-to-earth Yorkshireman, he would have laughed if a friend had told him that he had seen one. At around 5.05am on 29 November 1980 he was looking for an elusive herd of cows that had been reported running loose in a housing estate. He had already made several unsuccessful attempts to find the cows earlier in his shift, but decided to give it one last try before he finished work at 6am. He accordingly set off in his panda car up Burnley Road, which runs out of the town in a north-westerly direction. He had travelled only a few hundred yards when he saw a very bright light ahead. It could not have been the early morning bus because he had already seen that pass; something about it attracted him to it.

He drove to within about 100 feet (30.5m) of the light and then stopped, amazed. What he saw appeared to be a giant, fluorescent top in the middle of the road. There was a line of five dark windows about two-thirds of the way up. The domed, upper part of the craft was still, while the bottom was spinning in an anti-clockwise direction, making the nearby bushes and trees shake. Godfrey calmly observed the craft. The beam of his headlights was reflected from its surface, so he assumed that it was made out of metal. He tried to use the police radio to call for assistance, but neither the UHF nor the VHF channel worked. Although he later admitted that he was too terrified to get out of his car, he felt perfectly safe inside and picked up his clipboard and sketched the object on the back of a traffic-accident report. By comparing its size to the width of the road and the height of the street lamps he jotted down his assessment of its dimension estimating that it was 20 feet (6m) across and 14 feet (4m) high. The windows were oblong, 3 feet by 1 foot (91cm by 30cm), and the entire thing was hovering 5 feet (1.5m) above the surface of the road. Suddenly, without a sound having been made, Godfrey found himself 300 years (91m) further down the road. The UFO was gone.

He initially reported that he had turned his car around and had driven directly to the police station. But six weeks later he recalled that he had first got out of the car and looked upwards, into the sky, to see if the UFO had really gone.

Godfrey then picked up a colleague from the police station and together they went back to the place where he had seen the UFO in order to examine the scene. There they found that there was nothing that he could have mistaken for a UFO. Furthermore, the only remaining evidence of the UFO were dry, swirling-shaped patches on the road over which the UFO had hovered; the rest of the road was uniformly wet, as a result of the rain that had fallen earlier that night.

Back at the station, Godfrey's colleagues began teasing him and he decided not to make an official report of the incident. However, he became increasingly convinced that what he had seen had been real. Later that day he had one of his first flashback memories: he suddenly recalled that just before the object had disappeared he had heard a voice in his head saying 'You should not be seeing this. This is not for your eyes'.

When Godfrey turned up for work the next night he discovered that an anonymous man had reported seeing UFO answering the same general description as the one that Godfrey had observed hovering around the area that night. Godfrey then decided to file a report, only to find that three other policemen, who had been searching the moors for stolen motorbikes, had seen a similar UFO just twenty minutes before him. After they had radioed in their report of the sighting they had seen the UFO again, this time travelling in the direction of Todmorden. Even thought they were convinced that it was not a helicopter or plane, they checked with Leeds/Bradford Airport, which, however, could offer no instant solution to the UFO's identity. A special constable also reported a UFO sighting made at around that time.

All of this corroborating evidence helped Godfrey to reassure himself that he had not gone mad. He now noticed that his foot had been injured and that his sturdy, new boots had been damaged: the left one looked as if it had been dragged over a hard surface, but he could not recall any such incident. He also remembered that it had not been damaged before 29 November.

With the aid of UFO investigators Godfrey subsequently painstakingly reconstructed the events of the early morning of 29 November and discovered that 15 minutes could not be accounted for. The missing time left him totally flummoxed; he wanted an explanation for what had happened and eight months later agreed to undergo regressive hypnosis.

Under hypnosis, he again remembered getting out of the car, but then recalled having seen a light appearing under the UFO. He had got back into the car, but when he had attempted to drive off it would not move. Then, in middle of the session, the regressed Godfrey cried out and covered his eyes; he was connected to an electrocardio-gram (ECG) machine which showed that his heart beat was rocketing upwards at this point. The hypnotist, the psychiatrist Dr Joseph Jaffe, had to go to great lengths in order to calm him down.

Eventually continuing the session, Godfrey remembered that he had been blinded by the light when he had felt something take hold of him – he was being carried, float-ing. Suddenly he was in a room. There was a table in it, as well as a man wearing white sheet and a skull cap. He was about 6 feet (1.83m) tall, with a beard and a long, thin nose. Somehow Godfrey knew that his name was Joseph. There were eight small, metallic figures in the room, as well. They were about 3 feet 6 inches (1.6m) tall, with heads 'like a lamp' and eyes resembling single, vertical lines. Godfrey squirmed when they touched him, but Joseph told him telepathically not to he afraid: the robots were his. Furthermore there was a big, black dog in the corner.

Joseph led Godfrey to the table. A bright light was shin-ing and Godfrey recalled receiving the telepathic message. 'This is not for your eyes.' Telepathically again, he was told to get onto the table. He experienced a pain in his head, which Joseph soothed. The robots removed his left boot and examined his foot. The bracelets were put onto his right wrist and left ankle before being plugged into some type of machine. He smelt a sickly odour and saw flashing lights – similar sensations to those that many peo-ple experience when undergoing a general anaesthetic.

Details of his past life flashed before him and his mind was probed by a series of questions. He got the impression that Joseph was telling him that they had met somewhere before and he suddenly had a very vivid childhood memory of seeing a huge ball of light in his bedroom.

Godfrey remembered little of his sessions of hypnosis; the tapes of the sessions were played back to him afterwards and he was at a loss to explain them. Jaffe could not suggest any reason why Godfrey would have made up what he was saying and concluded that something very strange had happened to him.

Although PC Godfrey never claimed to have been an abductee his experience changed his life: West Yorkshire Police declared that he was mentally unfit for duty and sacked him. 'I know what I saw on the road that night. It was real', he said. 'As for what I said under hypnosis, I just don't know. It seemed real, but it might have been a dream.' But why would anyone dream of being abducted? As in the case of Zigmund Adamski, the mystery remains unsolved.

All in the Family

In 1988 the 45-year-old Carol Thomas and her 24-year-old daughter, Helen, were working in a Birmingham mill. Each morning Helen went round to her mother's house and together they made the 15-minute walk down a number of alleyways to the mill.

As they walked to the mill on the morning of 30 March 1988 it was still dark. They heard a distant, humming sound and suddenly a bright searchlight shone down on them from above, it was directly above their heads and appeared to be becoming bigger and bigger. The next thing that they knew was that they were stumbling dizzily down an alleyway. By the time that they reached the mill they discovered that they were late for work – even though they had left Carol's house at the same time as usual.

They reported suffering from a curious type of what appeared to be sunburn after the incident – the skin on their arms and faces was blistered and red. They also experienced discharge from their navels, as well as nosebleeds.

Under hypnosis, Carol recalled how the light – which had first been above them – had appeared below them. She saw the moon and seemed to be travelling towards it. Then she found herself in a white room, which had windows all around it. Naked, she was strapped to a table and a wet cloth was spread over legs. She was surrounded by little aliens with big, black eyes and three long fingers on each hand. They were led by a taller alien, which had blond hair and blue eyes and was wearing a silver suit with a badge on it. The aliens pushed a long, thin, glass tube through her navel; she thought that they were taking eggs from her ovaries. An odd-looking cup was put onto her head and she was shown a series of shapes on a screen and then a war film. After that she had found herself back in the alleyway again.

Carol said that Helen had been with her on the ship and that she had begged the aliens not to harm her daughter. Helen confirmed this under hypnosis. She, too, had found herself lying naked on a table, her legs having been covered with a wet cloth, next to her mother. She had cried. Helen had seen the little aliens, although she thought that they only had two fingers. What she took to be a camera was hovering over her. The aliens had pushed a thin rod up her nose and had implanted a small, silver ball into it. A wire was fed into her ear and a thin, glass tube into her navel. She also reported that two wires were inserted in her cervix; something was removed from her body, she believed.

Helen seems to have been given a more comprehensive physical examination than her mother. As with Carol, they had placed a metal cap on her head and had showed her images on a screen. She also remembered seeing her mother with a tall, blonde alien wearing a silver suit. The alien, she said, looked like a beautiful woman. The touch of the aliens, both women reported, was damp. Indeed, everything inside the ship seemed clammy. Helen recalled observing an alien feeling the texture of her leather coat. Back in the alleyway, her mother had asked why her coat was wet. Neither of them could remember why.

However, Helen's hypnosis revealed something even more interesting: she recalled having been abducted before, when she was five. She had been taken from a field full of buttercups before having been examined in a room by what she took to be weird-looking children. They had given her a strange stone, which Carol remembered her having when she was a child and which she had kept under her bed. Helen also began to remember numerous other occasions on which she had been abducted. Indeed, it seemed that aliens had been plaguing her for all of her adult life.

Working in the Woods

At 8pm one night in September 1996 Mary and Jane – two young mothers in their thirties living in Scotland – went out to buy some coffee in a nearby village. They took Peter, Mary's 10-year-old son with them; Jane's 14-year-old daughter, Susan, stayed at home watching television.

While driving on a remote stretch of road they saw two bright lights. In the sky above them was a huge, dark shape. They stopped and got out of the car for a better look. The UFO was 80 feet (24m) across and shaped like a triangle, with a red light at each corner. It then sped off across the sky at incredible speed. After they had bought the coffee they were returning down the same road when they saw the UFO again: this time it was hovering over the car, as if it was observing them. They were frightened, but then it shot off towards the horizon again.

When they reached home Mary called the UFO Research Association's hot line, which advised them to go back to the spot where they had seen the UFO and to take a camera and binoculars with them. They did not have either, but went back anyway. This time Susan came along, too.

There was no sign of the UFO when they got back to the place where they had originally seen it, but they became aware of a weird glow in a nearby field; through the gloom they saw a number of small creatures milling about. Terrified, they drove to Jane's home. Her brother was there and lent them a pair of binoculars to take with them before they again returned to the scene. Hiding behind a dry-stone wall, they observed the aliens carrying objects – which they took to be cocoons – from a small wood. The aliens seemed to be approaching, so the group fled again.

Jane subsequently began to get the impression that she had been taken aboard the UFO and had been examined – the examination had not hurt, but had made her feel happy. A few weeks later the other three also began to recover memories of having been on the ship. UFO investigators went to the site of the alleged abduction and found burn marks on the ground. They ascertained that a heavy object had certainly rested on the ground, although there was no road access to the site. The surrounding trees and bushes were furthermore covered in a strange, cobweb-like substance.

Alcoholic Aliens?

In July 1995 Steve and Annie asked their neighbours, Mike and Debbie, to their Derbyshire home for a barbecue. At around 11pm that night a flying saucer descended on to the garden; they spotted it when it was about 20 feet (6m) above their heads.

A door at the bottom of the spacecraft opened and bathed the garden with light. They then fell into a trance and later remembered only the light dimming to a pinpoint as the craft move away leaving them feeling nauseous and dizzy. The whole episode, they thought, had lasted a matter of minutes, but when Steve looked at his watch it was almost midnight.

They called the police, who took statements from them. Curiously, two of the four glasses of cider that had been on the table when they had spotted the UFO had disappeared and were never found again. Their physical reactions continued for the next few days and they found it hard to sleep. In an attempt to deal with the aftermath of the encounter they agreed to undergo regressive hypnosis.

Under hypnosis, Debbie recalled having suddenly found herself alone in a room, unable to move her arms or legs. She was surrounded by aliens with enormous, black eyes. They were doing something to her, hurting her. She became so distressed by her recollections that the hypnosis session had to be terminated.

Annie, too, remembered seeing big, black eyes, but they had belonged to a creature with a pale face and a pointed chin that had appeared in the garden. Another two had then materialised and had pulled her towards the illuminated door of the craft. They had dragged her towards a table and had started taking off her clothes. She recalled them examining her and doing something to her navel.

Under hypnosis, Steve similarly remembered having seen small aliens with large, black eyes. He seemed to have been restrained within a transparent tube. He had seen a map of the solar system and had noted that the aliens seemed to be carrying balls of white material in their hands. He got the impression that at one point he had been in a second, larger ship.

For his part, Mike was too traumatised to risk hypnotising him. All four were profoundly affected by what had happened to them, even though they had not believed in UFOs or alien abductions before the incident occurred.

A Girls' Night Out

On Thursday 16 July 1981 three young women had enjoyed their weekly night out at Tiffany's, in Shrewsbury. They left Tiffany's at 2am for the 20 minute drive back to their homes in Telford New Town. The 27-year-old Viv Hayward was driving; next to her in the front seat was the 20-year-old Rosemary Hawkings, a mother of three; and in the back was the 26-year-old Valerie Walter.

Their route took them down the A5, through the village of Atcham. While driving on the road to Norton their collective mood suddenly changed and all conversation ceased. Later they reported that they thought that this was part of a 'conditioning process'. Across the fields they saw some lights – two white, two red. Hawkings wound down the window in order to take a better look and saw that the lights were attached to some sort of object. They did not know what it was, but were not unduly concerned until it started to follow them. Walters peered out of the back window, and noticed that the craft had windows along its side, while the lights on the base seemed to be tilting towards the car.

It was then 2.10am. They were just ten minutes from home so Hayward put her foot down, but the engine did not respond and began to splutter. The car was losing power and the three women began to panic. The last thing that they wanted was their car to break down on a deserted road in the early hours of the morning in the presence of a UFO.

Suddenly, however, the spacecraft disappeared into the clouds. The car's engine regained power and they sped on down the road, convinced that they had just had a close encounter with an unidentified flying object. Racing on to Telford New Town, they went straight to Malinslee Police Station to report the sighting. The women said that they got there at 2.55am, but the police report stated that they had arrived at 2.40am. If the police report was true, they had lost around 20 minutes.

Under hypnosis, Hayward recalled that everything had gone very quiet when the engine failed. Then she had had the feeling that she was floating. The car seemed to have been lifted from the road, which had disappeared under a white cloud. She recalled having been alone in the car while it was lifted vertically into the UFO, through two big doors that had opened in the underside.

Small, hairless aliens, with big eyes and wearing green cloak, had taken her out of the car and had placed her in a reclining chair similar to a dentists' before holding her down. The aliens were very strong, she said. They had wanted to discover things from her, particularly 'how humans are made'. They had told her not to be afraid, but she felt that they were taking something from her body – it was as if they had put their hands inside her legs and were pulling on her bones. This hurt terribly and she cried out in pain, but their only response was again to tell her not to be afraid. They had then given her body a thorough examination and had scanned it with a 'big light', a procedure which had left her paralysed so that she had had to be carried back to the car. The next thing that she remembered was driving along the road again, with the UFO gone.

When regressed, Hawkings reported a similar tale. She, too, said that everything had gone quiet and that she had had the sensation of floating (but she had not been in the car, a statement which corroborated Hayward's recollection). She had found herself alone in a room which had a table in it and had been very frightened. She had no idea what she was doing there, but had a strange compulsion to get onto the table; it was something that she had seen in a film, she said, and it calmed her. Then a number of robots 4 feet (1.2m) tall – 'round on top with a round body and round leg' (like R2-D2 in the film *Star Wars*) had entered. Hawkings said that they had not given her a full medical examination – 'they just wanted to have a look'. She had the impression that she was in an area leading off a larger room waiting for her friends.

However, under hypnosis, Walters told a sightly different story. The UFO, she said, had come at them. Its lights were blinding and she had felt dizzy. She remembered having been in the car with Hayward, although Hawkings had gone – mysteriously having been 'beamed out' of the car.

She recalled that the car had then come to a halt; the doors were open, so she had got out to look for Hayward. Next she remembered having floated into a room. Two aliens had entered – one male, one female. They were different from the aliens that Hayward had seen, however, being 6 feet (1.8m) tall, with white skin, blue eyes and long, dark hair, but like Hayward's aliens they were wearing long, green cloaks.

The aliens seemed inordinately interested in fashion. While Walters was told that the women had been abducted for purposes of observation, Hayward was given a full medical and Hawking's reactions were being studied, it was Walters' clothes that they had been especially interested in. The female alien had even slipped her high-heeled shoes off Walters' feet and had then tried them on; she had had difficulty walking in them, though.

Walters was the only one who had questioned the aliens – not that it had got her very far, however. She had asked them who they were, but they had curtly replied that she was not to know and that she would not understand. Walters later came to believe that her encounter with the aliens had not been as benign as she had first recalled: they had done something to her, she thought, that was 'gynae-cological in nature'.

Abductions Worldwide

A Canadian Abduction

In what would later become Canada's most notorious alien-abduction case on record, the 14-year-old David Seewaldt revealed under clinical hypnosis what it was like to be seized by an alien spacecraft and then closely examined by repulsive-looking creatures.

The event that he described happened in Calgary, Alberta, on 19 November 1967. At around 5.45pm on that fateful Friday Seewaldt had just left a friend's house and was walking the two blocks home. It was late autumn and darkness had fallen. He took a short cut across a field; normally the walk took just a couple of minutes. Suddenly he heard a high-pitched sound. He turned round to see a silver-grey object flying in the sky. Dotted around the centre of the UFO were coloured lights that were flashing on and off.

The next thing that he knew he was running up to his front door, with the sinister craft still hovering overhead. He burst into the house and ran upstairs. His older sister, Angela, followed him and found him cowering behind his bed in a state of blind panic. 'What happened?', she asked, grabbing him, 'Why are you home so late?' It was 6.30pm by then; 45 minutes had passed since he had left his friend's house. 'I ... I was chased by flying saucer', said the clearly terrified boy. He managed to describe the spacecraft and to give a detailed account of its pursuit. But the ordeal that he described would have lasted only a minute or so and there was no way that he could account for the missing time.

Seewaldt was usually a calm and mild-mannered child, but he was a bundle of nerves for the rest of that weekend. His worried parents called the celebrated ufologist William K. Allan, who hosted a UFO-related show on the local radio station, CFCN-AM, in Calgary. Allan had a meeting with the boy, but could coax no more information out of him.

Seewaldt gradually forgot about the incident and the household settled back into its regular routine. Then, in April 1968, five months after the incident, Seewaldt awoke one night having had a nightmare. He had dreamt that he had been taken aboard the alien spacecraft and had been given a medical examination by creatures so repulsive that he could only describe them as 'monsters'. He was convinced that his dream was an accurate recollection of what had happened during the missing 45 minutes.

The nightmare had left the boy badly shaken and the Seewaldts again turned to Allan for help. Allan realised that he would have to use regressive hypnosis in order to get to the bottom of Seewaldt's tale. He therefore secured the services of a local dentist who used hypnosis in his practice, but when the dentist hypnotised Seewaldt it transpired that he had been so traumatised that he had mental block. When he was asked what had happened after he had first seen the craft he was unable to speak; his legs shook violently and he began to perspire.

Under the supervision of a psychologist at the University of Calgary, Seewaldt was subsequently videotaped while under hypnosis. In order to try to get through the mental block it was suggested to him that he recount what had happened as if he were seeing it unfold on television. This is the tale that Seewaldt then recounted.

After he had first seen the spacecraft he had turned and run. Then an orange beam had emerged from the bottom of the ship. When it reached him he had fallen into a trance, after which the beam had seized hold of him and had lift him into the ship. Once inside, he had seen two monsters, which had brown, scary skin, like a crocodile's, holes for their noses and ears and slits for their mouths. They were about 6 feet (1.8m) tall, with four fingers on their hands and feet and no thumbs.

Four of them had put him onto a cot and had then start-ed to examine him. They had stripped him of his clothes, including his underwear, which had upset him greatly. They had lifted up his head and had examined his hair, eyes and nose. One of them was talking to the others in a language like none that he had ever heard; he described it as a sort of buzzing, like the sound that emanates from a beehive or high-voltage, electrical spark. After they had let him put his clothes back on again, the aliens had taken him down a hallway to another room, in which there were all sorts of bright lights. They had laid him on a table. Then came what he described as being the most terrifying part of the ordeal: a grey blanket was thrown over him before a huge, orange light descended and was shone on to him; one of the monsters next took a grey needle and stuck it into his arm.

The next thing that he knew was that he was walking through a computer room before emerging into a hallway. The orange beam had then reappeared and Seewaldt had next found himself back on Earth. A high-pitched whine came from the spacecraft as he dashed for home; just as he arrived the spacecraft shot upwards and disappeared.

Not Fade Away

The multiple abductions of a Canadian rock musician named Jack were reported in *Flying Saucer Review* in 1984. He first recalled having been abducted at the age of two, when he was taken from his pushchair by aliens with big, black eyes. He had been put onto a table inside a room and felt that his thoughts were being siphoned from his head. The aliens told him that this was a test: they want-ed to know if he was suitable for their purpose (although he did not recall what the purpose was). When he was six he was abducted again, this time along with his father. Jack's father subsequently underwent regressive hypnosis and recalled the abductions as well. Then, at the age of ten, he and his friend, Jim Voss, were taken from Twelve Mile Creek, near their home.

By the time that he was 16 Jack had joined a rock-and-roll band. On the night of 16 October 1971 the band was returning from a party at around 1.30am when strange lights appeared on the road ahead. They thought that there had been a road accident until their van skidded out of control. It finally stopped at the side of the road, whereupon a small alien ordered them to get out. Some members of the band refused and there was struggle, during which a drum was kicked out of the van before rolling across the road. The alien asked – telepathically – what this odd instrument was.

The alien explained that they wanted to test the young men, but only had enough facilities for three; volunteers should therefore step forward for testing. Jack, the drummer and the bass-player stepped forward and were taken aboard the UFO. Jack recalled that they were in some sort of drugged, or hypnotic, state.

The bass-player, Calvin, was taken into a separate room. Sam, the drummer, and Jack were told to take off their clothes. They complied before lying down and being inspected with lights and instruments. Blood and hair samples were taken. The two young men asked some questions, including where the aliens came from. They replied 'A long way away ... If we told you you would not understand'. Before the band members were released Jack gave their abductors one of their records, which he was carrying in his bag, as a gesture of goodwill. To this day Jack believes that the inhabitants of some planet in a distant galaxy are grooving away to Canadian rock music.

European Encounters of the Fourth Kind

On 1 November 1954 an Italian woman named Rosa Lotti was walking through a wood on the way to Cennina when she came across a weird, egg-shaped craft. Two 3 foot (91cm) tall aliens then grabbed her. She did not have a very clear recollection of the incident, but remembered that she had been paralysed by a beam of light and that the aliens had been talking in what sounded to her to be Chinese. She did not remember whether she had been medically examined, but during the encounter she lost a stocking and a bunch of flowers that she had picked. Some village boys had come to her rescue, who saw both the spaceship and the aliens.

In September 1955 the 27-year-old Josef Wanderka was riding his moped down a road in Austria when he inad-vertently rode straight up the ramp of a flying saucer. He apologised profusely to the occupants, who explained – in perfect German, naturally – that they were from the 'top point of Cassiopea'. Even though they were plainly adept at interstellar travel, they were fascinated that they might be harbouring totalitarian tendencies (Austria had been de-Nazified relatively recently), so he launched into an anti-fascist diatribe. They evidently found this so boring that they kicked him out of their flying saucer without subjecting him to an invasive medical examination.

In July 1965 the 41-year-old Maurice Masse was tending the crops on his farm near Valensole, in France, when he heard a strange, whistling sound. An egg-shaped craft then descended from the sky and landed nearby. At first Masse thought that it was a new helicopter that the French Air Force was trying out, but then a number of what

133

appeared to be small boys got out. On closer inspection they turned out not to be boys at all: they had big, pumpkin-shaped heads, thin mouths and big eyes – classic greys. Masse would have left it at that had the aliens not begun picking his lavender. He tried to remonstrate with them, but was zapped by a beam of light. It was subsequently suggested that Masse had made up the last part of his story and that he had actually been taken on board the spacecraft. He later admitted that he had not told the whole story because something had happened that he found too embarrassing to talk about. He particularly did not want his wife to know about it (so perhaps it is not just Brazilian men that aliens abduct for the purposes of sex).

In July 1968 a 25-year-old man was approached by some tall aliens with domed heads and huge, oriental-shaped eyes on the Grodner Pass in the Dolomites. They had a small robot with them. They communicated telepathically with him and passed on some useful information, including that they came from a 'far galaxy' and that 'everything is God, man'. More pertinently, they told him that the Earth's poles were about to shift, that its crust would crack and that humankind was about encounter a lot of trouble.

A 31-year-old Belgian businessman had a narrow escape on the evening of 7 January 1974, when he was travelling down a deserted border road near Warneton. He was trying out his new radio-cassette recorder when the car's electrics suddenly packed up. When the car had finally cruised to a halt he got out, intending to replace a fuse. Then, in the road in front of him, he saw a domed craft. Two aliens got out of it; they were about 4 feet 6 inches (1.35m) tall, with pear-shaped heads, big eyes, narrow mouths, small noses and long arms. As they approached him one pulled out a tube-shaped instrument and pointed it at him, whereupon he heard a painfully high-pitched whistle and felt a shock at the base of his skull. At that moment another car drove down the road, causing the aliens to run back to their ship and speed off. The driver of the second car saw the flying saucer take off; he then stopped and ran over to help the businessman, who was still temporarily paralysed. The second driver, who lived nearby, next went to find some friends to help him to search the area. 'If I come up with nothing', he said, 'I will keep quiet. Without proof, no one would believe a word of this.'

On 5 February 1978 in Soria, Spain, a 30-year-old vet named Julio was driving in his car with his dog, a pointer named Mus, in the early hours of the morning. He did not remember what had happened to him during the drive, but his eyes later hurt and he seemed to have lost some time.

Under hypnosis, he recalled having been suddenly blinded by a light. Then he had been led into a room and a soothing voice had reassured him that they only wanted to borrow his dog for examination. But the frightened dog had lost control of its bladder – with unpleasant consequences. The aliens had then decided that if the dog was going to behave in that way they may as well experiment on the human instead, and had accordingly taken samples of Julio's semen, blood and gastric juices. The aliens had told him that our sun hurt their eyes, which was why they usually only appeared at night. Their world was a dark and polluted place, they had said, but humanity's was a beautiful jewel and a rare oasis of fertility in an otherwise sterile universe. That is what had brought them to Earth, they had explained: they wanted to study the humans' world before they destroyed it, just as the aliens had destroyed their own.

But Julio was warned to be careful because not all of the aliens who came to Earth were so benign: there was a bunch of ugly little brutes, for instance, which was intent on reprogramming human beings biologically. After that Julio had blacked out and had next found himself back in

135

his car, where he had sat in a daze for a long time.

On 2 April 1980, in the early hours of the morning, Aino Ivanoff was driving her car near Pudasjarvi, in Finland. Suddenly a mist came down. The next thing that she remembered was that she was in a room, lying on a table being examined by small aliens. They seemed rather worried and told her that war was bad and that she should join the peace movement. Then they explained why they were so concerned about the fate of humankind – they could not have their own children (which explains a lot).

Hairy Giants

On the night of 6 December 1978 Fortunato Zanfretta, a 26-year-old security guard in Genoa, Italy, saw four lights moving about in the courtyard of a deserted house in the Marzano district. He tried to call his headquarters to report the intruders, but his two-way radio failed. He could not fetch help either, as his car's electrics would not work. So he therefore went to investigate himself.

As he walked up to the front gate the lights in the courtyard moved towards him, before disappearing around the back of the house. Emboldened, Zanfretta followed them and stuck his head around the corner at the back of the house. He was suddenly pushed to the ground and caught a glimpse of his assailant, who quickly vanished. Zanfretta scrambled to his feet; as he ran back to his car he heard a loud, whistling sound behind him, followed by a blast of heat. He called for help on the radio, which was now working. It was then midnight, much later than when he had arrived at the house. After that he blacked out, and an hour later his friends found him in a field. Nearby they discovered a horseshoe-shaped depression about 24 feet (7.3m) wide.

Under hypnotic regression, Zanfretta recalled having seen a triangular spacecraft. He had been abducted by aliens who were 10 feet (3m) tall and covered with green hair. Their ears were pointed and they had two narrow, triangular-shaped eyes that sloped upwards; in the middle of their foreheads, among the folds of skin, was a third, more human-shaped, eye. The hairy giants had taken him into a circular room, where they had put something onto his head that had caused him a great deal of pain.

The Strange Case of 'Dr X'

The 'Dr X' case is one of the most famous French alleged alien abductions on record. Although the abduction itself did not last long, its recollection did not rely on regressive hypnosis and there was also a large amount of evidence supporting it.

'Dr X' was a well-known and respected biologist. He allowed the French ufologist Aimé Michel to publish an account of his abduction provided that Michel did not use his name – hence his pseudonym, 'Dr X'.

In 1968 the 38-year-old Dr X was living with his wife and 14-month-old son in a house that overlooked a valley in south-eastern France. On the night of 2 November 1968 he awoke to hear the cries of his son; there was also a thunderstorm raging outside. His wife was sound asleep, so Dr X got up to attend to the child. He did so with some difficulty because a couple of days before he had been chopping wood and had slipped; the axe had struck his left leg, puncturing an artery and causing extensive internal bleeding. The wound had been treated by a doctor, who had examined it again on the previous afternoon. He already had problem with his right leg. During the Algerian War a mine had exploded and had fractured his skull. This injury had damaged the left hemisphere of his brain, as a result paralysing the right side of his body. Although the paralysis had passed after a couple of months the muscles on his right side were left permanently wasted. His disability had cost him his career as a musician and he still could not stand on his right led properly.

Despite his walking difficulties, he managed to totter into his son's room. The boy was standing up in his cot shouting 'rho, rho' (this was the word that the child used to describe a fire burning in the hearth or, indeed, any bright light). He was pointing to the window and Dr X assumed that he was indicating the lightning flashes that were visible through the cracks in the shutters. He got the boy some water and settled him down. While he was doing this he could hear a shutter on a window in an upstairs room blowing back and forth in the breeze. Half asleep, he went up and closed it, noticing as he did so that the room was bathed in a pulsating light.

After closing the shutter Dr X felt thirsty and therefore went downstairs for a glass of water. Still puzzled by the intermittent light, he went out on to the terrace to investigate it. It was 3.55am by the kitchen clock, he noted. Once outside, he immediately saw the source of the light: it was being emitted by two silver, disc-shaped UFOs that hovered over the valley. They had long antennae sprouting from them, which seemed to be collecting electricity from the storm clouds. The furthest end of the antenna would begin to glow, the light would build up along its length and would then suddenly be discharged onto the ground in the form of a lightning bolt. The build-up happened rhythmically and the emission illuminated the whole valley with a flashing light.

The two craft then merged into one and the pulsating light ceased. After that the unified object moved up the valley towards him. Dr X noticed its underside, which was covered with dark, rotating bands, causing patterns that defied the laws of science and logic. When the craft got within 500 yards (457m) of Dr X he had the feeling that it had noticed him. It then turned a bright beam of light onto him, bathing the whole house in an intense glow. He raised his hands to protect his eyes, but suddenly there was a loud bang and the craft shot skywards, so fast that it looked like a single streak of light.

It was 4.05am when Dr X went back inside, which surprised him as he did not think that he had been outside for one minute, let alone ten. He returned upstairs, woke his wife and told her what had happened. As he talked excitedly to her he paced up and down, stopping every so often in order to make notes and draw sketches of what he had seen. His wife suddenly observed that he was walking normally. He pulled up his left pyjama leg and saw that the axe wound had healed completely, which would normally have been impossible is such a short period of time. What was more, his withered right leg was now functioning perfectly, too.

Eventually Dr X went back to bed. Later on that night his wife was disturbed by him talking in his sleep. She noted down what he was saying; one of the things that he repeated was 'Contact will be re-established by falling downstairs on 2 November. When he awoke at 2pm the next day she did not tell him about this, but suggested that he write to their friend, the ufologist Aimé Michel. Dr X asked why he should do so, whereupon his wife discovered that he had ho recollection of his UFO sighting on the previous night. And when she showed him the notes and sketches that he had made he became alarmed.

Later that afternoon Dr X tripped and fell down the stairs; it was as if something had grabbed his leg, he said. He hit his head during the fall and suddenly memories of his experiences of the previous night flooded back to him. Twelve days afterwards he dreamt about seeing another UFO. It was not like the ones that he had observed before, being bright, luminous and triangular. Three days after that he felt an itching sensation on his stomach and on the following day a red triangle appeared around his navel. The dermatologist was baffled and wanted to write a scientific paper about it, but Dr X prevented him from doing so.

Dr X then contacted his friend, Michel, who discovered that there had been a rash of UFO sightings around the area where Dr X lived on the same night that he had seen his flying saucers. Michel suggested that the red triangle might be psychosomatic in origin. Dr X agreed, only to find that a similar red mark then appeared on his son's stomach the next day.

The experience left Dr X feeling depressed and confused. He gradually began to take an interest in ecology. The triangle disappeared, but later reappeared occasionally; other injuries that he sustained healed miraculously. But his house seemed to be constantly plagued by poltergeists and aliens, who would take him on journeys over impossible distances.

A Good Communist is Kidnapped

Alien abductions even happened in the former Soviet Union. A good communist named Anatoly was walking along the shores of Lake Pyrogovskoye in May 1978 when he was grabbed by two aliens who wore dark suits made out of Cellophane. They took him for a spin in their spacecraft, but it appears that they really just wanted a chat.

He asked them to help the Soviet Union to fight the evils of the world – that is, capitalism, which Anatoly believed caused world-wide poverty. Although they acknowledged that helping the poor was a noble aim, the aliens found the idea impractical: 'If we helped the poor,' they argued, 'then we would have to help the not so poor, then we would end up helping everyone'. They gave him a drink which tasted like lemonade laced with salt. If their civilisation was so advanced, he asked, why did they not drink vodka? 'Perhaps if we drank vodka we would not be such an advanced civilisation', the aliens astutely replied.

The aliens then kindly dropped him off by the lake. When he got home and revealed what had happened to his wife, she told him to keep quiet about seeing aliens in case he found himself doing hard labour in a gulag. However, Anatoly felt that by not reporting the abduction he had not fulfilled his duty to the state and therefore told the local commissar about it. The commissar believed that Anatoly was fabricating the story in order to avoid a court martial. Nevertheless, the proper procedures had to be followed and Anatoly was examined by a psychologist, given a lie-detector test and put under hypnosis. They could find no evidence that he was making it up – in fact, when he described events more fully his story even became convincing. And as he had tried to recruit the aliens to the communist cause they let him off.

Brazil Nuts?

Per capita, there have been more UFO sightings in Brazil than in any country apart from the USA. Furthermore, as we have already seen, alien abductions practically began there.

Ten years after Anôtnio Villas-Boas was abducted for sexual purposes on 1957, another abductee wrote to a leading Brazilian weekly magazine in response to a piece that it had published called 'I Saw a Flying Saucer'. The anonymous letter purported to come from a leading citizen of the city of Fortaleza, the capital of the province of

Ceara. He had withheld his name, he explained, because of his prominent position: if he disclosed his identity people would think that it was either a publicity stunt or that he was a lunatic or liar. He also said that he had been with four other people at the time in question. Two of them were well-known doctors, another was a local bank manager and the fourth was a metallurgical engineer from São Paulo, who spent his holidays in Ceara.

All five were keen fishermen, and in June 1967 they had spent a weekend fishing in the beach resort of Morro Branco, in the municipality of Cacavel, not for from Fortaleza. They were out fishing at around 3am on a Sunday morning. There was a full moon, but the sky was cloudy and dark. Suddenly they heard a low, humming sound, so low that it became painful to the ears. Then it stopped as abruptly as it had begun.

The engineer went off to investigate the source of the noise while the rest went on fishing. Five or six minutes later they heard the humming again, only this time it was louder and more intense. It appeared to be coming from a hill around 500 yards (457m) behind them, and when they turned around they saw pulsating beams of smoky-coloured light spreading across the sky. Frightened, they frantically started to pack up their gear, but then the engineer suddenly appeared behind them and told them to stop. Gesticulating with a gun in his hand, he ordered them to climb up the hill, towards the lights. The others thought that he was joking and laughed – that is, until he fired a bullet and said that he would kill all if they did not do what he said. His eyes looked strange and they decided to obey him, thinking that he had gone insane.

They accordingly staggered up the hill at gunpoint. The lights had now disappeared, but they heard weird sounds in the darkness. The engineer responded by shouting 'We're coming'. Then he urged on his fishing companions, who, by this time, were all suffering from splitting headaches. At the top of the hill they saw a huge UFO hovering 30 feet (9.1m) above the ground; it was 15 feet (4.6m) tall, round, finned and glowed in a phosphorescent-green hue. The top was covered with funnels (from which the beams of light had come) and there was ramp leading up to a square hatch which they climbed, into the UFO.

They found themselves in a room that was about 6 feet (1.8m) tall and had no window or doors. The light that illuminated it was soft and gentle on their eyes. A strange-looking grille was situated in one corner, near the ceiling. That was all that there was to see. They were very frightened, but a voice told them to be calm: nothing would happen to them, it said; they had only been summoned to answer some questions. The voice explained that their answers would have important implications for the universe in general, and then went on to say that it had come from very far away, but could not explain from where because they would not be able to understand.

The first question that it asked was how they had been made (it did not know what a woman was). It also wanted to discover what humans were made of, what they ate, how many people lived together on the planet and what it was called. Did they live with other species and, if so, did the species also talk on 'frequencies'? The alien was interested in what the men had been doing when they were abducted. Fishing turned out to be a concept that it could not grasp, no matter how clearly the fishermen tried to explain it. The alien furthermore wanted to know whether humans had travelled to other planets, whether they lived on them and also whether they lived under the Earth's crust. When he asked how long a human life span was the men had to explain what a day, a month and a year were. The alien laughed when he computed just how short a human life was

The abductees found it very stuffy in the windowless room and plucked up the courage to point out to the alien that they would die if they did not get some air. Miraculously, a triangular window appeared in the smooth, seamless walls and a light breeze blew into the room. The alien then considerately asked if there was anything else that they needed. One of the doctors replied that they needed an explanation for their abduction and the others said that they would not answer another question until they were told what was going on. The alien retorted that there was no point in concurring since they would not be able to understand, but it finally conceded that it would answer their questions. They then fired so many at it that it had to stop and ask them to talk one at a time.

The alien explained that it and its five companions would not make themselves visible to the men as they were very different from humans, although made of similar substances. Their life span was 300 years. It said that they had come from Goi to study Tonk, which is what they called the Earth. This was their first visit, but aliens from other planets had visited Earth before. The humming sound that the craft made was used to attract humans and lights had been used to hypnotise the engineer. The alien was interested to know what the engineer was carrying in his hand. After they had explained what a gun was the alien realised that this was how four of his companions had been killed. He said that it had been their own fault, however, because they had been following 'something that flew in the air'.

The alien revealed that there were other inhabited planets nearby, but that those who lived on them were much more advanced than the people of Earth. Aliens would be living among humans within a year, it predicted; creatures from Goi already dwelt on Earth, but in places to which humans never went and, indeed, where they would not be able to live – indicating the polar icecaps. The alien also said that two people from Earth lived on Goi; they had travelled there may years previously and had now forgotten about the Earth. Before the alien went, it said that it needed some of their things – one item from each man. 'We heard nothing more', said the letter, 'and woke up.'

The five men then found themselves back on the beach. It was 5.20am; two hours and twenty minutes had elapsed, but they had no subjective idea of how much time had passed. Puzzled, they trekked back to the hill, but although they could see the footprints that they had made when they had ascended it there were none to show that they had come down. None of them has any recollection of how they had returned to the beach. They climbed the hill and found a shallow crater at the top which seemed to have been caused by a blast of air. Back on the beach they found that some of their possessions were missing – one item from each man.

The letter concluded with the writer making the following statement: 'The purpose of writing this narrative is solely that it should be used as data for research purposes'.

Alien Doppelgänger

A prominent Brazilian industrialist reported a similar incident to the fishermen's. It had happened when he was 18. At 7.30pm on 28 February 1974 he had been at home when he had suddenly found himself elsewhere: in a square room – with not windows or doors – the colour of smoked glass.

'Keep calm, we are your friends', said a voice, which came from a box with flickering lights on it in the corner; it appeared to be a tape recorder or a translation device of some sort. A door mysteriously opened in the wall and an egg-shaped machine appeared. An intense, yellow light shone down on him from the ceiling before an exact replica of himself stepped out of the machine. His *doppelgänger* patted him on the back and said, in a voice identical to his own, 'keep calm and wish me a good journey, for I do not intend to harm anyone'. He then stepped through the wall.

At that exact moment a large screen appeared in the room, which showed images of a replicant in the abductee's home. However, there were some significant differences between the abductee and his *doppelgänger*. The abductee was affectionate by nature, for instance, but saw the *doppelgänger* treat his younger sister very tersely, which surprised and upset her. Indeed, he was rough and thoughtless with everyone around him. Even though he was only 18, the abductee was already working for the family firm and watched as the replicant went out on business in his place. It become clear that he did not know anything about money, or even how to write.

The screen prompted the abductee to help his *doppelgänger*. Watching the screen, the abductee saw himself hovering above the replicant and whispering advice. Another thing was that he could see through the replicant's eyes. On one occasion, however, the screen was switched off: this was when the replicant had left the apartment at night. It was switched on again when he returned in the morning. The screen was also switched off when the replicant visited a rainforest (he seemed particularly concerned about the welfare of plants and trees). The abductee furthermore watched as the alien camera that he used was not the abductee's own.

All this time the young business was able to talk to his abductors via the box in the corner. He asked them what they wanted with him. 'We need to conclude some tests', they replied. They assured him that he would not be harmed: they were friendly and only wished to help, they said, but added that 'the less you say about this, the better'.

They said that they could not tell him who they were,

but revealed that the world from which they came was very different from Earth. They wanted his help. He asked why he should assist them if they were invading Earth, whereupon they protested that they were not invaders but had come for the benefit of the human race. A good Catholic, he asked whether they knew about God; everyone did, they responded. And he also asked what the thing that had replaced him was. 'He is an image that can be touched', they said. Inside him there is one of ourselves and a recording of your mind.'

The abductee said that he was held for 24 hours and that he never once felt hungry, thirsty or tired during that time. In the room was a stool for him to sit on and a bed that was fixed to the wall, with a shelf beside it. When the screen went blank the abductee expressed a wish to have something to read – photographic images of a newspaper, books and a *Spider Man* magazine then appeared on it. (He later discovered that the issue of *Spider Man* that he had read in the room was not yet available on the news stands.) For his part, the replicant took a lot of his books on Brazilian history, geography and science. He also helped himself to newspapers, flowers, leaves, photographs and card, as well as promotional material from the family firm.

The young businessman's watch had stopped during his abduction, but started again when he was returned, exactly 24 hours later. When he tried to explain to his family that he had been abducted and then replaced by a *doppelgänger* they not unsurprisingly found it hard to believe, although they conceded that he had been acting strangely for the past day and could offer no alternative explanation for his behaviour. What is more, one of the family-firm's employees said that he had walked through a wall, although he had felt like flesh and blood to touch.

Guided By the Light

In 1972 the leading Brazilian ufologist Irene Granchi was contracted by an army general – had also trained as a doctor – who claimed to have been abducted by a UFO.

The incident had happened in March 1969. On the night is question he had been out for a meal at Bara da Tijuca, an isolated spot. After dining on shrimps and beer, the general – who was alone in his car – set off for home at around 2am. As he drove along a section of dirt track the engine suddenly cut out and the car trundled to halt near a place called Morro do Chapeu ('Halt Hill' in English). He got out and inspected the engine, but could find nothing wrong. The road was deserted and by 2.45am he had given up all hope of getting help. It was at that point that he saw a light approaching him around the hill. At first he thought that it was a lorry, but then he realised that the light was brighter than anything that he had seen before. As it came closer he noticed that it was hovering about 1 foot (30cm) above the ground.

He walked towards it as if in a trance, following it for 300 or 400 yards (274 or 366m); the ground felt unusually smooth and he appeared to be treading on 'solid light'. When he reached the source of the light he saw it was a spacecraft which had a translucent wall about 10 feet (3m) square, with, metal attachments at the side. The wall then tipped up, just like a garage door. Inside, he saw a chamber, whose far end seemed rounded.

There were already three people in the chamber: an engineer, a woman who was a law graduate and a doctor. The doctor – who appeared to be very much at home – told them that they had been brought there to see something that they had never seen before. He also said that they were in no danger and that he would be grateful if their response was peaceful.

The rounded end of the chamber then turned into a screen and he noted that they appeared to be travelling across the landscape at an amazingly fast speed. Next, the city of São Paulo hove into view and they seemed to hover above it at a height of about 100 feet (30m). Before he knew it, they were above another large, Brazilian city, which he did not recognise, but thought that the others did. The spaceship then moved on again at an apparently incredible rate until they reached Rio de Janeiro. Afterwards, the doctor asked them whether they had enjoyed what they had seen. He then said 'This will be useful to you, to each in his own way'. (What the general himself gained from the experience was a sensation that time was not as linear as he had once thought.)

After the doctor had said goodbye the general left the chamber and crossed the 'solid light' to his car, which now worked. He drove a mile down the road and then stopped at a bar to attempt to make sense of what he had seen; he ordered a coffee and tried to convince himself that he had had a hallucination. Then he went home and slept.

Over the next few months he studied the available literature on UFOs and tried to evaluate his experience. He was sure that he had been awake throughout and that he had not been dreaming. Using his medical experience, he concluded that he had not been hallucinating: the experience had been unique and isolated and there was nothing that he could identify that would have induced it. His own behaviour at the time had been stable, while routine medical tests could reveal no physical problems. He wanted to unravel the philosophical and intellectual implications of his abduction and also reported the case in order to furnish others with scientific data. However, he balked at undergoing hypnosis.

An Alien Hitch-hiker

Onilson Pattero was abducted not just once, but twice. On the first occasion, on 22 May 1977, he was driving home to Catanduva from São José to Rio Preto in his blue Opala. He was 41, married with two children and worked as a library administrator in the state of São Paulo.

At around 2.55am he crossed the Tiete river and the Anhanduva Falls. Soon afterwards he saw a hitchhiker at a petrol station and stopped to pick him up. The young man had short, fair hair and deep, penetrating, blue eyes. He said that his name was Alex. He mentioned that he was 'not from here' and was carrying a silvery object that appeared to be a cigarette case, although he said that he did not smoke. It was still 25 miles (40km) to Catanduva and Pattero stopped for a coffee; Alex had only a sip of mineral water. He said that he was on his way to Itajobi, which lay a few miles beyond Catanduva, to find a job. Pattero offered to drive him there, but Alex did not want to be taken all the way to the village and jumped out of the car in a deserted spot.

After Pattero had turned around to drive back to Catanduva the radio in the car packed up; then the engine spluttered and died. Suddenly a circle of blue light moved slowly across the car. It seemed to make everything transparent, and as it passed over the dashboard Pattero could see the engine beyond it. He initially thought that is was a strange optical illusion caused by the moonlight before realising that it was overcast and raining and that there was, in fact, no moonlight. A large circle of bright, blue light then shone down the road at him. He feared that it might be an oncoming lorry and tried to move the car on to the hard shoulder, also flashing the car's lights to signal its presence. The light only grew brighter, however, and he lowered his head onto the steering wheel to try to protect his eyes.

When he heard no vehicle approaching he looked up and saw a UFO hovering about 30 feet (9m) above the road, some 15 yards (14m) away. For a moment he thought that it might be a helicopter, but soon realised that this was no helicopter, but a flying saucer. It looked, he said like two metal soup plates that had been welded together rim to rim. The object itself was dull, with no noticeable features; it was around 20 feet (6m) high by 30 feet (9m) wide. There was a halo of light around it, but he could locate no specific source for this.

It had grown hot and airless inside the car, so Pattero got out; it was intensely hot outside, too. He then noticed a tube extending from the bottom of the craft towards him and tried to make a run for it. He had got less than 40 yards (37m) before he felt what he thought was a lasso ensnaring him. He turned around, but could see nothing holding him. After that he saw the beam of blue light running over the car, rendering it transparent. He had not yet finished paying for the vehicle and was worried that he would face financial difficulties if it were damaged. Then he fainted.

About an hour later two young men from Itajobi were driving past when they saw an unconscious Pattero lying face down on the road; next to him was his car, with the driver's door flung open and the headlights on. They immediately went to call the police. When the police arrived they found a road map of northern Brazil lying beside the car. Inside, Pattero's suitcase had been opened and ransacked, which surprised Pattero after he had come round because he had locked the suitcase and the key was still in his pocked. The map had been inside the car. Nothing was missing and the vehicle worked perfectly.

Pattero was taken to hospital, where he was passed fit. Strangely enough, however, his hair had turned jet black, but after a few days it returned to its normal, chestnut colour. He developed an itching feeling around his abdominal region and blue spots appeared on his hips and buttocks, which turned yellow and then disappeared. No cause could be found for these physical symptoms and he was in perfect health otherwise. Regressive hypnosis later revealed that he had been abducted.

Nearly a year after his first abduction, on 26 April 1978 Pattero was travelling home at night again when his car's engine cut out and a beam of light passed over him that appeared to make his car transparent.

On this occasion he was subsequently taken into a spaceship, through what he described as a sort of 'curtain'. Inside was Alex, the erstwhile hitchhiker, who told Pattero not to be afraid. He was shown a fabulous laboratory and Alex then told him to take off his clothes. There followed a detailed physical examination by what seemed to be humans, who did not say a word. After that his hands and feet were secured with steel rings and he was put into a long case. While this was happening he saw three 'men' walking by – one of them was an exact replica of himself and he was dressed in the clothes that Pattero had been wearing when he was first abducted

He remembered nothing else until he awoke on the top of a hill; it was reputed to be a spooky spot, where neither cattle nor men dared venture. Pattero shouted for help; fortunately a gaucho heard him and his boss, the ranch-owner, came to his rescue. Pattero then discovered that it was 3am on 2 May. He had been missing for seven days, but was still clean-shaven. He was also over 500 miles (805km) from home, in a different state. His car had been found – with the driver's door wide open – not far from home and his family had practically given up hope of ever hearing from him again.

Eminent ufologists later endorsed his case and he was regressed again. However, there were some inconsistencies in his story. After giving a series of interviews to the newspapers he then told a reporter that the authorities had forbidden him to speak to the press. He was subsequently denounced at a UFO conference by the ufologists who had originally supported his story on the grounds that he had claimed that he had not watched the TV series *The Invaders*, although his wife revealed that he had. This omission made him an unreliable witness.

Hundreds of Miles From Home

On 4 January 1975 the 28-year-old Carlos Diaz was returning home form his place of work in Bahía Blanca, Argentina. When he got off the bus, just 100 yards (91m) from his home, a powerful beam of light shone down on him from the sky and he found that he could not move. The air around him then began buzzing quietly, whereupon a gust of wind lifted him off his feet. He guessed that he was about 8 feet (2.4m) above the ground when he lost consciousness.

When he came round he found himself inside a plastic sphere around 10 feet (3m) in diameter. He tried to move, but his body was paralysed. There were small holes in the plastic which allowed him to breathe and he discovered that if he turned away from them he felt ill. The sphere seemed to he travelling at a rate of knots.

After about 15 minutes three aliens approached him from behind. They were humanoid, about 5 feet (1.5m) tall and moved very slowly. Their skin was greenish and felt like a rubber sponge. They were naked and had no mouth, nose, ears or hair of any sort; instead of hands they had suckers on the ends of their arms. One of the aliens grabbed him, while another attempted too pull out his hair. Diaz tried to struggle, but they were very strong and he eventually passed out.

Some time later he was awoken by a man, who took him to hospital. There he discovered that he was in Buenos Aires, more than 200 miles (321km) from his home. Numerous doctors interviewed him and the military authorities wanted to know what had happened to him, too. It did not seem possible for him to have travelled by conventional means from Bahía Blanca to Buenos Aires in the time available. Some researchers claimed that a paper that he had bought in Bahía Blanca bore his story out, but others said that his tale was riddled with discrepancies.

Hopping Continents

In May 1968 Dr Gerardo Vidal and his wife were driving the 100 miles (161km) from Chascomus – 80 miles (129km) away from Buenos Aires – to Maipu, in the south of Argentina. They were following another couple, but somehow their friends lost sight of them. Although they waited for them when they got Maipu, the Vidals never arrived.

Two days later they had a phone call from the Argentinian consulate in Mexico City asking them to collect the Vidals from the airport and warning them that Mrs Vidal was ill and would have to be hospitalised. On return to Argentina Dr Vidal explained that while they had been driving along the road to Maipu they had suddenly found themselves surrounded by a dense fog that seemed to have appeared from nowhere. The next thing that they knew was that they were sitting in their car on some unknown side road, suffering from terrible headaches and general feeling of fatigue. When they asked passers-by where they were they were shocked to discover that they were in Mexico – over 4,000 miles from Buenos Aires.

Abductions That Failed

On 28 November 1954 Gustave Gonzales and José Ponce were driving to Petare, about 20 minutes away from their home in Caracas, Venezuela. Suddenly, they saw a glowing, spherical object, about 10 feet (3m) across, blocking the road in front of them.

They stopped the car and got out to investigate, whereupon a small, hairy alien approached them. Gonzales grabbed it, but the creature fought back. While the two of them were struggling Ponce ran to the nearest police station, which was only a couple of streets away. As he ran he saw two more of the aliens, who had apparently been collecting rock and soil samples. On seeing him they fled to the sphere and jumped into it.

The creature that was attacking Gonzales now unsheathed its claws. In response, Gonzales pulled out a knife and stabbed it. The knife hit the alien on the back, but the blade just bounced off, as if the creature was made of metal. Another alien then stepped from the sphere to give his companion some assistance and shot a beam of light at Gonzales, which left him temporarily blinded. The two aliens then jumped aboard the craft, whereupon it blasted off into the skies.

Gonzales took to his heels and sprinted to the police station, where he and Ponce were accused of being drunk. However, a subsequent examination of Gonzales revealed claw marks on his back. Furthermore, a few days later a doctor came forward and said that he had seen Gonzales fighting with the alien, but had not wanted to become involved.

Two weeks later the hairy aliens again tried to abduct some humans. This time their intended abductees were Lorenzo Flores and Jesus Gomez. Both aged 19, they had been out hunting on the night of 10 December 1954 and were on their way home, travelling down the trans-Andean highway in Venezuela, when they stopped for a rest. A flying saucer then landed near them and four aliens got out of it. They were only about 3 feet (91cm) tall and were covered with hair.

They seized Gomez and tried to drag him aboard their craft, but the youth was not eager to comply. Flores grabbed his rifle and hit one of the aliens with it. 'It felt like I had hit a rock', said Lorenzo. 'My gun broke.' Gomez, who had been screaming, now fainted. Flores rushed to his assistance and was so concerned about his friend's welfare that he did not see what had happened to the aliens (who had presumably thought better of it and had fled back to their ship).

Gomez and Flores made their way to the local police station, where they reported the incident. When they were examined they were found to be covered in scratches, as if they had been attacked by animals.

However, it seemed that the hairy aliens had not given up, for six days later another Venezuelan named Jesus was grabbed by a hairy alien after he had slipped off the road to relieve himself in the parkland near San Carlos. Again a friend came to the rescue and took Jesus to hospital in a state of severe shock.

After 1954 all the small aliens who appeared on Earth seemed to have moulted. And why hairy dwarf aliens only appeared in Venezuela has not yet been explained.

Gaining Time

Missing time, it seems, can work both ways – it does in Chile, at least. On the night of 25 April 1977 Armando Valdes, a corporal in the Chilean Army, was on exercise on a desert plateau 12,000 feet (3,658m) above the town of Putre, in Aric province. At around 3.50am Pedro Rosales, who was on watch, spotted two violet lights, which were illuminating the ground as they descended from the sky.

Rosales informed Valdes of the phenomenon, who ordered his men to screen the fire that they had lit with blankets and then to take cover behind a low, stone wall. Next they saw a large flying object a little way ahead of them, down a slope. They observed a central light, with a smaller light on either side. Whatever it was, it was Valdes' job to find out. Praying for God's protection, he ordered his men to cover him before stepping over the wall and setting off into the darkness. After a few minutes the object disappeared.

Fifteen minutes later Valdes reappeared, stumbling into the camp from the opposite direction from which he had left as if in a trance. He was mumbling the words 'You did not know who we are or where we come from, but I tell you we shall return'. Then he blacked out.

The men tended to him overnight and in the morning were shocked to see that he had grown a full beard – it was as if he had not shaved for days. His digital watch had stopped at 4.30am, but the date display said that it was 30 April – five days later.

Australasian Abductions

On the night of 30 October 1967 a wool-grader named Harris, a married man with three children, had paid off his shearers at an outpost near Mayanup, Western Australia, and was driving to another station at Boyup Brook. Suddenly, on a deserted stretch of the road, the electrics in his car just cut out. 'I had no feeling of deceleration', he said. 'The car just came to a halt.'

Everything around him became quiet and then a beam of blue light hit the car. It came from an oval object in the sky above and Harris had the sensation that he was being watched. Then the light disappeared and the car started working again. Again, Harris noted no sensation of acceleration: the car was just speeding down the road again. It was as if he had been momentarily frozen in time.

He later noticed that his watch, which was usually extremely accurate, had lost several minutes. Furthermore, for two weeks after the event he had a constant headache. Thinking that he was ill, Harris went to the doctor, who referred him to a psychiatrist in Perth.

A medical team searched for evidence of temporal-lobe epilepsy, which can cause brief lapses of consciousness, but found none. Regressive hypnosis was not used in Australia in 1967, but even in the absence of recovered memory the Harris case bears all the hallmarks of a classic alien abduction.

Maoris and Aborigines also had similar abduction experiences. On 22 February 1969, for example, at Awanui in New Zealand, one Maori encountered several tall, white-skinned, blond-haired non-humans who had emerged from a big, glowing object in the woods. Before they could grab him he ran off and performed the rites used locally with which to ward off evil spirit (which apparently mainly consisted of running around in a circle while urinating).

In 1971 there had been a rash of UFO sighting in Kempsey, New South Wales, when, at around 10pm on 2 April, a 34-year-old Aborigine went into his kitchen to get a glass of water and suddenly felt himself being sucked up into the air. He passed out and later awoke outside the window, which was broken, although the bar across it remained in place. It seems that the unconscious abductee would have had to have been squeezed through a hole just 10 inches (25cm) wide.

On 27 September 1974 a 19-year-old musician and an 11-year-old boy were hunting in the Snowy Mountain when they saw a large, white light on the horizon which was giving off a low, humming noise; everything else was deadly silent. They remembered nothing more.

Although no regressive hypnosis was used on him, nine years later the boy – who was now twenty – became psychic and began having dreams that relived the experience. In them he saw himself being pulled towards the humming object.

Floating inside it, he was then laid onto a table and tall, thin, grey aliens attached instruments to him to see how electromagnetic fields affected his body. (For his part, the musician resisted the aliens' advances and was consequently drugged.) The aliens showed the boy a devastated landscape, which he took to be their home planet. Strangely enough, he did not recall having been afraid.

At 6.19pm on 21 October 1978 the 20-year-old Frederick Valentich flew out of Moorabbin Field, Melbourne, in a Cessna 182 (call sign Delta Sierra Juliet). According to his flight plan, he was flying to King Island, in the Bass Straits, midway between Tasmania and the coast of Victoria. He was on his way to pick up crayfish and had collected A$200 from friends with which to buy them.

At 7pm Valentich passed the coast and began his long descent. Everything was going to plan when, at 7.06pm, he radioed air-traffic control in Melbourne and asked if there were any other aircraft in the area. The air-controller, Steve Robey, replied in the negative, whereupon Valentich asked if Robey had radar contact with the peculiar aircraft that he could see in his vicinity, but Robey replied that it was out range.

It was now twilight, and Valentich said that he could see a large, dark mass with four bright landing lights on it. Robey later said that Valentich had sounded scared; the craft, he reported, had flown closely overhead at great speed. Robey checked again and ascertained that no civilian or military air-craft was scheduled to be in the area.

Robey then heard Velntich say 'It's not aircraft … it's a …', before his voice trailed off. A minute later Valentich said that the object was circling directly above him. Then, at 7.12pm, six minutes into the encounter, Valentich reported that the Cessna engine was failing and that he was going to try to glide to King Island. The UFO was still in the vicinity: 'It is hovering and it's not an aircraft', said the panic-stricken Valentich.

The beleaguered pilot then repeated his call sign, before falling silent. His radio continued transmitting for another 17 seconds, during which time Robey heard strange noises, like metal scraping. Neither Valentich nor the Cessna were ever seen again. No wreckage was found and the accident enquiry could venture no explanation for the loss of the plane.

However, there had been numerous reports of UFO activity over the Bass Straits shortly before Valentich's flight and several UFO contactees have reported that he is alive and well and living on another planet. One even claims to have seen him.

On 8 August 1993 the 27-year-old Kelly Cahill and her husband were driving through the Dandenong foothills on their way home to Belgrave, Victoria, when they saw an alien spacecraft hovering a short distance in front of them above the road. The light that it gave off was so bright that Cahill had to shield her eyes with her hand. What seemed like just moments later, she asked her husband what had happened. He did not know. When they got home they found that they had lost an hour and Cahill also discovered a strange, triangular mark around her navel.

Memories of the experience later came back to her. They had been driving around a long curve in the road when the spaceship had landed to the side of it. Cahill had told her husband to stop the car. Both of them had then got out, only to see that the spacecraft was not a spacecraft at all, but a black, 7 foot (2.1m) tall, alien creature. Although it had arms and legs like a human, its eyes were complex, like those of an insect, and glowed red. And it was not on its own; there were others nearby. A group of the aliens had approached the couple, while a second group blocked their retreat to the car. Cahill had felt that these creatures were evil and her husband had begged them to let them go. Then Cahill had been sick. The next thing that she remembered was being back in the car.

Afterwards, like other abductees, Cahill was plagued by nightmares about the encounter. She did not undergo regressive hypnosis to find out what had happened during the missing hour, but believed that one of the creatures had bent over and kissed her on the stomach, hence the mark.

African Abductions

During the spring of 1951 a British engineer working in Paarl, Near Cape Town in South Africa, had fixed his troublesome car himself and had then decided to take it for a test drive, even though it was late.

He was in a deserted spot halfway up Drakensteen

Mountain when he decided to stop, whereupon a very short man dressed in brown laboratory coat emerged from the shadows. He had no hair, a smooth face and a domed head. 'We need water' he said, in a strange accent. The only water that the engineer had was in the car's radiator, so he took the stranger to a nearby stream. It was then that he saw a huge flying saucer half hidden in the lee of the mountain. The alien was grateful for the engineer's help and invited him aboard. Inside, he was shown a bed, on which another alien was laying.

The engineer was not subjected to a humiliating examination, nor was he offered sex, but the alien said that as recompense for the man's kindness he would answer any questions that he had.

Naturally, being an engineer, he wanted to know how the spaceship worked. In response, the alien explained 'We nullify gravity by means of a fluid magnet'. He also asked the alien where he came from, but the alien rather unhelpfully simply pointed to the sky and said 'Up there'.

Also in South Africa, in 1956 the 56-year-old Jean Lafitte was abducted by alien with large heads and no hair. They paralysed him, before taking him into the room in which he was put on a table and medically examined. During this procedure they implanted something into his brain which would active his psychic powers, they explained. They said that they were from the Pleiades star cluster and also told him that there was another species of alien visiting Earth from Alpha Centauri.

The aliens were not hostile, they abducted Lafitte regularly, and in 1986, in the wake of the Chernobyl disaster, told him that they had mopped up the excess radiation spilled by the nuclear-power station.

A Rhodesian named Peter met another helpful alien in 1974, when he was driving overnight with his wife Frances from what was then Salisbury, Rhodesia – now Harare, Zimbabwe – to Durban, in South Africa.

All of a sudden, on a deserted stretch of road between Umvuma and Beit Bridge, he saw lights in the sky and had the feeling that the car had been taken over by a strange force. Everything fell silent, the scenery looked unreal; and the car seemed to be gliding along without touching the road. Peter then fell into a trance and lost all track of time, while Frances slept through the whole experience (maybe her sleep had been induced).

Later, they realised that they had lost some time. Furthermore, the car had not used as much petrol as it should have done.

Peter was something of a veteran of close encounters. At the age of 13 he had been on a delivery run with his father, a truck-driver, when they had seen a UFO at Shabini. When they had dropped off the electrical equipment that they had been carrying it was found that all the circuits in the equipment had been destroyed by a power surge. Peter also reported having had dreams of floating, as well as out-of-body experiences.

Under hypnosis, he recalled that during the 1974 experience an alien had been beamed down into the car. Using telepathy, he had shown Peter a laboratory on an alien spacecraft in which aliens experimented on the humans that they abducted. The aliens, he said, were able to appear in any form that the witness found acceptable, and on this occasion – as on so many others – the alien had appeared as a small, hairless human with no reproductive organs.

The alien revealed that there were thousands of them living among humankind. They never interfered directly in human affairs, he claimed, but instead tried to influence individuals. He said that they had used this influence to charge the world in the past and would do so again – perhaps to end war or to introduce 'their way of doing things'.

Yet the alien's various explanations about where he

came from were not consistent. At one point he said he and his fellows had not travelled across space, but across time: they had 'come back in time to get to the earth'. He also said that they came from the 'outer galaxies', as well as from the '12 planets of the milky way'.

In the early evening of 15 August 1981, in a heavily forested area around Mutuare, on the border between Zimbabwe and Mozambique, 20 workmen were returning to the village when they saw a ball of light drifting across their path, lighting up the entire La Rochelle estate. All of the men ran for it except for Clifford Muchena, the head men, who watched as the ball turned into a glowing disc and travelled rapidly over the estate.

The light that the disc gave off was so bright that Muchena feared that it might set the forest on fire and therefore raised the fire alarm. The glowing disc then seemed to set down and three tall beings approached him: they were silhouetted in the glow and he thought that they might be estate workers, so he called out to them. This was the wrong thing to do, because they were aliens. They promptly turned towards him and zapped him with a brilliant flash that hurt his eyes. Then he was assaulted with mind-jarring force and lay dazed and paralysed for some time. When he recovered his senses the aliens and their glowing craft were gone.

Muchena seems to have experienced some missing time, but his native language did not contain the words to describe some of the key concepts needed to explained an alien abduction. (In fact, this deficiency may also be true of the English language). Furthermore, his tribe knew nothing about space travel and did not even believe that men had walked on the moon.

Asian Abductions

In the late evening of 3 October 1978 Hideichi Amano, a 29 year-old snack-bar-owner from Sayama City, Japan, drove to the top of a nearby mountain, where the reception was good, to contact his brother by Citizens' Band (CB) radio. In the back of the car was his two-year-old daughter, Juri.

At the top of the mountain the car's engine cut out and the radio stopped working; next the interior began to glow. Amano stuck his head out of the window, but could see nothing; then he looked over his shoulder to check that his daughter was alright. To his horror, he saw a strange, orange light playing across her body. Suddenly he had the sensation of metal pressing against his head and looked up to see a short alien with a tiny nose. Amano was paralysed; weird images and a hideous, screaming noise flashed through his brain.

When the glow disappeared the car's electrical system began to work again, but the dashboard clock had stopped and Amano had no idea how much time had passed. Terrified, he started the car and drove away.

Although his daughter was none the worse for the experience, Amano was left with an excruciating headache. He later recalled that the alien had abducted him and had implanted something into his head, which he understood would vibrate the next time that they came to get him.

An even more disturbing tale comes from India. In rural area in 1958, an Indian businessman and his companion saw a flying saucer land in broad daylight and four aliens 3 feet (91cm) tall get out; they seemed to have some difficulty in walking.

It was later discovered that two boys who had been playing on the rock on which the alien craft had landed were missing. One was subsequently found dead and an autopsy revealed that several of his organs had been removed, as if by an expert surgeon. The other boy was found in a catatonic trance and taken to hospital, where he survived for five days. He never regained consciousness and was therefore unable to explain what had happened to him.

Not all alien abductors in Asia are dangerous, however. In June 1969, for example, the 27 year-old Machpud met a beautiful female alien in Bandjar, West Java. She took him back to her spaceship, where, in a brilliantly lit room, she indicated that she wanted to make love to him; he obliged. Apparently, the sex was so good that he lost consciousness, later to awake in the Gunung Babakar Forest to find his clothes distributed across a tree. Other than acute embarrassment at being found in this way by a passer-by, he suffered no ill effect.

In Malaysia, people are regularly kidnapped by the 'Bunian people' (no known human tribe). According to the Malaysian UFO investigator Ahmad Jamaludin, the Bunian are smaller versions of humans who have a unique property: during abduction only the abductee can see them.

At 10am on one morning in June 1982 the 12 year-old Masweti Pilus was going to wash some clothes in the river behind her house when she bumped into a female Bunian around her own size. The sounds of the village faded and Masweti said that it was as if only she and the Bunian woman existed. The Bunian told her that she was going to take her to see a strange land. Masweti had no option but to go with her, yet she was not afraid and the Bunian took her to a beautiful place. She lost all sense of time, which seemed to fly by.

Her relative later discovered her lying unconscious on the ground not far from her home. Two days had passed.

Chinese people do not seem to be abducted, and perhaps with good reason. The truck-driver Wand Jian Min was driving on the road near Lan Xi, in the Cheking province, at 4am on 13 October 1979 when he almost ran into a parked car. The car-driver told Wand that he had seen a flying saucer on the road ahead and was now too scared to drive on, whereupon the fearless Wang announced that he would lead the way.

The road wound to the top of a hill and Wang accordingly drove up the slope slowly. At the top was a dome-shaped craft that gave off an odd, blue glow; two silver-suited aliens about 5 feet (1.5m) tall clad were standing beside it; they were wearing bright lights – like miners' safety lamps – on their heads. Wang wondered whether he might be witnessing an optical illusion and switched off his headlights, but the alien and the craft were still there.

Determined to resolve the situation, Wang then rooted around in his truck's cab and pulled out a crowbar, but when he turned to confront the aliens with it they, along with their craft, had gone.

Heavenly
Messengers

Preaching to the Converted

In 1974 Betty Andreasson went public with her experience of having been abducted by aliens. The incident had supposedly taken place seven years before, on 25 January 1967, a few months after *Look* magazine had made the abduction of Betty and Barney Hill internationally notorious.

Andreasson claimed that she had known nothing about the Hill's abduction until after she had been abducted herself (although her letters to her mother revealed that she had read a report of the incident, which had appeared in the Boston newspapers in 1964).

Many abductees are reticent about coming forward because they believe that others will brand them either mad or liars. For her part, Andreasson overcome any reticence that she may have had when she heard about the $100,000 that the *National Enquirer* was offering for irrefutable evidence an extraterrestrial visitor. Sad to say, she did not snatch the prize – nor did she win the $5,000 'best-case' award that went to Travis Walton in 1975. Undaunted, however, in 1979 she published the book *The Andreasson Affair: the Amazing Documented Account of One Woman's Terrifying Encounter With Alien Beings*, in collaboration with ufologist Raymond Fowler.

The book told the story of Andreasson' abduction by aliens. The details of the abduction were later recovered during 14 regressive hypnosis sessions held between April and July 1977. The incident had taken place after one of a series of electrical black-outs that had plagued the northeastern United States during the 1960s and 1970s. When the lights went out on 25 January 1967 Andreasson's six children came scurrying into the kitchen to ask what was going on. Then Andreasson saw a pinkish light outside the kitchen window; it was getting brighter, as well as pulsating. Her eldest daughter, 11 year-old Becky, saw the glow, which was now a reddish-orange, too. Everything fell silent: even the normal noises of the night disappeared 'like the whole house had a vacuum over it', Andreasson said. She became frightened and shooed the children into the living room.

Next, Andreasson's father hurried into the kitchen to find out what was happening. 'The creatures I saw through the window of Betty's house were just like Hallowe'en freaks', he said, in a sworn statement. 'I thought they had put on a funny kind of head-dress imitating a Moon man. It was funny the way they jumped one after the other – just like grasshoppers. When they saw me looking at them, they stopped ... the one in front looked at me and I felt kind of queer.' Becky also saw the creature silhouetted against the light; after that she, along with all of the members of the family apart from her mother, found themselves unable to move. They were unaware of anything else.

The lights then came back on and four 'entities' marched into the house, passing straight through the solid, wooden door. They wore dark-blue, tight-fitting uniforms, which had an insignia on the left sleeve that resembled a bird with outstretched wings. They were about 4 feet (1.2m) tall, with large, pear-shaped, hairless heads and tiny noses and mouths. Their skin was grey and clay-like, they had only three fingers. Andreasson was an accomplished artist and her sketches of the aliens accompanied the book. She was also a fundamentalist Christian and at first took her visitors to be angels. After all, she said, 'Jesus was able to walk through doors and walk on water'. Admittedly, the creatures that she saw did not fit any of the descriptions of angles given in the Bible, but the scriptures urged 'Entertain the stranger, for it may by an angel unaware'.

The leading alien introduced himself as Quazgaa; he

already knew her name. Their conversation took place by telepathic means. Quazgaa was slightly different from the others; one of his eyes was white, the other black, and above his eyes he had feelers, 'like a bee'. He held out his hand. Andreasson then hospitably asked if the aliens wanted something to eat. They nodded, so she went to the fridge, got out some meat and started to cook it. Quazgaa then told her that they could not eat food unless it was burned, so she accordingly turned up the heat. When smoke started to come off the meat, however, the aliens stepped back in astonishment. 'That's not our kind of food', explained Quazgaa. 'Our food is tried by fire. Knowledge is tried by fire. Do you have any food like that?'

The devout Andreasson knew exactly what he was talking about and gave him the family Bible, whereupon Quazgaa passed his hand over it and suddenly 'other Bibles appeared, thicker than the original'. He handed out the volumes to the other aliens, who began flipping through them. She saw that each page was pure, luminous white.

In return, Quazgaa gave Andreasson a thin, blue book. (This would have given her the incontrovertible proof of alien visitation that she needed in order to win the *National Enquirer*'s $100,000 prize had she not mislaid it. However, under hypnosis, Becky managed to confirm this part of the story.) Before she lost it, Andreasson said that she had had a chance to skim through the blue book. It was full of symbols that she did not understand, but one of the aliens told her that it was about the importance of love.

Andreasson then asked the aliens why they had come. 'We have come to help', they replied. 'Will you help us?' She wanted to know whether they had come from God; they answered that they had come because the world was trying to destroy itself. Andreasson then asked how she could help. In response, they told her to follow them and reassured her that her parents and children would be alright: although they appeared paralysed, they were, in fact, just 'resting'.

Convinced, she followed the aliens, whereupon all of them whooshed through the door again, without opening it. Outside, she saw an oval craft standing on struts. (When later asked why her neighbours had not seen the huge spaceship, she said that there had been a 'haze'. Weather reports confirmed that it had indeed been misty on that night. Other witnesses, however, saw a glowing, hovering object. One man said that his car's engine had died and that when he had got out he had found himself immobilised. A lot of UFO activity was reported in the area at that time, some of it correlating with Andreasson's story.)

Andreasson was frightened of entering the strange craft, so Quazgaa made the hull transparent so that she could see inside the spaceship. She noticed an odd-looking apparatus, which, she said, looked like glass balls suspended on arms that could rotate in an inner tube. Quazgaa then transformed the bottom of the ship back into a silver-gold colour. A door opened and they entered the craft. Once inside, she felt weightless and nauseous.

The aliens cleansed her under a shower of intense light, after which she put on a flimsy, white gown. Then she was strapped down onto a table in a dome-shaped room and was given an intimate examination by silver-clad creatures. Like Betty Hill, a long needle was inserted into her navel. The aliens mentioned the word 'creation', but were disappointed when they discovered that she had had a hysterectomy. Worse still, another needle was pushed up her nose, a procedure that was excruciatingly painful. When it was removed there was a 'little ball with little prickly things on it' on the end of the needle. (Ufologists believe that this was an alien device that had been implanted during an earlier, unremembered abduction.)

After that she was scanned with an instrument that looked like an eye. Other things may have happened, too – Andreasson had a severe emotional reaction while under hypnosis at this point in her tale, but she was relaxed and relieved when she related that she had been allowed to get dressed again.

Later she sat down on a chair that had plainly been made to suit human dimensions. There were a number of these chairs, each of which was covered with a plastic bubble. Air tubes were connected to her nose and mouth and the bubble was filled with grey liquid. She kept her eyes closed while she was fed some syrupy liquid through the tube. The liquid around her vibrated and she thought that she was being transported somewhere. When the fluid had drained from the bubble she felt as though the aliens had taken control of her body. Then her eyes opened and she saw a number of creatures with black hoods over their heads.

Released from the chair, she followed the creatures out of the ship through a labyrinth of dark tunnels. Their silver suits glowed in the gloom, but their black hoods made them look as though they had no heads. There was a mirror at the end of one of the tunnels and they walked through it into a world of vibrating redness. As they floated past concrete buildings along a black track, weird, lemur-like beings clambered all over them. These had no heads – instead giant eyeballs sat on long stalks that sprouted from their skinny bodies. They climbed like monkeys and their eyes swivelled round to stare at Andreasson.

Andreasson's party moved on into a green world with lush vegetation and the creatures took off their black hoods. It was very beautiful and Andreasson saw odd creatures that she described as a cross between fish and birds. In a green city she saw a white pyramid crowned with a male head that looked like that of the Sphinx, albeit thinner. Next, she saw a giant, crystalline structure whose prisms gave off beautiful rainbows of light. In front of it stood a gigantic bird that resembled the emblem that the aliens wore on their suits; it looked like an eagle and was 15 feet (4.6m) tall. As she approached it she became hot, but then it suddenly disappeared, leaving only ashes, which then turned into a worm.

After that, she heard a voice that told her that she had been chosen to fulfil a mission. She asked whether it was the voice of God, but was told that its owner would only be revealed to her 'as time goes by'. From the things that it said, however, she soon began to believe that it was indeed the voice of God and started to cry. Andreasson was given an important message for the world. Unlike the earlier messages, which had been communicated telepathically in English, this important directive was in the aliens' own language. It was 'Oh-tookûrah bohûttah mawhûlah dûh dûwa ma her dûh okaht tûraht nûwrlahantûtrah aw-hoe-noe marikoto tûtrah etrah meekohtûtrah etro indra ûkreeahlah.' (That, at least, was Fowler's phonetic rendition of it. When she was later asked what it meant Andreasson said that she did not know.)

Quazgaa took pity on her and gave her a message in English to convey to humankind when she returned to Earth. The aliens loved the human race, he explained, and had come to help. But unless humans could learn acceptance they would not be saved. 'All things have been planned', Quazgaa said. 'Love is the greatest of all ...We have the technology that man could use ... It is through the spirit, but man will not search out that portion.' He also warned her that the human race would not believe her message until much time had passed.

Andreasson then retraced her steps though the strange, alien world. Thereafter the spacecraft transported her back to Earth while she sat in the immersion chair. Another of the aliens – Joohap, who was carrying two glowing balls –

accompanied her into the house. He kindly put the children to bed before re-entering the spacecraft, which zoomed off.

That was not the end of Anreasson's contact with aliens, however. A being about 4 or 5 feet (1.2 or 1.5m) tall that appeared to be made entirely of light appeared in her house soon afterwards and leapt down the stairs. Then, several months after her abduction, while she was washing the dishes, something took control of her mind which enabled her to see into the future. She said that she could envisage inventions made thousand of years hence, but did not reveal any details.

During the hypnotic sessions Andreasson claimed that she was still engaged in telepathic communication with one of aliens, whose name was Antonio. The ufologist present asked if they could communicate directly with him. 'You would worship him if he was to come here', she responded. 'But that is not his way. He is just a servant and a messenger.' They asked if Antonio could supply proof that he actually existed and was talking through her, to which Andreasson replied. 'The world seeks proof. They cannot see with spiritual eye. Only those worthy will see.' She then confirmed that the aliens' visitation of Earth heralded the second coming of Christ.

After the hypnotic sessions were over Andreasson divorced her husband and moved to Florida, where she married again (her new husband had also been abducted by aliens, in June 1967). They heard angry voices talking in unidentifiable languages on the telephone and Andreasson's daughter awoke one night to see a huge ball of light swoosh over her head; on the following night Andreasson's two sons were killed in a car accident.

Renouncing the World

On the night of 14 December 1983 Antonio Taqsca, a radio announcer from Chapeco, Brazil, was abducted by a pale-skinned beauty with oriental eyes. She told him that her name was Cabal and that she came from the planet Agali. She said that like others who had cosmic minds he had been specially chosen to 'spread the word'.

Most abductees believe that their alien abductors have hypnotically implanted the suggestion that they must forget everything that they have experienced, but not Tasca: indeed, Cabala used a machine with which to implant the message 'you will never forget' into his unconscious mind. She also gave him another token that he would not forget – burn marks on his back. 'The mysterious burns are inexplicable', said his doctor. 'They cause no pain, erythema, fever or other symptoms of first or second-degree burning.' After his abduction Tasca lost all interest in material things.

Keeping the Secret

In 1976 the 50 year-old farm-manager Ted Pratt and the 42-year-old British Rail washroom-attendant Joyce Bowles, both from Winchester, in Hampshire, were regularly visited by aliens. They were driving on the first occasion when they saw an orange glow in the sky before suddenly being pulled across the road by a strange force. The car's engine then cut out, the lights dimmed and they found themselves on the grass verge next to an egg-shaped spaceship. A tall figure, with pink eyes, blond hair and beard and wearing a silver suit, got out of it. He came over to them and leant on the car, but when Pratt and Bowles looked away he simply vanished, along with his spacecraft. Then the car's light came on, the engine started and they drove home.

Bowles had red marks on her face for several days thereafter. She had to remove her wedding ring because the skin underneath it was sore. She also found that her watch had been magnetised and had begun to gain time.

On the night of 30 December in the same year they were driving near Winchester when they again noticed an orange glow in the sky. Although they seemed to have remembered everything about the previous alien encounter, this time there were gaps in their recollection. The next thing that they remembered was being in a room with three tall beings who spoke in a strange language. The only word that they remembered was 'mi-ee-ga', which the aliens seemed to use rather a lot.

Pratt was asked to walk up and down and to say if he felt hot or cold. The couple was then shown odd, transparent images on the walls. 'This is our field', they were told. Pratt, a farmer, assumed that they meant that it was pasture land, a misunderstanding which seemed to annoy them. Nevertheless, in order to put the abductees' mind at rest the aliens explained that they had not come as invaders. 'That's what Hitler said', commented Bowles. The aliens were offended by this, too, but still continued to impart more information to the hapless couple; sadly, most of it went over their heads.

After that there was a sudden flash of light and they found themselves back in their car on an unfamiliar road; about an hour seemed to have gone by. Bowles was psychic and said she was sure that the aliens would return. They did, but this time they appeared to her alone, giving her a message of a religious nature which she refused to divulge – even to Pratt.

The People From Janos

Apparently there was once a planet called Janos, which was located several thousand light years from Earth. It was inhabited by humans who had left Earth many thousand of years ago. It was like Earth, only nicer, in that there were no wars, work or similarly unpleasant things. Yet it seems that such an untroubled life could not go on in this paradise for ever.

Janos had two moons, both smaller than the Earth's. The one nearest to the planet was called Saton. For some reason Saton's movement slowed down imperceptibly and it gradually moved closer and closer to the surface of the planet until the tidal effect caused by the gravity of Janos ripped the moon apart and huge rocks showered down onto the surface of the planet, destroying everything.

The people of Janos had seen this disaster coming, however, and had built huge, orbiting spaceships, which were large enough to accommodate the entire population of the planet. Other, smaller ships were built in enormous, underground factories and were used to ferry the population to the vast, orbiting cities. Despite their hugely advanced knowledge of science and technology the people of Janos were caught unawares when Saton began to break up ahead of schedule. Those who were not killed outright by boulders the size of houses falling from the sky made for the underground shipyards. Ships carrying as many survivors as they could hold took them up to the orbiting spacecraft through showers of rock. Many made repeated trips, but in the end the rock showers became too severe and many Janosians had to be left to die.

The survivors on board the enormous spacecraft watched helplessly as the nuclear-power station on the planet blew up, encircling Janos with a shroud off radioactive dust.

When it was safe for rescue ships to travel to the planet's surface again they found that those who had survived the catastrophe were terminally ill with radiation sickness. They did what they could for them and the giant spaceships remained in orbit until the last of those on the surface had died. Then the traumatised survivors of this planetary disaster set out across the vastness of space – their destination a planet that they had read about in their history books: Earth.

They accelerated to something approaching the speed of light, but still the journey took thousand of years (in earthly time). When they arrived above the Earth they abducted an unassuming English family and showed it videos of the catastrophe that had befallen their planet.

The family in question consisted of John, his wife, Gloria, their daughters, five-year-old Natasha and three-year-old Tanya, as well as John's sister, Frances. On the evening of 19 June 1978 they were travelling home to Gloucester after attending a family funeral in Reading, taking the A417 via Faringdon and Cirencester, a route that they were all thoroughly familiar with.

Shortly after passing Standford-in-the Vale they noticed a bright light in the sky which seemed to be keeping pace with the car. But John was driving at less than 50 miles (80km) an hour, which would have been impossibly slow for an aircraft. John remarked, half jokingly, that it was UFO; he wanted to stop and have a good look at it, but they kept finding reasons not to do so. Then things seemed to become somewhat spooky: for example, they saw an illuminated house which was not there when they travelled along the same route on the following day.

John slowed down and watched the light moving ahead of them before turning back. He stopped by a hedge, but when he got out of the car the light was no longer there. Suddenly a huge, bowl-shaped craft appeared directly in front of the car, thereafter moving away to the right and sinking down behind a line of trees. He could now see a line of coloured lights along its rim. Gloria and Frances, who had stayed in the car, saw it too, and described it as looking like a flying-saucer-shaped craft: round and flat, with a domed bulge in the middle. Apart from the lights along the rim it was dull, black and featureless. Similar craft had been seen before in the Cirencester area. They next heard a weird, swishing sound and the spacecraft moved up and down several times as if it was having trouble landing. Gloria become frightened and shouted to John, telling him to get back into the car. He did so and they drove off.

From then on things become even more strange. The road – which they had travelled along many times before – seemed unfamiliar: it was narrower than they remembered and flanked by tall hedges at points where they knew that there were broad views of the countryside. The adults were uneasy. Yet there was no possibility that they had taken the wrong road: there were no turn-offs along that stretch. Their journey down this tunnel-like section of hedged-in road seemed endless; details seemed to repeat themselves and the two sides of the road looked like mirror images. The road, which they knew to be straight and flat, now curved and undulated. They had the sensation that they were floating. It was a hypnotic, dream-like experience. John felt that he had taken his hands off the steering wheel and his foot off the accelerator; he believed that it would have continued under its own volition.

The 1 mile (1.6km) to Faringdon took between 30 and 45 minutes, yet John had been driving at a speed of between 35 and 40 miles an hour all the way. All the time the light in the sky had followed them. Suddenly they were in Faringdon. It was only a small town and after they had left it Frances noticed that although the light was still accompanying them it now seemed to blink out every time they passed through a town or village before reappearing again. After they had passed Cirencester, 18 miles (29km) down the road, however, it disappeared completely.

They arrived home nearly an hour later than they had expected. John phoned the RAF and asked whether any experimental aircraft had been flying in the area. Extensive checks were made with military and civilian airports, but nothing airborne on that night could be found that could have accounted for the family's UFO sighting.

When Frances was leaving to go to her own home Natasha told her that she should keep her windows tightly shut 'or you might get sucked up into a spaceship'. John and Gloria went to bed immediately, feeling rather unwell. An hour later they awoke to hear the same swishing sound that the UFO had made. For the next few days John and Gloria suffered form an unexplained itching of the skin, and when Frances went to the hairdresser's four days later the shampoo stung her scalp. All three adults found strange marks on their bodies: they were like bruises, only more sharply defined and they did not hurt when they were prodded.

A week after the incident, when John was in bed with flu, he dreamt that he had actually entered the flying saucer. In his dream John and the others had got out of the car and had mounted a sloping, hazy beam to a doorway in the side of the ship. Once inside, they had walked along a corridor that followed the curved contours of the ship. Some inner compulsion had told John to enter a room off the corridor; the two women, who were carrying the children, also entered separate rooms.

The room that John entered was full of electronic instruments covering the walls. What appeared to be a black dentist's chair was standing in the middle of the room. Telepathic instruction told him to sit in it and restraints gripped his legs when he complied. (The position of these restraints matched the marks found on the adults' legs.)

then a thin beam of light shot out of the panel in front of him and scanned his body. When it had finished he was released from the chair. He got up and returned to the corridor, where he met up with the others, and by way of the beam the five walked down the corridor, out of the doorway and back to the car.

A few days later Frances also had a dream, which had numerous parallels with John's except that she recalled that they entered the ship via a moving ramp. She also said she had seen aliens watching them as they entered the craft; they seemed to have been wearing tight-fitting, silver suits.

Natasha had dreams, too, in which she was examined in a room by a number of silver-clad aliens. What had disturbed her most was their eyes, which she said were 'funny'. She also began to have waking recollections of what had happened. She said that when they were leaving the ship the aliens had given them a fizzy drink which they said would help them to forget their experience; she had refused to take it, however. (Frances later confirmed under hypnosis that she had accepted a fizzy drink.)

Natasha remembered that a lady in a gold suit had come to fetch them from the car; it was she who had switched off the lights, because Daddy had forgotten to do so. The UFO had had retractable legs, Natasha said, which went up into the spaceship in a lift. When they entered the craft she had seen a big screen that showed the road and their car below them. She said that she had been in a room full of aliens of both sexes. They had all sat down and had been gripped around the middle by some sort of safety belt. Then she had felt the sensation of the spaceship taking off.

Natasha furthermore revealed that a lady named Akilias had taken her into another room and had shown her a video of a flying saucer exactly the same as the one that they were on. The film showed the saucer flying over, and landing on, several different planets. There were 'monsters' on one of the planets: huge, naked, hairy men (which sound rather like the North American sasquatch) who were covered in mud and lived with their families in caves, sleeping on straw mattresses. She recalled having seen four types of monsters: their names, she had been told, were saunuses, vonasons, fains and phusantheases. When she sketched them they all looked distinctly humanoid. The first three were decidedly unfriendly, but the fourth, the phusantheases, appeared to be the little green men of classic UFO yore.

University lecturer and UFO investigator Frank Johnson was convinced that the family had had a CEIV – a close encounter of the fourth (IV) kind – and persuaded all of them to undergo regressive hypnosis. Over a number of sessions Frances, John and Natasha came up with a fairly consistent recollection of their abduction. What had happened was as follows.

After the car had stopped it was surrounded by fog or mist. The spaceship straddled the road, but as it was 350 feet (107m) in diameter they were still 50 feet (15m) from the bowl-shaped centre of the hull. Several shadowy figures dressed in silver suits surrounded the car – John said that there were seven. They opened the car door and when the family got out a beam of light shone down from the spaceship above. They then stood in the light and floated up the beam into the ship. Natasha saw one of the silver-suited creatures below lean into the car and switch off the ignition and lights.

They floated into the craft through an inner hatch and found themselves in the middle of a large, circular room some 150 feet (46m) in diameter. An alien ushered them to a moving ramp that led up to a balcony on which were three or four silver-suited men. One of the reception committee made a short speech of welcome. He explained that the family would first be examined to see if earthlings were the same as them, after which they would be pleased to answer any questions that the family may have. After a guided tour of the ship they would be taken back to their car and everything would be exactly the same as if they had never stopped.

After this introduction the adults were taken into separate rooms – the children remained with their mother. As in John's dream, Frances found herself in a room which contained what seemed to be a dentist's chair. She was told telepathically to sit down and was assured that she would not be harmed. However, when she sat down she found herself being thrust back into the chair as if being pushed by a heavy weight. Then she was dazzled by a bright light. She began to panic, but was told to be calm; a soothing image appeared in front of her which seemed to help. It was only at that point that she noticed that there were two thin, silver-suited men in the room who were attending to the instruments that lined the walls. They were about 6 feet (1.8m) tall and had blue eyes and blond, crew-cut hair.

She was in the chair for about 20 minutes. After the examination was over she was told that there were very few differences between humans and their abductors. The human pulse rate, however, was higher (which is perhaps hardly surprising under the circumstances). The aliens said that if they settled on Earth they thought that their pulse rates would adjust themselves automatically.

Outside the examination room Frances met a bald alien who was two or three inches taller than the others. She was taken to a café to an area where there were around 20 aliens sitting around drinking. An alien called Uxiaulia then introduced himself to her. He was a pilot from Janos, he said, explaining that although they had come from a planet that was several thousand light years away the journey had taken them only two Earth years. (If they were travelling close to the speed of light, according to the theory of relativity, time would slow down, so this would be perfectly possible.)

Next Uxiaulia showed her a film of people dying of radiation sickness on Janos while he explained how their planet had been destroyed. Then he showed her stills and films of how it used to be, including images of children laughing, women wearing colourful clothes and happy people boating on the lake and enjoying barbecues in the garden. Frances was particularly interested in the odd-looking fruit and vegetables that they were eating (Janosians apparently ate little meat). Uxiaulia told her that their technology was very advanced – they even had cars that floated above the ground. He also told her that they powered their ship with static electricity; for them, coming to Earth was like stepping back in time.

Suddenly he said that they would have to go – someone was coming – and the tall alien escorted her through the maze of corridors. She could hear a deep humming – like that of a turbine engine – somewhere below; then the deck pitched, knocking her off her feet. The tall alien took her into a room where she was reunited with Gloria and the children. They sat down and waited for John, noticing that all the furniture was bolted down, like that on a sea-going ship. All around them the aliens were tapping out instructions on buttons.

John had been accompanied to the examination room by an alien called Anouxia. He seemed to be the captain, but it appeared that no one was really in charge – it was not necessary, because the ship was run by computer. Anouxia was about 6 feet (1.8m) tall, with blue eyes and a blond crew-cut. After he left John two slim, silver-clad figures entered – female aliens about 5 feet 3 inches (1.6m) tall. John was aware of the shapes of their breasts showing through their silver uniforms; their skin was soft and young-looking and they wore flimsy, silver helmets over their hair. Most of the aliens wore round badges on their chests, but the two female aliens wore similar badges, carrying flying-saucer emblems, on their belts.

Their names were Serkilias and Cosentia, they said; they already knew his name. They explained that the purpose of the examination was to see whether their race could adapt to life on Earth, from which John inferred that they planned to settle on the planet permanently. Then they strapped him into the chair and performed a number of psychological tests on him, which seemed to involve hypnosis. John also recalled the sensation of movement, as if the spaceship was taking off. When the examination was over the females told him that they had taken blood samples. After that Anouxia returned and talked to the females in their own language, which was completely incomprehensible to John. Thereafter he took John on a tour of the ship. At one point they passed a porthole; looking outside, John could see only darkness. Anouxia reassured John that the members of his family were alright – they were still undergoing their medicals.

After that Anouxia and John took one of the ship's floating lifts down to the balcony where they had been greeted. In the middle of the a large, circular room, John saw their car. There were also a number of pillars in the room, with square units roughly the size of a filing cabinet at the bottom of each; when Anouxia shouted into a microphone around 50 aliens ran to man these units. John looked upwards and saw that the columns supported a transparent deck, above which was a huge rotor which now began to move. Anouxia explained that if it span fast enough it acted as an antigravity machine, revealing that the rotor generated electricity at a 'very very high voltage'. Huge cables emerged from the bottoms of the columns and inside one of the cabinets John saw a black box which could have been a capacitor. He later noticed what he took to be a giant transformer – indeed, it seemed that the aliens powered their ship with nothing more complicated than a huge Van der Graaff generator.

Suddenly Anouxia said that John and his family had to go – someone was coming and they were frightened of being captured. Soon the rotor had sped up so fast that its blades looked blurred, causing considerable vibration and making a loud, humming nose. Like Frances, John was knocked off his feet by a jolt, whereupon Anouzia laughed and showed him the special shoes that he was wearing that kept him anchored to the floor, even when there was no gravity. John was then given an extended tour of the engine room, but it appeared to him to be nothing more than a jumble of wires and monitors. Then they took the floating lift to the navigation room, which contained a horseshoe-shaped desk from which the ship was controlled. Anouxia pushed buttons and turned knobs at the desk, but nothing that he was doing made any sense to John.

John asked Anouxia where he had come from and with the help of images shown on a screen Anouxia ran through the journey in reverse. Starting from Earth, they shot out through the solar system until Janos and its two moons eventually came into view. Then Anouxia told John about the destruction of his planetary home: John was shown how green and fertile it had once been before seeing images of the surface of Janos strewn with huge boulders. The video probed beneath the surface of the planet down long tunnels to underground chambers that were crammed with people who were clearly near to death, their bodily disfigurements being hidden by monks' cowls.

After showing him the video Anouxia took John to the room where the others were waiting. They were told that the spaceship was now in position and that they were about to be returned to their car. Anouxia said that they would meet again. 'When you see us again, you will know us', he promised. Then they were offered a drink to help

189

them to forget their experience, the aliens explaining that if they remembered everything they would make their knowledge public and would consequently be exploited. With a little dose of the elixir of forgetfulness the memories would take some time to come back to them. When Natasha refused to take her drink Akilias, her minder, said that it did not matter – because she was little no one would believe her anyway.

Anouxia shook John's hand and kissed the women goodbye. They were then shown to the circular room, but their car was no longer there: instead they could see it 30 feet (9m)below them. One of the silver-clad men said that he would accompany them to their vehicle and walked out into what appeared to be thin air. The family followed him and began to descend slowly; once they were clear of the ship they could feel the breeze on their faces. They landed so gently that their knees did not even bend. The alien then told them that they would remember none of what they had seen, before floating back up to the ship, which ascended and sped away.

From his regressive-hypnosis debriefing of these five abductees Johnson put together a history of Janos and its people. The message was clear: humankind's brothers and sisters from outer space needed somewhere to live. Johnson made a rough estimate of the number off survivors of the planetary catastrophe and suggested that an island the size of either New Zealand's North or South Island could sustain them. However, as the Janosians were clearly Nordic-type aliens – space-Vikings, as Johnson called them – perhaps Iceland would be better. Either way, they have not come back to explore the possibility.

190

Men in Black

At 5.15am on 23 January 1976 a 17-year-old receptionist named Shelley got off the bus in Bolton, Lancashire. She was on her way home from work and had just a short walk ahead of her, across a quiet housing estate, when she saw two coloured lights in the sky which appeared to collide. Then a flying saucer the size of a house swooped down to rooftop height. The wind that it generated flattened her, after which the terrified Shelley picked herself up and ran home.

So hysterical that she could not speak, she dragged her mother into the street to show her the UFO. It had gone, however. Her mother feared that she had ben raped and called the police. By the time that they arrived Shelley was calm enough to tell her story; the police concluded that she had been startled by a low-flying aircraft and left.

The day after her experience Shelley was having a bath when she noticed burn marks on her body. For the next few days she felt dizzy and sick and her eyes were sore. Her doctor diagnosed flu, but this did not explain what had happened to her fillings. During the encounter she had sensed a vibration in her mouth; it later transpired that her fillings had turned to dust and that she required emergency dental treatment.

A number of other young women reported having seen lights in the sky on the same night and the local paper decided to run a UFO story. It called the police, who told it of Shelley's encounter. Reporters beat a path to her door, but Shelley was young and unable to cope with public pressure and therefore did not tell the whole story. She crucially omitted to mention that she had lost around three-quarters of an hour: although she had seen the UFO at 5.20pm – when she was less than a minute from home – she had not arrived at her house until 6.05pm

During the following years Shelley had a number of psychic experiences, which often included the sensation of floating; her mother and sister also claimed to have seen her levitate. Eight years after the encounter Shelley contacted a UFO investigator, who organised hypnostic-regression sessions for her. Under hypnosis, Shelley recalled having been in a room with a female alien who had platinum-blonde hair and was dressed in a surgical gown. The word 'Babinki' then sprang into Shelleys mind. (She had no reason to know that Babinki was the name of a French neurologist who had developed a neurological test to check for spinal damage. In a healthy patient the big toe automatically extends when the sole of the foot is stimulated; this reflex is checked on new-born babies and accident victims.)

The female alien poured all sorts of ideas into Shelley's head. At home some time in the future, she was led to believe, these memories would be triggered and she would be used as a messenger. She was also told that the aliens would return to see her again when 'man rises against man and nation against nation'.

The hypnosis sparked nightmares concerning a holocaust and Shelley developed a morbid fear of atomic energy. She said that she was also visited by men in black suits, who tried to persuade her that the UFO that she had seen was, in fact, an experimental aircraft.

Swedish Conversation

In April 1969 a Swedish hippie named Kathryn Howard was enjoying a trip to the fjords with two friends, Martin and Harvey. Although it was a beautiful, sunny day, the conversation turned to world problems, particularly the wars in Biafra and Vietnam, which were under way at the time. Kathryn became upset when she thought of all the cruelty in the world and began to cry, whereupon she looked up to see a huge, oval UFO hovering the sky with its landing gear deployed. Martin saw it too, but before Harvey could look skywards the UFO had vanished. The next thing that they knew was that they were at home on the sofa. It was 11pm and they had 'lost' hours. Martin and Kathryn had had some sort of vision. They thought that they had seen the moon at close quarters – although the first lunar landing would not be until July, Apollo 9 had orbited the moon in March and had beamed back pictures of it. In her vision Kathryn felt herself floating and saw the Earth below her. She had the feeling that she now understood the forces of nature; hearing a slow, steady pulse, she believed that she was hearing the heartbeat of the universe – somehow it seemed that she had tapped into the 'cosmic consciousness'. She saw the whole history of the human race flash before her eyes and was overwhelmed by an intense feeling of joy. Wars, famine and diseases, she felt, were just part of the evolution of the planet, but if everyone had the insight that she had been given there would be no more human conflict.

In 1986 Kathryn underwent regressive hypnosis in an attempt to relive her vision. Under hypnosis, she recalled having seen the UFO retracting its landing gear. Then it had seemed to suck her inside it before whisking her into space – she had been terrified that she was being taken away permanently. The aliens whom she had seen on board were transparent. They had taken her to another planet, where she had seen a rocket being launched. She had been allowed revisit her past lives and had had a terrible vision of the world being destroyed in a holocaust – it was a warning to humankind.

The Anti-gravity Machine

In 1979 Graham Allen, a painter and decorator, was living in the wilds of Staffordshire while his fiancée, Charlotte, stayed in Maidenhead, Berkshire. At weekends they took it in turns to drive to each other's house. On Friday 17 June – a sunny day – it was Graham's turn to drive to Charlotte's and he was heading along the A34 as usual. He was nearing the point at which he usually turned off on to the A423 when he heard the DJ on the radio say that the time was 5.55pm. Then the radio crackled and fell silent; he looked down at it and saw that the power light was still on. When he looked up again he could hardly believe his eyes: it was no longer bright and sunny, but overcast and teeming with rain and he had to switch on his windscreen-wipers.

Although he had driven up and down this road twice every other weekend he no longer recognised anything that he saw. He discovered why when he saw a sign saying that it was 3 miles (4.8km) to Newbury – he was therefore 20 miles (32km) past the turn-off to the A423, yet the clock on the dashboard said that it was only 6.05pm. Somehow he had covered 20 miles in 10 minutes, which meant that he would have had to have been travelling at 120 miles (193km) an hour. He certainly hadn't been driving that fast.

He pulled in to the entrance to a farm in order to turn around. The next thing that he knew was that although he was sitting in the car in the farm entrance the car was facing towards the road, which meant that he had already turned round. And the radio was now working perfectly. He was three-quarters of an hour late when he finally got to Maidenhead, but on his arrival he could not get out of the car: his legs were temporarily paralysed. This 'missing-time' experience heralded a series of bizarre dreams, which gradually subsided, however.

Over Christmas in 1987 Graham awoke at 3am and suddenly had a vision of being in his car on that day eight years earlier. He heard a low, humming sound that frightened him. There was a man on the other side of the road and Graham screamed at him for help. But it was too late: the humming intensified and the car was bathed in a golden light, whereupon it began to rise upwards. He passed out and awoke to find himself lying on his back and unable to move because three aliens seemed to be operating on him. Since then he has received telepathic messages from the aliens, including details of how to build an antigravity machine, which, sadly, he has not yet done,

How Not To Be Abducted

Jason Howard was an insurance salesman who had gone back to university to work for a PhD English literature when he became involved in Professor Jacob's abduction study. Under hypnosis he recalled having been abducted in 1976, when he was 17.

The aliens had shown him a picture of a nuclear explosion which had produced a huge, white cloud that enveloped much of the world. They had discussed the atomic bombs that had been dropped on Japan at the end of World War II and the alien had been surprised that Howard knew about it because it had happened before he was born. But the explosion that he was being shown on the screen was not something that had happened in the past, he was told – it would happen in the future. He was being given a terrifying preview of a nuclear holocaust and he must warn the world about it. It would happen one month before his fortieth birthday – in other words, in

1999.

Passing on one such dramatic message to the world via Howard was evidently not enough, for the aliens repeatedly abducted him, which became rather tiresome. Later, however, he stumbled upon one of the few effective defences against alien abduction: he discovered alcohol. When the aliens grabbed him from his college room one night he was in no mood to co-operate. Although he was happy enough to lie on the examination table, when they started fiddling about with him he got up and protested. The aliens tried to calm him down, but he would have none of it and stood unsteadily in the middle of the room making wild karate movements. The smaller aliens cowered against the wall while a tall alien tried to reason with him, but with no success because Howard was fighting drunk. The tall alien then stared deep into his eyes and the next thing that he remembered was that he was standing on the college lawn, about a mile away from his room, in his underpants.

Alien
Experiments

Alien Evidence

Sometimes when alien abductions are performed carelessly, the aliens leave evidence behind them. This was the case in the abduction of two teenage girls who were taken more than a decade apart.

On the night of 7 October 1955 a young girl named Jennie was tucked up in bed in Nebraska when something strange happened. She had to wait for nearly 30 years to find out what, until, in 1984, she was hypnotically regressed and recalled that a creature whom she called 'The Explorer' had come to visit her. He had been hovering outside her bedroom window telepathically willing her to come to him. She had tried to resist and had pretended that she was dreaming, but the power of his suggestion had been too strong and she had eventually asked how she could follow him. She had received the answer telepathically: the Explorer told her about his laboratory, which, she said, was like two 'dessert bowls stuck together'. She could not really envisage it, but he had nevertheless placed the concept in her mind.

Jennie's resistance was now weakened and she began to float towards the UFO; drifting through the walls of her house she even glimpsed the dirt and cobwebs in the cavity between the inner and outer walls. Once outside, she saw the UFO clearly, although the vision pulsed. As it faded she could see inside the craft and then right through it, to the car park below.

It was chilly outside, but even colder inside the UFO, like being inside a freezer. The Explorer was waiting for her in his laboratory wearing a white, surgeon's cap, while smaller creatures milled around him. He was between 3 and 4 feet (91cm and 1.2m) tall and had a head that was shaped like an egg. His face was waxy and greyish – it looked sensitive, as if she would hurt him if she touched it. His nose was a tiny bump with two slits for nostrils and his mouth was another slit. Jennie said that he was stern, but not angry – he clearly meant business.

The Explorer did not speak to her, but instead implanted thoughts into her head. She was told to jump up on a silver table. She then asked him where they were going, to which he replied 'Nowhere'. Clamps grabbed her and held her down. Along with a helper, the alien collected samples of her hair. A blood sample then was taken using a capillary tube. She complained that this hurt, and although he pretended that he did not care Jennie knew that he did. When the examination was over her returned to her bedroom. The sole purpose of her abduction seemed to have been the collection of samples, as if human beings were something to be studied and experimented on. In the morning she awoke and remembered her abduction as having been a vivid dream. When she looked out of her window however, she saw the elm tree outside had been burnt. Her father told her that it had been struck by lightning, but Jennie knew better.

The aliens who abducted Shane Kurz, a 19-year-old nurse's assistant, did not set a tree on fire, but instead left muddy footprints that led from outside up to her room. At around 4am on the night in question, 3 May 1968, Shane recalled having seen a UFO. She remembered nothing more until her mother woke her in the morning; she could not explain the footprints.

Six years later she approached Professor Hanz Holzer, a parapsychologist, with her story. She was desperate: since that night she had suffered from nightmares and migraines; strange, red rings had appeared on her abdomen and she had stopped menstruating for nearly a year. Doctors were at a loss to explain her condition. Then, in February 1973, her symptoms suddenly disappeared. She still wanted to get to the bottom of what had happened to her, however, and she felt that her unexplained illness was in some way connected to both the UFO and the mysterious, muddy footprints.

When Professor Holzer hypnotised her Shane remembered having been abducted from her bedroom by aliens. They were small, with grey skins, probing eyes and no hair. Having taken her to a strange room, they had then insisted that she get on to a table; she had not wanted to, but had had no choice. One of the aliens had seemed to know her, but she could not work out how. He had stuck a long needle into her abdomen and had taken samples of her ova. She was told that she had been chosen to give them a baby and that they were experimenting on her to see if she could. After that she was raped.

Professional Courtesy

Given the aliens' predilection for examining human women intimately, who better to find out what they are up to than a professional gynaecologist? On 7 August 1965 a highly respected Venezuelan gynaecologist and two businessmen were visiting a study farm at San Pedro de los Altos, 30 miles (48km) south of Caracas. During the late afternoon they were discussing horses and investments when suddenly there was a blinding flash. The three men looked up to see a glowing sphere, bathed in yellow light, drifting down from the sky. They sensed a soft humming, which seemed to be coming from inside their heads. One of the men turned to run away, but the gynaecologist grabbed him. 'Stay and watch', he said.

By that time the sphere was hovering just above the ground, but it was still a safe distance away. Suddenly a beam shot out of the side of the globe; it was angled to the ground like a ramp and two aliens then floated down it.

They were 7 feet (2.1m) tall, with blond hair and huge, round eyes and their suits were made of tinfoil. The three men were terrified, but fear rooted them to the spot and they could not run. The aliens casually walked over to them. 'Don't be afraid', they said. 'Calm yourselves.'

They communicated with the men telepathically, the men hearing the voices inside their heads. The gynaecologist asked the aliens what they were doing on Earth. 'We come from Orion', said the aliens, explaining that they had come to Earth to study the psyche of humans in order to adapt them to their own species. They also wanted to experiment with the possibility of interbreeding with humans to create a new, hybrid species.

The aliens also helpfully unravelled another mystery – why some people report encountering tall, blond, blue-eyed, 'Nordic' aliens, while others see small, black-eyed 'greys'. They explained that there were other extraterrestrial visitors to Earth besides themselves. The short aliens were from the 'outer dipper', but their purpose was unclear. The Nordics said that their own mission was peaceful. For example, they could have brought a 'wave compressor' big enough to disintegrate the moon, but instead they had only brought a few small ones with them. Although puny, these were 'powerful enough to halt an atomic explosion' (they presumably carried them for self-protection).

The Venezuelans and the aliens then had a long, philosophical discussion, but the men's memories of what had been said later became hazy.

Mysterious Bleeding

On the night of 29 April 1995 a company director named Malcolm, his wife, Samantha – a charity organiser – and their daughter, Lizzie, were driving across the south of

France. They were travelling down the motor way between Epagny and Caignes Cordon, two hours from Dijon, where they intended to spend the night, when they saw a UFO, which appeared to be following them. They tried to shake it off by speeding up and then slowing down, but the only time that it disappeared was when other traffic appeared on the road. Neither Malcolm nor Samantha believed in UFOs, but after it had followed them for several hours they were forced to come to the conclusion that not only did they exist, they were not necessarily friendly.

Although they drove for hours they did not reach Dijon that night, and at around 2am, after checking that the UFO was gone, they slept in the car. When they awoke Samantha had a nosebleed and Lizzie was bleeding from the anus. It was only when they got back to England that they realised that it had taken them three-and-a-half-hours to cover the 40 miles (64km) from Epagny to Caignes Cordon. (They later drove along the route again and ascertained that the could not have taken the wrong turning.)

After seeing the UFO all three were plagued by disturbing dreams and saw humanoid creatures with black eyes. Nine months after the encounter Samantha awoke to hear Lizzie crying, but found that she was temporarily paralysed. Malcolm went to see what was wrong with Lizzie, but before he did so he first stopped by the window – there was someone digging up the road outside, he said. Malcolm said that he had then found a crying Lizzie looking out of the window. When Samantha could move again she also went to the window and observed people digging in the garden. In the morning, however, there was no sign either of any road works having been carried out or of the earth having been disturbed in the garden. Worse still, both Samantha and Lizzie were bleeding from the rectum. They believed that the bleeding was connected to the UFO and that aliens were carrying out a bizarre experiment on them.

A Seminal Experience

The 19-year-old computer-programmer Will Parker was driving his wife, Ginny, through Virginia late one night in 1974 when he pulled into a petrol station that was closed. For some reason that he could not explain he turned off the car's engine and lights and waited expectantly in the darkness.

Under hypnosis, he remembered that they had chatted nervously while they waited, but had had no idea what they were waiting for. Then Ginny had told him to be quiet because she had heard something; someone was out there. Indeed there was, and a small alien now appeared beside the car. Ginny was shocked and started praying, but Will was calm: he suddenly realised that he had seen aliens before. Ginny soon fell quiet and Will saw that she was asleep. Once Ginny was fully unconscious the aliens took Will out of the car. As they took him away he could only think about going back to lock the car – he did not think it safe to leave Ginny alone in there.

Four or five aliens were waiting at the back of the service station; 'Where are the rest?' asked Will. Telepathically, the aliens told him not to be afraid: they were not going to hurt him. They promised that they would bring him back and told him not to worry about Ginny – she would remember nothing. They were standing together in a tight group, waiting, when an alien spacecraft about the size of a building appeared. They were next lifted up as a group, although Will did not know how, and suddenly found themselves inside the spacecraft, with the Earth beneath them.

An alien pushed its face into Will's and he had the impression that it was scanning his mind. Next, they fitted a machine over his genitals, whereupon it vibrated and Will felt semen being sucked out of him; he had no orgasm or sense of pleasurable release, however.

Harvesting Foetuses

In 1979 Tracy Knapp, a 21-year-old musician, was driving from Los Angeles to Las Vegas with two girlfriends when they saw a light swooping down at them from the sky.

Under hypnosis, she remembered that as the light had whizzed past the car had started spinning, whereupon all three of them had started screaming and crying. The car was being lifted up into the sky when Knapp remembered seeing hands coming in through the window. When they touched her she became limp and they lifted her out of the car. From then on she lost sight of the other women and did not see them again until they were back on the ground.

She recalled lying down with her legs pointing upwards; two creatures were pressing on her and one was cutting her internally with long-handled scissors that had very small blades. They then dowsed the wound with a fluid that burned her. The procedure continued for a long time. The aliens seemed to have been cutting threads before pulling out their instruments and removing a sac containing a tiny foetus. This was put into a small, silver cylinder about 3 inches (7.6cm) wide, which was in turn put into a drawer in the wall, along with numerous other live foetuses.

Sisters in Space

Janet Demerest, a secretary, and her sister, Karen Morgan, the owner of a public-relations firm, suffered numerous abductions. The first was in 1963, when Demerest was nine. On the occasion in question she remembered having been playing near her house with some of her fellow Brownies; the other girls formed a circle, but Demerest wandered away on her own. Under hypnosis, she recalled having seen a man with grey skin; he was not very tall – only about the same height as she. Together they walked into the woods holding hands, which reassured her. She then saw a UFO landing in a clearing and she and the alien walked into it up a ramp.

Inside the craft she was led to a large room in which there were a human woman and an odd-looking girl, who had greyish skin, thin arms and long, slender fingers; she seemed to have no bone structure and no ears. The man wanted Demerest to play with her, so they sat down together on the floor. While Demerest found her stare riveting and could not pull her eyes away from it; all of the time she was aware of the man and woman watching her. Eventually the girl hugged her and told her that it was time to go. The man then took Demerest back.

When Karen Morgan was 28 she also recalled having had an abduction experience, 6 years earlier, in 1981. Under hypnotic regression, she remembered having entered a UFO and having been taken to a waiting area where there were a number of benches in arched alcoves. She sat in one and saw that there were other men and women waiting in the others. Some wore night clothes; one young man was slumped as if he was not well; another woman looked very frightened. They were then strapped in and Morgan had to tell herself not to panic; she had the curious feeling that she had been through all this before. Then the aliens came, two per human. The first woman was stripped and they were all herded into an examination room; the sick man had to be helped.

Morgan tried to resist, but the aliens pushed her along anyway. She was the last into the examination room, in which there were four operating tables and a shelf with instruments on it running around it. She was stripped and strapped to the table. (Demerest recalled having had a similar experience, but said that she had not really had a sense of what was happening or of who she was). Morgan had braces on her teeth at the time, which fascinated the aliens, who asked her to take them out. She refused, but when she awoke the next morning she found them lying on her stomach.

The aliens also cut out a sample of her gum for analysis, which infuriated her. She asked how much more of her they were going to take and how long it took to study someone. Their answer was that it could take years. A tall alien then asked her to look into his eyes; she did so and immediately felt that she was being overwhelmed, as if she was falling into them. Her will-power had been sapped and she found that she could not look away or fight the alien in any way – it was as if she no longer had a mind of her own.

Morgan later became angry about the gynaecological procedures that the alien was performing on her and cursed the creature in her mind. The alien read her thoughts and reassured her that she would come to no harm. He then carried out what seemed to be a smear test, but she believed that they were inserting an embryo into her, implanting it into her womb. She found this idea repulsive and told the alien that he could not do this, but he replied that she had no choice: it was part of a very important programme. Still she protested, however, saying that once she was back on Earth she would have an abortion, to which the alien countered that she would not because she would not remember the embryo having been implanted into her. Although she kept insisting that she would, she nevertheless sensed the alien's hypnotic suggestion that she would forget the incident overwhelming her mind. There was nothing to worry about, the alien reassured her: they had done this many times before.

Morgan then remembered that she had indeed been through this procedure many times before. She felt sick, comparing herself to an animal that was being experimented upon. The embryos, she knew, were hybrids: part human, part alien. Sometimes the procedure was quick, but this time it had taken longer because she had resisted. When the alien had finished he pulled his instruments out of her body and patted her on the stomach. Morgan was disgusted and told the alien to take his hands off her; he reluctantly did so, shaking his head as if bewildered by her unco-operative attitude. In the morning, Morgan awoke in her own bed. On feeling a mysterious, gooey substance between her legs she took a shower and washed it off.

Demerest recalled having undergone a similar procedure in 1987, when she was 33. A long needle had been inserted into her vagina and the aliens had implanted a 'little round thing' into her womb. Afterwards a female alien had helped her off the table. She had been left with the over-whelming feeling that she wanted to have a baby.

Morgan was given a glimpse of the results of her labours when she was 32, when she was shown a large number of babies – 50 or even 100 – lined up in boxes behind a glass panel. They were not moving and looked as though they were dead, although Morgan knew that they were alive. They seemed to have been suspended in some sort of liquid and were being fed through tubes by a machine. Although some were only foetuses they appeared to be in all stages of development. It seemed clear to Morgan that the aliens were running some sort of breeding programme and that these were foetuses that had been taken from hundreds of women. She was also shown nurseries, in which babies were being tended by aliens; some of them she was told, were hers.

Morgan was sometimes told to wash a baby, or else to play with it, and saw other women (who were naked) doing the same thing. Morgan had the impression that the intention was for the babies to be touched as much as possible, but they were very unresponsive and did not even laugh or smile when they were tickled. When the women made baby noises to them the children did not make a sound.

In 1987 Morgan, along with the other humans whom she had seen tending the children, was shown in a film what would happen to the babies. It depicted an idyllic land-scape – like a park, with a stream running through it – and she was told that this was where the aliens were taking the babies. At first she thought that the place was on Earth, but as the camera panned the panorama she realised that although the trees, rivers and canyons were like those on Earth they were not quite the same. The narrator explained that this was how the world would be some time in the future, when their programme was complete. Morgan suddenly felt that she did not want to look at the film any more – it was as though the aliens were trying to implant something into her mind – but she was unable to avert her eyes from the screen.

On another occasion when Morgan was abducted – when she was 30 – two aliens had lifted her up from the examination table and had held her at a strange angle. They had told her to look at a picture, but there was no screen or picture in the conventional sense – the picture was inside her head. She saw mother dying of cancer in her uncle's home, with all her relatives standing around the bed saying the rosary. It was as if the aliens had filmed it; she could even see herself standing at the back. Morgan asked why she had to watch this and one of the aliens said that they wanted her to feel it again. But reliving the expe-rience of watching her mother die was too painful and Morgan became angry and tried to fight it – it was as if the aliens wanted to see her suffer.

During another abduction she saw an attractive man whom she knew and had the distinct impression that she was going to make love to him before gradually realising that the man was, in fact, an alien who had staged the erot-ic scene by means of hypnotic suggestion.

Abducted by a Grey Deer

Virginia Horton was a corporate lawyer who lived on the USA's east coast. She was married, with a family, and had an ordinary, everyday life. In 1979 friends told her that there had been an item on the NBC nightly news about the Bershad abduction. It was then that she remembered strange things happening to her in the past.

In the summer of 1950, when she was six, she had been on her grandfather's farm in Manitoba. She remembered that she had gone to the barn to collect eggs. The next thing that she remembered was that she had been in the yard again. She had felt an itch on the back of her calf and had pulled up the leg of her jeans see what was causing it. It turned out to be a neat and surgical wound, as if it had been cut with a scalpel, but there was no hole in her jeans and she did not understand how it could have happened.

She now asked her mother about the incident, but she did not remember it. However, she did recall another strange event that had occurred when Horton was 16. They had been holidaying in France and were having a picnic in the countryside when Horton and her brother had gone off to explore a nearby forest. They had become separated and her brother had been searching for her for about half an hour when she had suddenly reappeared, with her dress splattered with blood. She had had no explanation for this and had claimed that she had been gone for no time at all. When pressed, however, she had remembered meeting a beautiful deer.

Horton then contacted the ufologist Budd Hopkins, who arranged a regressive-hypnosis session for her. It was then discovered that when she was in Manitoba she had been taken into a bright room in a spacecraft by grey aliens. They had said that they were from far away among the stars and that they just wanted to take a little piece of her back with them. She remembered that some type of apparatus on the end of a retractable arm had been used to make an incision in the back of her leg.

In the woods in France she had seen a deer – a grey deer, which had large, black, hypnotic eyes. (Hopkins concluded that the deer was a cover story that had been implanted into her mind by the aliens in order to cover their tracks.) She again remembered having been on a spaceship; the same aliens whom she had seen when she was six were also there. They had shown her a star map and had told her that they were from another galaxy. Then they had given her a lecture on the importance of ecology, bio-diversity and the preservation of endangered species, including the human species – all advanced ideas in 1960. They had also taken her for a short trip in their flying saucer, in which they had ascended upwards for 100 miles (161km) or so. She had been examined on a table and they had stuck an instrument up her nose which had given her a nosebleed, thereby explaining the blood on her dress.

Put to the Test

In 1965, when attorney George Kenniston was 16, he recalled having been abducted in order to act as the navigator of an alien spacecraft. On the bridge, he had followed orders, but had had no idea how he knew what to do. He remembered that he had been undergoing some sort of test: he had been taken to a complex control panel and had been told to operate it. The aliens had given him a destination and he had been supposed to fly the ship there; they had scrutinised him closely while he did so. It had been an almost impossible task, like driving for hundreds of miles in a straight line.

When the test was over Kenniston had floated down from the spacecraft, flying over the town and landing on a hill behind his home. After walking home he had let himself in through the kitchen door and was making his way to his bedroom when he bumped into his father. Kenniston told him that he had got up for a glass of water and then went back to bed.

Multiple
Abductions

Abduction at Copley Wood

Debbie Tomey was a mother of two in her early thirties. Following a divorce she had moved into her parent's house in the Indianapolis suburb of Copley Wood. After having closed the pool-house door on the night of 30 June 1983 she noticed that it had opened again and that the light was on inside – when she looked later the door was closed and the light had been switched off, although none of the family had been near the place. Furthermore, a bright light seemed to be hovering over the bird-feeder in the garden. Fearing burglars, and carrying her father's 22 firearm, she went out to investigate, but found nothing untoward, except that the dog was cowering the back of the car and had to be coaxed out. There was a patch of burnt grass in the garden around 8 feet (2.4m) in diameter, from which a straight path ran, ending in an arc about 50 feet (15m) away. The dog would not go near the area and the birds avoided the bird-feeder.

Like Virginia Horton, Tomey had a scar on the back of her calf – in fact, two; one of them had appeared when she was about 13. Her sister, Laura, and mother each had identical scars, but none of them knew how they had got them. Her sister had reported seeing a UFO when she was a teenager and also had the feeling that she was missing some time. When Laura was hypnotised in order to help her to lose weight she had emerged traumatised and had had to be sedated. And both Tomey and her mother suffered from the same, recurring nightmare, in which they were being terrorised by small, grey aliens. Tomey furthermore recalled that when pregnant she had been plagued by weird phone calls, in which the caller had made odd, guttural sound, as well as clicks and moans. The calls had persisted, even though she had changed her number and had gone ex-directory. Other people – including her mother and best friend – had heard them, too, but they had suddenly ceased when the baby was born.

Under hypnosis, Tomey remembered having seen a light and having gone into the garden on the night in question. Aliens had then abducted her and had performed a number of medical experiments on her, including implanting a small device into her body. She recalled having met a child which she believed was hers – the result of an earlier abduction. When the experience was over she had woken up in the garden, bleeding.

Regressive hypnosis revealed that she had been abducted on numerous other occasions. When she was a child, her mother – who was a victim of alien abduction, too – would hide her in a cupboard in an attempt to protect her.

Tomey recalled having been abducted from a car in December 1977 (the other people in the car had been 'switched off' and therefore could not corroborate her account). She then remembered having undergone a gynaecological examination (abduction expert Budd Hopkins believed that she was impregnated by the aliens on this occasion). When she was a few months pregnant she had been abducted again, and during the subsequent examination had felt a terrible pressure inside her. Then, under hypnosis, she screamed out 'It's not fair. It's mine'.

She later began dreaming that she had given birth to an odd-looking, super-intelligent, hybrid baby. (Similar dreams of abductees giving birth to super-intelligent children – so-called 'wise babies' – soon began appearing in UFO-related literature.) Tomey revealed that all of the female members of her family had had been abducted on several occasions. During these abductions they would be stripped and forced to lie on an examination table with their feet in stirrups while small, grey aliens examined them internally either manually or by means of a needle-shaped laparoscope inserted through the navel. The aliens would sometimes implant something into the women or, at other times, remove something. Tomey also believed that her son would become another victim of repeated alien abduction.

In November 1983 Tomey reported having been abducted once more; this time the aliens had removed some of her ova. When she was subsequently abducted, in April 1986, she was shown two elf-like infants, which she was allowed to hold and name. These were her children, the aliens said, who were presumably trying to encourage her to forge some sort of mother-child bond. The aliens told her that they had nine of her offspring altogether. Tomey realised that the aliens had been impregnating her with alien genes and had then been harvesting the foetuses, which were incubated elsewhere. During one of her abductions she met another of her half-alien offspring – a slim girl, with a huge head and no eyebrows or eyelashes. Even though she looked strange and unearthly, Tomey could not help but feel a twinge of maternal affection for her.

Tomey believed that alien impregnation of human women was quite a common phenomenon. In one case a 13-year-old girl had become pregnant, even though she had protested that she had never slept with a man. Her parents, not unsurprisingly, did not believe her, but when they sent her to the doctor for an abortion the doctor reported that her hymen was still intact.

Abducted to an Alien Base

Christa Tilton was another victim of multiple abductions. He first remembered being taken at the age of ten, when she was visiting her aunt's house in Tucson, Arizona. Walking down a road there, she noticed a huge, orange ball of fire falling to the ground. Directly afterwards, she saw a small, grey creature, with a big head, large eyes and a thin body. She began playing with it, exchanging rocks, before blacking out. She woke up on a table with someone standing over her. This, she was told, was the 'doctor' and she would always remember him. Although he was an alien, he was very human in appearance. He gave her a thorough examination; skin samples were taken and her abdomen was probed with a long needle; a sharp instrument was also inserted into her ear.

This was the first of many subsequent abductions, during the course of one of which she believed that she had been impregnated by aliens. (The foetus was then removed from her womb during an abduction from New Orleans in 1971.)

Under hypnosis, she recalled an abduction from Arizona in July 1987 that stood out from all of the rest. She had been driving through the desert when she had seen an alien sitting on a hill. She had stopped the car, whereupon two other aliens had attacked her. Although she had locked the car doors they had still managed to unlock them, and as she struggled they had manhandled her into their waiting spacecraft. Inside the craft, she was given something to drink which knocked her out.

When she woke up she found herself being led out of the ship, which had landed at the entrance to a cavern. This, she later discovered, was an underground alien base. There were numerous security checks; on one level her alien escort had had an argument with one of the guards, but she did not understand what it was about. As usual, she was given a thorough examination by the aliens. When it was completed she was taken down to level six, but she was not allowed into it because, as her escort explained, there were things on that level that she would not be able to understand and that might upset her.

After this seemingly fruitless trip the aliens returned her to ground level before transporting her to her car. It was now late, and she drove to her aunt's house and went to bed without disturbing her aunt or her best friend, who was staying with her. On the following morning her friend pointed to the long, red, scratch marks on Tilton's back. These, Tilton maintained, were the result of struggling with the aliens when they were trying to abduct her.

Tilton lectured regularly on her abduction experiences, which continued unabated.

'Why Do You Keep Taking Me?'

A 32-year-old nurse called Alex, who had no interest in UFOs or science fiction, reported a bizarre experience on 2 July 1997. It had been her day off and she had been relaxing in a field in the sunshine when she had suddenly felt herself being sucked up into the sky. Above her, she had seen a huge, triangular craft. The she had blacked out.

When she awoke she was lying naked on a table in an operating theatre. She could not move, although there seemed to be nothing restraining her. She saw a number of small creatures with big heads – 'like babies before they are born' – around her. She noticed a teenage girl, who was also naked, lying on another table. Then a taller creature, with black eyes, came in. When he bent over the table she had the feeling that she had been seeing things that she should not have seen; she then went back to sleep. The next thing that she knew was that she was back in the field, but some yards from where she had been sitting. The book that she had been reading had been flung aside, into a bush. There were strange marks on her arms. Her watch had stopped, and when she got home she discovered that five hours had passed.

Thereafter she began to have a weird effect on electrical appliances. She found it difficult to sleep, but when she eventually did so her dreams were full of aliens and space-ships. She also had nosebleeds. Up until then she had been a country girl, but she soon moved to a city in order to be near people.

Under hypnosis, she related the story of her abduction, this time adding some more details. The tall creature, she said, had told her to be calm: everything that they were doing was for her own good. But she still protested. 'Why do you keep taking me?' she asked. The UFO investigator then plumbed her memory for earlier encounters. She said that she had first been abducted at the age of five, when a band of ugly children whom she had met in the woods took her to a circular house to play, but she did not want to play with them because they were too ugly.

In the following year they had come to get her again; this time she had been examined by a doctor. Then, when she was 14, she had been out with her dog when she had been sucked up through the air into an alien spacecraft. This time she had found herself naked, lying on a table; samples had been taken from her and she had been told that she would not be able to bear children. When she was returned to the wood her dog had gone; it turned up two days later, but refused to go near the wood.

She was again abducted at the age of 18; that time she had seen a large room in which lots of naked women were lying on tables. At the age if 27 she found herself unex-pectedly pregnant; when she was abducted shortly after-wards she had found herself in a room containing lots of foetuses in large jars full of liquid. She had later lost the child and claimed, under hypnosis, that the aliens had taken it away from her. The aliens abducted her again when she was 30, but this time they only took samples of her hair, skin and fingernails.

Phantom Pregnancy

Sharon, a woman in her thirties, heard the UFO investigator Tony Dodd talking on the radio about alien abduction and phoned in. She told him that although she had no recollection of having been abducted by aliens she had experienced many of the strange phenomena that plague abductees. For example, electrical equipment malfunctioned around her. Her car inexplicably stalled on lonely roads, only to restart again. She had had 'missing-time' experiences and her clocks seemed to run backwards. She had suffered from mysterious discharges from her navel and, after exhibiting all the symptoms of pregnancy for three months, had found that she was not pregnant at all. She had found blood on her pillow on several occasions, although she had no visible wounds. She had woken up in different parts of the house, although she had no history of sleepwalking. Mud and grass were once found in her bedroom. And one morning she had woken to find that the nightdress that she had put on the night before had gone – it was nowhere to be found.

Under hypnosis, she revealed that she had seen huge UFOs outside her window which had terrified her. Inside them, she said, had been big, white rooms filled with tall aliens with round heads and big eyes. They had taken her aboard regularly and had told her that they had come to help humankind look after the Earth.

During one abduction, when she was a small child, Sharon had found herself in a room with a number of other people. She could see the distant Earth outside. They were given a lecture about ecology and pollution before being marched off into separate rooms. She then found herself lying down, possibly floating. There were small creatures around her, with four, long fingers, who gave her some sort of injection; a circle of gold light had made her feel elated and happy. Then she was allowed to dance and play until a big man told them that it was time to go home. When she was eight the big man again came to take her to the spaceship. Afterwards she thought that she could fly and broke her arm trying to launch herself off a slide.

At the age of 14 she was abducted from the back garden of her aunt's house after she had slipped out to have an illicit cigarette. On this occasion she was taken to a different ship, which was covered in Egyptian hieroglyphs. She remembered having been undressed and examined. Later she was again told about respecting the environment, as well as about the pyramids and spirituality. Energy, she said, was pumped into her body; it gave her pins and needles but made her feel good. Indeed, her abductors always made her feel good, but they also made her feel alone – isolated from everyone else – as if she were something special or different.

They came again when she was 22 and pregnant. The aliens placed the naked Sharon on a table, saying that they were going to check the baby which made her anxious. They stuck a probe into her and pronounced the baby to be fine – they said that it was a girl and that it was going to help them just as she had. When Sharon was 24 they abducted both her and her daughter, Louise. Sharon was given the usual physical examination while Louise was taken to another room. They than performed a strange procedure on Sharon, which, they said, would cure her. When Louise returned she called one of the aliens 'Mummy'.

The next time that Sharon recalled having been abducted was when she was 33. By then she had had a hysterectomy. Her husband was asleep beside her in bed when the aliens took her onto the balcony, where there was a light that she had to walk into. She told them that she had nothing to give them – no ovaries or womb, which, she felt, were the only things that they valued her for. They replied that that was not what they had come for. When she tried to resist them they pushed her on to the examination table and shone the golden light on her, which, for the first time, made her feel guilty.

Lifetime Abductee

Patti Layne, a high-school teacher, was abducted several times during her life. The first time appeared to have been when she was 15 years old. She was on a camping trip with her school when she and eight other students went skinny-dipping in a reservoir. A light approached them from the sky and then shone directly onto them, where-upon they were lifted into the UFO.

Layne was abducted again in the following year. This time the aliens told her that they needed some parts from her that would help everyone on the planet. They took her to a small room; she could see the stars and space from its window. The aliens then sat her on a chair and attached some kind of apparatus to her head which played terrible images of nuclear war, the destruction of cities, devastating earthquakes, mountains collapsing, the sun turning black, people starving and her own family struggling to survive. All this would happen, the aliens said, because humankind could not stop being greedy. Afterwards one of the aliens looked deep into her eyes – it was as if he was looking into her soul, she said. Then he told her to forget everything that she had seen; the next day she would think that she had just had a bad dream. At school the next day she was haunted by the awful fear that there was going to be a nuclear war. And for many years afterwards she dreamt of a nuclear holocaust.

Layne was next abducted in 1982, on her twentieth

birthday, when she was studying at a college in a small town in Pennsylvania. She had decided to go for a drive in the mountains and on reaching a wild spot she turned off the road and began following a dirt track. All the time she was thinking that she should not be there – she had a class that night and was supposed to be studying. Then the car stopped; she waited for a while and it started again. Or at least that's what she thought had happened – under hypnosis, she recalled the car door having been opened and aliens taking her out.

A year later, when she was still at college, she and her friends went into the mountains for a picnic. They spread out blankets to sit on and started drinking wine. Shortly thereafter Layne wandered off into the woods to relieve herself and was squatting down in a clearing when a light was shone at her. She thought that it was Freddy, one of her friends, who was always playing practical jokes. 'Cut it out', she yelled. To her surprise, however, she could not hear her own voice; nor could she hear the others, even though they were quite close, just down the trail. She then became scared and walked back to the picnic spot, only to see a large band of aliens grabbing at her friends. One of them, James, looked ill; the others were being examined with some kind of instrument.

The aliens were small and had no genitals; they seemed to be wearing some sort of military insignia, which resembled a bird. One of them came over to her – he looked friendly and took her by the arm, like a boyfriend would. With the rest of their friends apparently having been frozen, Layne and James were marched off. Two of the aliens practically had to carry James, who looked as though he was going to vomit. They were then taken deep into the woods, where a craft was waiting; it was not a conventional flying saucer, but looked more like a black bubble, with a hatch opening into it and a row of windows around the top. It was light inside and they walked into it up a small ramp. Layne was taken into one room, James into another. One alien pressed its face to hers, as if scanning her mind. She felt a deep sexual bond with him and empathised with his mission, even though she did not know what it was. The alien, she thought, was tapping into her mind, but this was not an unpleasant sensation – in fact, it was rather pleasurable.

Layne was abducted from her bedroom in 1985, when she was 23. On that occasion she recalled having had strong feelings for a tall alien and had been saddened when he had told her that it was time to go home – she had wanted to stay with him. He had assisted her off the table and his colleagues had picked up her clothes and had helped her to dress. The tall alien had then led her down a corridor and shortly there after they had entered a wood. The alien had walked Layne home in the moonlight. When they reached her house Layne had floated through the bedroom window and had then found herself back in bed with her husband. She had shaken him awake and had told him that she had had a strange dream, but he had just fallen asleep again. In the morning she found that the aliens had put on her nightdress inside out.

Bedroom Visitors

Communion

The book – and subsequent film – *Communion* made Whitley Strieber the world's most famous abductee. In it, he claimed to have been the victim of multiple abductions. In fact, during his whole life, he said, he had been plagued by little, bug-eyed aliens stomping around his home at night and whisking him off for proctology practice.

Strieber, the son of a wealthy lawyer, was born in San Antonio, Texas. As a child he had been fascinated by the 'space race' and once, as a joke, his school friends had painted him green because he was always talking about little green men. He later graduated from the University of Texas and went to film school in London. Then he moved to New York, where he worked in advertising, although he spent much of his free time studying witch-craft and mysticism.

In his thirties he decided to become a writer, and in 1978 published the book (which was later made into a film) *The Wolfen*, which was about a pack of super-wolves who roamed Manhattan ripping people's throats out. *The Wolfen* was followed by *The Hunger*, which told of a romance between a couple of high-school vampires. He continued writing horror-thrillers with supernatural over-tones. In one book, *Cat Magic*, the female protagonist was abducted by 'fairies' – small humanoids who had 'sharp faces with pointed noses and large eyes' – who took great delight in examining her, both physically and mentally. Another common trend in his fiction was his claim that his books were true stories related by fictional characters. But in the futuristic book *Warday*, which Strieber co-wrote with James Kunetka, the two authors wrote autobiographically, fictionalising only their future experiences. The line between fact and fiction had therefore become well and truly blurred.

Then it happened. On 4 October 1985 Strieber, his wife, Anne, and their son, Andrew, along with their friends, Jacques Sandulescu and Annie Gottlieb – who were both writers – went to stay in the Striebers' cabin in upstate New York for the weekend. It was foggy that night and Strieber lit the stove before going to sleep. He awoke to see a blue light shining on the ceiling. He became frightened; the party was in the woods and therefore could not see the headlights of passing traffic and in his sleepy state it occurred to him that the chimney might be on fire. Then he fell into a deep sleep.

He was later awoken by a loud bang, which also woke his wife; he could hear his son shouting downstairs. When he opened his eyes he saw the cabin was shrouded in a glow that extended into the fog. He cursed himself for having fallen asleep, thinking that the roof was on fire. He told Anne that he would get Andrew and that she should wake the others. Before he could go downstairs, however, the glow suddenly disappeared. In the morning Sandulescu mentioned that he had been bothered by a light during the night, but nothing more was said about it.

During the course of the following week Strieber found himself becoming increasingly disturbed by the incident. Then he remembered having seen a huge crystal, hundreds of feet tall, standing on end over the house. It was this that had emitted the strange, blue glow. Anne thought that he was crazy.

The cabin gradually became a dark and terrible place in Strieber's mind. New York City also seemed dangerous, so he decided to move back to Texas. Strieber and his wife accordingly went house-hunting in Austin in November 1985, but when he saw the house that they intended to buy he became paranoid: the huge Texas sky, he thought, was a living thing which was furthermore watching him. He therefore cancelled their plans to move to Texas, to the fury of his wife; if he did not pull himself together, Anne said, she would leave him.

Strieber managed to put aside his fears and they visited the cabin again at Christmas. On the night of 26 December 1985 he awoke in the middle of the night to hear a peculiar sound, as if lots of people were running around downstairs. Suddenly a small figure rushed at him and he lost consciousness. After that he had the impression that he was being carried and woke to find himself lying in a small depression in the woods. He was paralysed, and someone was doing something to the side of his head. The next thing that he knew was that he was travelling upwards, high above the forest. The terrified Strieber then found himself in a messy, circular room, with small creatures scurrying about around him. They first inserted a needle into his brain and then they examined his rectum with a probe, possibly taking samples of faecal matter, but leaving Strieber with the impression that he was being anally raped. After that they cut his forefinger.

Strieber awoke in his bed in the cabin feeling distinctly uneasy. He read a report in the newspapers about a UFO having been sighted over upstate New York that night. His brother had given him a book about UFOs for Christmas which he tried to read, but which inexplicably frightened him. He ploughed on, however, until he finally got to the chapter about alien abductions, and then everything began to make sense.

Back in New York City, Strieber looked up the UFO expert Budd Hopkins' phone number in the telephone directory. On calling him, Hopkins suggested that he investigate the events of the night of 4 October 1985, when there had been other witnesses. Strieber did so. His wife remembered having been woken by the bang, but did not recall the glow. Sandulescu remembered the light, but neither of them could come up with an explanation for it. Strieber's son had the most interesting recollection: when he had heard the bang he had been told that it was alright because his father had just thrown a shoe at a fly. Strieber then asked his son who had told him this. 'A bunch of little doctors', Andrew replied. He had dreamt that the diminutive doctors had carried him to the porch and had told him telepathically that they were not going to hurt him. The boy said that it was a strange dream because it was 'just like real'.

At Hopkins' suggestion Strieber, who had been feeling suicidal, went to see a therapist, who suggested that they try regressive hypnosis, as well as therapy. After a couple of weeks Strieber dropped the therapy, but kept on with the hypnosis.

During the first hypnotic session Strieber was regressed to the night of 4 October. He recalled having noticed something flash past of the window before seeing the light. In the corner of the bedroom he had seen a goblin, wearing a cloak, which had rushed at him and had struck him on the forehead with a wand. At this point Strieber screamed so loudly that he came out of the hypnotic trance; when he was put under again he saw images of the world blowing up. 'That's your home', said the goblin.

After that he saw his son in a park, an image that Strieber associated with death. The goblin said that he would not hurt Strieber and then took a needle and lit the end. It exploded, whereupon Strieber began to think of the house burning down. It was as if the goblin had implanted the image into his mind. After that he came out of the trance again, but was hypnotised once more in order to discover what had happened. This time he saw his son lying dead in the park. Then he watched his father dying: he was sitting in an armchair, choking, while his mother looked on – but this was not the deathbed scene that his mother had described.

In a later session Strieber was regressed to the night of 26 December. He recalled having been naked while being pulled from his bedroom by aliens wearing blue overalls. They had dragged him into the woods and had seated him on a chair, which had then propelled him hundreds of feet upward, into the air. Next he was sitting on a bench in a room in the company of a female alien dressed in a tan-coloured suit. She then pushed something resembling a penis up Strieber's rectum and told him – in a flat, Midwestern accent – that he was the 'chosen one', but he scoffed at her. After that she tried to encourage him to have an erection, but he could not comply (perhaps because she had leathery, yellow skin and a mouth like an insect). The next thing that he knew was that he was lying naked on the living-room sofa; he went upstairs, put on his pyjamas and went to bed.

Further hypnosis revealed more bizarre abductions. On one occasion he had been in a spaceship and had seen his sister, wearing a night dress, sprawled on an examination table. His paralysed father had stood beside her. Lying on other tables were unconscious soldiers in uniform. He had even been invited to give a lecture on the evils of the British Empire. Another time he had woken up to find a group of hybrids standing around his bed. The aliens seemed to be tinkering with his mind, showing him symbols and images that evoked thoughts and memories in his mind.

Outside hypnosis, Strieber's life became more and more bizarre. He woke up one night to find himself paralysed and convinced that a probe had been shoved into his brain via his nose. He developed nosebleeds, as did his wife and son. He began to smell the odour of the aliens around his home. Symbols that the aliens had shown him appeared on his arm. He regularly suffered the sensation of missing time. He was plagued by weird memories and strange phone calls and his stereo began speaking to him. He eventually became so afraid of his flat that he moved to Connecticut, but when Connecticut proved frightening, too, he returned to New York.

Numerous psychological and physical tests were carried out, but no one could find anything wrong with Strieber. The answer, he concluded, must lie with UFOs. Hopkins had introduced him to the UFO network, but he had managed to alienate Hopkins' contacts with his eccentricities. He became paranoid about reporters, fearing that if news of his abduction experiences were made public he would be held up to ridicule, which would damage his career.

However, Strieber began to feel that he was all written out in any case: his last book had not done well and his abduction experiences were making it impossible for him to work. He therefore decided to confront his fears and to write about the abductions themselves, although he was not sure whether his experiences were real or whether they were instead memories of a former life. He nevertheless produced a manuscript for a book that he had provisionally called *Body Terror*, but one night, when his wife was asleep beside him, she spoke to him in a strange, deep voice and warned him he should change the title to *Communion*, otherwise he would frighten people.

Strieber then circulated the manuscript. According to Hopkins, it was consistent with Strieber's horror-thriller genre rather than reading like a factual account – in one section of the manuscript, for example, the female alien led Strieber around by the penis, like a dog on a lead. Hopkins persuaded Strieber to tone it down a bit and after he had done so Strieber secured a $1,000,000 advance on the book. It was published in 1987 as *Communion: a True Story* and sold millions world-wide (Hopkins' own book, *Intruders: the Incredible Visitations at Copley Woods*, was swamped in its wake). Strieber later published four more books in the *Communion* series (Hopkins only managed on investigative work).

234

The Brooklyn Bridge Abduction

After *Intruders*, Budd Hopkins' book about the Copley Wood abductions, was published Hopkins received a letter from a 41-year-old mother called Linda Napolitano. Although she did not believe in UFOs, she wrote, she had bought a copy of his book, but when she had tried to read it she had been overcome with fear and had had to put it down. She was terrified that the same cycle of abduction had been happening to her for 22 years.

When she was 19 she had been living with her parents in New York City. On one night, when she was in bed, her body had become numb and she could not move. She had had the impression that someone was in the room, but had been so frightened that she had kept her eyes tightly shut; even so, the image of a hooded figure had appeared in her mind. She knew that something terrible was happening to her and had struggled mentally against it. Sometimes she had been able to summon up the energy with which to open her lips wide enough to scream. She thought that she had screamed loudly enough to be heard a continent away – certainly her screams should have woken her parents, who were sleeping in the next room, but no help had come. In the morning her parents had said that they had heard nothing and Napolitano had written it off as a bad dream. Two days later, however, the same thing had happened. She had gone to the doctor, who could find nothing wrong wit her. But still these strange experiences carried on.

Five years later Napolitano got married. Although the experiences became less frequent they did not stop altogether. Even though her husband was sleeping next to her he could not hear her screams – all he could do in the morning was comfort her. Three years after her marriage

Napolitano had had a baby. It was then that she had noticed a lump on the side of her nose, which made her nose crooked. She had gone to see a specialist, who had asked her why she had had surgery on her nose. She insisted that she had never had surgery on her nose, nor on any other part of her body for that matter. He replied, however, there was a scar inside her nostril that had clearly been made by a scalpel. She checked with her mother and older sister, who both confirmed that she had never had any surgery on her nose as a child.

She felt that the scar had something to do with her night-time visitations and decided to protect herself by taking the only action available to her: she stopped sleeping at night and rested during the day. The incidents did indeed cease and she also stopped talking about them, fearing that they might come back again. But her experiences still disturbed her deeply and were always something that she wanted to get to the bottom of so that she could go back to sleeping at night. After reading the opening pages of *Intruders* she therefore contacted Hopkins.

Hopkins told Napolitano that her case showed all the signs of alien abduction. Under hypnotic regressions she first recalled having seen a UFO festooned with coloured lights outside her window when she was eight, and then gradually a picture of regular UFO abductions was built up.

Then, on 30 November 1989, seven months after she had first written to Hopkins, she phoned him to tell him about an experience that she had had the night before. At around 3am, just as she was about to fall asleep, she had felt herself becoming numb. As usual, she had sensed a strange presence in the room, but this time she had kept her eyes open and had seen a small creature, with a huge head and eyes, coming at her. She then remembered floating out of the living-room window (she lived in a flat 12 stories up, two streets from FDR Drive, which runs the length of Manhattan along the East river.) Once outside,

she had floated into a waiting UFO. This was the extent of
her conscious memory.

Under hypnotic regression, she recalled that four or five
creatures had manhandled her out of the window. They
had examined her aboard the UFO and had spent some
time rummaging up her nose before removing a small,
metal ball. The thought had come into her mind that she
did not want to have any more children. After she had
been dressed and returned to her flat she had been overcome by the feeling that her children were dead – that the
aliens had killed them. She had rushed into their bedroom, but could not rouse them. They were fast asleep and
she had held a mirror under their noses to satisfy herself
that they were still breathing.

Hopkins had taken the investigation as far as he could
when he discovered that there had been witnesses to
Napolitano's abduction. In February 1991 two New York
policemen wrote to him to tell him that early in the morning of 30 November 1989 they had been sitting in their
patrol car under an elevated section of FDR Drive when
they had observed a UFO hovering above a nearby building. They had then seen a woman dressed in a nightgown,
along with three ugly, little creatures, floating out of a
window and into the spacecraft. Once they were all inside
the UFO it had sped off towards the East river before
plunging into the water near Brooklyn Bridge. It had not
resurfaced.

Witnessing this incident had put them in a dilemma. It
was such an outlandish tale that they did not feel that they
could report it; on the other hand, they were policemen
and were supposed to protect New York tax-payers like
the woman concerned. For months they debated what to
do about it. One of them even got to the point at which he
would sit outside the building in question trying to pluck
up the courage to go inside and find out whether the
woman whom they had seen really existed.

Not knowing that Napolitano had already contacted him, they eventually wrote to Hopkins, asking his advice as a ufologist. It emerged that soon after the incident the policemen had indeed gone to Napolitano's flat. Even more intriguingly, there had been a third witness to the abduction: the police officers had been acting as bodyguards for a famous politician whom they had been ferrying to the New York heliport at the time. In due course Hopkins received a guarded letter from the politician concerned, who said that it was not time to tell the world about the incident. (In ufology circles the man is thought to have been the then secretary general of the United Nations, Javier Pérez de Cuéllar, although he denied it when asked directly.)

As luck would have it, another witness came forward. She had been driving over Brooklyn Bridge that night when the street lights had gone out and her car had stalled. Then she had noticed that all the cars on the bridge had come to a halt – all had had a grandstand view of the abduction of Napolitano. Sadly, she was the only one to contact Hopkins; none of the other witnesses contacted the police and the media carried no report of the abduction. Yet it was considered by some to be undoubtedly the most important alien abduction to date, and Napolitano became the darling of the UFO lobby.

The Invasion of Black Brook Farm

One evening early in 1979, Joyce Bond was watching *Coronation Street* in her family's 200-year-old farmhouse home, Black Brook Farm, with her three daughters – 10-year-old Susan, 23-year-old Laura and 14-year-old Jayne. At 7.30pm, just as the episode was beginning, the phone rang. It was Jayne's school friend, Sandra Streech, who lived on a farm about a quarter of a mile (40m) away. Sandra said that she and her mother had observed strange lights in the sky circling Black Brook Farm. They could hear no aircraft noise and the lights were spooky enough to set their dogs barking. (Strangely enough, the Bonds' dog remained silent.)

Joyce and the girls looked out of the window and located a large, red light hovering over the field. But Sandra had said that the lights were directly over the farmhouse, so they went outside and then saw a huge object, as big as their house, hovering over the stables. It appeared to be watching them. The UFO was covered in flashing lights and for a moment they were mesmerised, but then Joyce broke away, grabbed her daughters and they all ran back inside. They locked the doors and switched off the lights. The terrified Jayne tried to escape from the house through a small window at the back, but the others pulled her in again. Joyce phoned her husband at work and told him to come home quickly. Then the family tried to make a more organised attempt to flee – to a neighbour's house, 200 yards (183m) away – but something prevented them and they turned back.

When Mr Bond reached the village he was stopped at a police road-block and told that there had been an accident. The sky was now full of aircraft from the nearby airforce base, RAF Finningley, although the planes did not usually fly at night. A smallholder who lived down the road from Black Brook Farm said that he had also seen the UFO, while friends of the Bonds who lived 2 miles (3.2km) away said that they had observed a bright-red light streak upwards, into the sky, at about 9pm. Mr Bond had to take another, longer route home and as he neared the farm he saw that the top of a fence post was on fire. Subsequently, however, he found no report in the local newspaper about any accident having occurred in the village.

For years afterwards Joyce and her daughters felt that

they had an incomplete memory of the event, and in 1994 they submitted to regressive hypnosis. It was a long time after their experience, however, and the hypnosis was not very successful, partly because the women were terrified of what they might recollect. But Jayne recalled having floated high above the farm, having seen bright lights and having been manhandled by a mysterious creature. The stress of the hypnotic sessions caused her hair to begin to fall out, but when the sessions were stopped it grew back.

The UFO investigators continued to keep in touch with them and all four began to report strange dreams, often involving small, grey creatures, with big, black eyes. The women told of other dreams, in which they were paralysed or naked in empty, metal rooms. The girls associated dreams of flying or being sucked up into the sky with Black Brook Farm, even though they had since grownup and moved away. Laura often dreamt of the UFO just as she was going to sleep, seeing a door open in it, but then she tended to open her eyes, being too frightened to see what happened next.

They all had the impression that Black Brook Farm was haunted. A floating, white figure had been seen nearby on several occasions, but the girls had never associated it with the UFO. When sleeping at the farm in 1990, Laura saw a glowing, white apparition approach the back door; even the dog was spooked. For some reason she switched off all the lights and went upstairs in the darkness to bed. That night she felt that she had been assaulted by an alien and the next morning the dog was too frightened to go out. Jayne's husband also saw a ghost at the farm. And from 1979 it seems to have been inhabited by a poltergeist. The Bond girls' children were frightened to stay there and told of weird experiences of their own. For her part, Joyce felt a presence in the bedrooms at night; she also reported having heard laughter and sniffing, as well as anonymous assaults on the back door, and once her earring was pulled by someone whom she could not identify. Susan saw the dog levitate in 1992.

The girls also had the waking sensation of being paralysed while something in the room was holding them down and touching them; these attacks also occurred when they were in their own homes, that is, not just at Black Brook Farm. Furthermore, they sometimes heard voices and saw strange lights, as well as aliens, in places other than the farm. Such encounters seemed to have been contagious: Susan's friend, Jo, for example, began to report similar experiences, while Jo's boyfriend, Martin, saw an alien while he was in Australia. The whole family lived in constant state of fear that the Bonds' grandchildren would become affected and that the poltergeist activity would spread to their homes. The Bonds, it seems, had also been visited by 'men in black' – mysterious officials who are thought to ensure that alien abductions are kept secret by intimidating the victims.

Long-time Lover

Tom's wife, Nancy, was abducted from their bedroom by alien visitors while the couple was locked in the act of sexual intercourse. They were making love when Nancy suddenly complained of an electric shock passing through her hips. Instead of thinking that he must be doing something right, Tom turned to the alarm clock and noticed that 45 minutes had passed; as far a he could remember they had only been making love for a couple of minutes.

This 'missing time' disturbed him and Tom therefore consulted a hypnotist. During a regressive-hypnosis session he recalled having seen two small aliens come into the room and then 'switch off' the couple, the two of them had frozen as if they had been hit by a tranquilliser dart. Then the aliens had lifted Tom off his wife and had carried her off. Some time later they had returned her and had placed Tom on top of her again. Fortunately, he had not lost his erection in the meantime and they had carried on where they had left off. Suddenly she had exclaimed 'Ouch', and when Tom had asked her where she felt the pain she had pointed to her hip. After that he had looked at the clock.

Nancy also consented to being regressed. Under hypnosis, she was taken back to the love-making experience and recalled a beam of blue light pointing directly at her; where it came through the window it formed a sort of entrance. It was then that she noticed that Tom was no longer on top of her, but lying beside her in a trance. Aliens were in the room. She was angry with them and told them that she wanted another baby and that by interrupting her they were messing up her fertile period. They took her away anyway. When they returned her to the bedroom they put her back in the position that she had been lying before they appeared, with Tom lying on top of her.

Weird and
Wacky

Double Trouble

At 9am on 4 January 1975 a UFO crashed into the wall of a house in São Luis, Brazil, where the teenager Antonio Ferreira lived. The incident seems to have annoyed the aliens on board, because they came back later and zapped him with a beam of light. The aliens then emerged from the UFO and dragged Ferreira aboard. They were about 3 feet (91cm) tall, with dark hair and skin.

The aliens' chief interrogator, Croris, asked Ferreira about earthly food and cars; when the teenager's answers were not up to scratch Croris punched him in the chest. Ferreira then offered the aliens a deal: in exchange for rides to distant planets he would bring them earthly subjects for vivisection. The aliens agreed and they were content to receive a cat, a dog and a parrot. In fact, they were so pleased with Ferreira that they gave him a powerful ray gun.

Ultimately, however, Ferreira found himself the subject of an experiment: he was put into a glass case along with an alien who then miraculously transformed into his double. Ferreira then had to watch while the aliens tested whether the *doppelgänger* fooled his father.

Alien abductions are very common in Brazil, but they are not always as successful as Ferreira's. José Nobre Uchoa, a farmhand from Balem, was invited by aliens to go for a trip on the following night. He was told to walk down the middle of a motorway and that a UFO would then pick him up. Unfortunately, before the spaceship could collect Uchoa a car had flattened him.

Alien Error

In 1979 the 19-year-old Franck Fontaine disappeared in France. He was happily married, with a six-month-old baby. On 26 November 1979 he had been helping his two friends, Saloman N'Diaye and Jean-Pierre Prévost, to take a van-load of clothes to the market at Gisors. Before the set off they saw a twirling lamp descending, which they took to be a UFO. N'Diaye and Prévost ran to get a camera while Fontaine drove towards the spot where the light appeared to be landing. When the two men returned they saw Fontaine's van about 200 metres (656ft) away, shrouded in light, with three equally bright lights converging on to it. The halo then rose into the sky and disappeared.

The two men ran towards the van and threw open the door, but Fontaine had gone. Even though they had no driving licences they then drove to the local police station at Cergy Pontoise, Vale D'Oise, to report Fontaine's disappearance. The police were sceptical and tested the men for drink and drugs; subsequent interviews with their families showed that they were neither inveterate hoaxers nor practical-jokers. For a while the police believed that they had murdered Fontaine, but gruelling interrogations of the two men failed to shake their stories. Detectives then circulated Fontaine's pictures to other French police forces. Tests carried out on the van for radiation proved negative, while nearby air bases reported no UFO activity in the area on the day in question. The police were baffled.

Then, a week after he had disappeared, Fontaine walked into N'Diaye's flat. As far as he knew it was still 26 November. He recalled having blacked out in the light and the next thing he knew was that he had woken up in a cabbage field near the place where he had lost consciousness. His van had gone and he assumed that it had been stolen.

When they found Prévost – who was being wined and dined by a local journalist – the three men went to the police and explained the situation. By now the local constabulary was losing patience. Convinced that the abduction was some sort of prank, the police interrogated Fontaine, but found that he could say a little to enlighten them. All that he could remember was that a sphere of light the size of a tennis ball had landed on the bonnet of his van. That was all. The police were forced to release him and to close the case, unsolved.

However, due to the publicity stirred up by Prévost, the story was attracting considerable interest across France. The French science-fiction writer Jimmy Guieu contacted the three men in the hope of writing a book about the abduction. Although Fontaine refused to undergo regressive hypnosis, Prévost volunteered to do so. And according to Prévost's recollections under hypnosis, it was he – and not Fontaine – who had been the object of the aliens' attention.

Soon after Fontaine's return Prévost said that he had been visited by a strange man called the 'travelling salesman' who had taken him to a secret, alien base in a railway tunnel on a Swiss border near the village of Bourg-de-Sirod, where Prévost had been brought up. There was a railway carriage of World War II vintage in the alien base, and people were brought there from all over the world to imbibe the wisdom of the 'intelligences from beyond', as the aliens called themselves. They sat enthralled around a campfire while an alien called Hurrio lectured them on the dangers of technology, pollution and so on.

It was later discovered that although Prévost had denied any interest in aliens and UFOs he had been reading a sci-fi story in a magazine at the time of the incident which told of a remarkable similar 'abduction'. There were also several inconsistencies in his story.

A Floating Bullet

The oil-field-driller Carl Hignon was a keen hunter. On 25 October 1974 he was in Medicine Bow National Forest, in Wyoming, when he saw five elks standing in a group. He raised his rifle and fired, but did not hear a report – in fact he did not hear anything at all. Furthermore, there was no kick from the high-powered weapon and the bullet appeared to float from the barrel before falling harmlessly to the ground, 20 yards (18m) in front of him.

Hignon then turned around and saw an alien standing behind him. The creature was a humanoid, some 6 feet (1.8m) tall and weighing about 180 pounds (82kg). He had oriental skin tone and his face seemed to blend into his neck. He wore a one-piece jump suit and his belt bore a star shaped clasp. The alien seemed friendly enough, as well as polite: 'How you doin'?' it enquired. 'Pretty good', replied Hignon. The alien then asked whether he was hungry, but before Hignon could answer a small package floated towards him. This contained four pills, which, the alien said, should last him about four days. He took one and put the other three into his pocket.

In the distance Hignon noticed a metal box. Even though it had no landing gear, no hatches and no external features of any kind he guessed that this must be the alien's ship. The alien asked him whether he wanted a ride but before Hignon could answer he found himself on board the spacecraft. It then took off and they flew to a planet that the alien said was '163,000 light miles' away. On the surface of the planet he saw buildings, including one that was tall and needle shaped; around it were lights. He put his hands over his face and complained that the lights were hurting him. 'Your sun burns us', the alien commented, somewhat inconsequentially.

The next thing that he knew was that he was stumbling down the road, still clutching his rifle. On seeing a lorry he got into it, even though he did not recognise it as his own. Using its CB radio he called for help, and when it arrived he was taken to hospital, where he kept calling for the pills that the alien had given him. Under hypnosis, Hignon filled in some of the details of his abduction. The alien's name, for example, was Ausso One. Unusually, there was some corroborating evidence for Hignon's story. Not only had a UFO had been reported in the area that night, but Hignon had also had the foresight to pick up the floating bullet and put it into his canteen before setting off for Ausso One's planet. Although it was badly damaged, the Carbon County Sheriff's Office identified it as coming from a 7mm Magnum rifle – the type that Hignon used. The deputy who examined it said that the bullet appeared to have been turned inside out, but could not explain how this could have happened. On the strength of this evidence alone most UFO investigators believe Hignon's story.

The Gulf Breeze Abduction

Ed Walters was a UFO buff who lived in Gulf Breeze, Florida, which has long been renowned as a centre of UFO activity. Walters observed and photographed numerous alien spacecraft, but it was not until 2 December 1987 that he first saw the creatures who flew them – or so he thought.

He awoke during the Wednesday night in question to hear a baby cry. This was strange because he and his wife, Frances, did not have a baby at that time, neither were there any babies in his house nor, as far as he knew were there any in his neighbours' houses. Then he heard voices discussing the crying baby in Spanish. Walters grabbed his pistol and, along with his wife, checked the house for Cubans. When they were sure that the house was free of illegal immigrants they went out of the French windows into the back garden to see a UFO swooping down from the skies and halt about 100 feet (30.5m) above the pool. Walters and his wife ran back into the house; as they cowered inside the Walters heard a voice commanding them to 'step forward'.

A veteran ufologist, Walters grabbed his Polaroid camera before he stepped out with his gun. When he reached the pool he pulled out the camera and took a photograph. The flash went off and suddenly he felt terribly exposed, so he fled back into the house and locked the French windows. Peeping through the kitchen window, Walter and his wife watched as the UFO sped off into the distance. During the encounter, Walters had felt a strange buzzing in his head, but now it faded.

Walters and his wife returned to bed, but then the dog barked – just once. This was unusual, so Walters again grabbed his gun and camera and went downstairs to the French windows. An experienced UFO-watcher, he was certain that he would see the alien spacecraft. But when he pulled back the curtain he was instead confronted by some thing beyond his wildest imaginings: just inches in front of him was a humanoid. Although it was about 4 feet (1.2m) tall, with big, black eyes, it was not like the greys that most people see: this one was wearing a helmet with a transparent visor pulled down over his eyes (this apparently enabled it to see on Earth).

Walters, usually a pillar of steel when it came to snapping UFOs, was so taken aback that he forgot to photograph the alien. He screamed, backed away, tripped and fell backwards, still brandishing the gun. He was determined to shoot the alien if it tried to get into the house. The creature seemed to get the message, for when Walters picked himself up he could see it retreating. It was still only 20 feet (15m) above the patio, but he again failed to take a picture of the alien. The alien spacecraft then sent out another blue beam, which Walters believed picked up the alien that he had seen and beamed it aboard.

As a UFO enthusiast, Walters naturally consented to undergo regressive hypnosis. Once in a hypnotic trance he revealed that he had undergone a series of abductions. Locked away among his repressed memories was the recollection of an encounter that had happened when he was 17 years old. He remembered having been followed by a big, black dog when he was cycling to a shop. The dog had waited for him and had then followed him home – no matter how fast he had cycled it had still kept up with him. That night he had been tormented by the thought that there was something in the house and that it might jump on to the bed. The dog, apparently, was an alien.

There seems to have been a deep memory block regarding what happened next, because Walters then fast-forwarded to 1 May 1988, four-and-a-half months after he had seen the alien through the French windows. This time he remembered that he had been abducted. A rod had been forced up his nose and he recalled lying on the floor covered with a sticky residue. There were four aliens with him; three wore protective shields, while the other was dressed in a tight, pink body suit. They were all carrying silver sticks.

Like many other abductees, he had then found himself in a large room which had a table in it. Underneath the table he spotted what he thought was a serial number. (He later produced the symbols that he noted, but they were not publicly released because they were being kept secret by senior UFO investigators to check the veracity of future sighting, Walters said.) While Walters was under the table, another creature, which had white hair, entered the room. Walters was then zapped by blue beam, which instantly transported him onto the table. After that a gang of aliens came in; their skin-tight body suits matched the colour of their skins so perfectly that they might as well have been nude. They carried a smaller creature – maybe only 2 feet (60cm) tall – with them, who was darker grey in colour. The aliens seem to have probed Walter's mind, stirring up all sorts of unwanted memories.

As the hypnotic sessions continue Walter reviewed memories of long-forgotten abductions which had begun when he was a teenager. They all involved creatures with strange, long-fingered hands which resembled the alien hands described by Leonard Stringfield. (Unfortunately, Stringfield later revealed that these hands were part of a hoax – a hoax in fact, that long predicted alien abduction: a showman had once displayed them claiming that they were the hands of a mermaid.)

Abducted in the Bath

British abductions are often more prosaic. In February 1976, for example, in Keighley, West Yorkshire, two tall aliens, with cat-like eyes and grey faces, appeared in the bedroom of an ambulance-driver named Reg. Although they observed him as though he were some sort of specimen, they seemed to need his talents. He was a dab hand at fixing things and the aliens showed him a piece of rubber tubing and asked him to come with them.

They told him to lie on the bed and then they paralysed him. Soon he found himself floating through the ceiling and up to a waiting UFO, which Reg said looked like a bathtub. In common with other abductees, once inside the UFO he was laid on a bed and given a full medical examination. A large, purple device, which looked like an eye, was scanned across his body. The aliens who had abducted Reg kept quoting the Bible at him, but when he asked them who they were they replied rather rudely 'An insignificant being such as a worm should not ask such questions'. However, they did volunteer that 'a thousand of your years are but a day to us'.

Suddenly Reg found himself back in his bed: he was still paralysed, but the numbness soon wore off. Although there were large chunks of recollection missing from his memory he nevertheless immediately contacted the British UFO Research Association and told his story.

A Persian-carpet Ride

During the late 1970s Dr Simon Taylor lived and worked in Tehran as an instructor for the Iranian Air Force. On 16 September 1976 he went with an Iranian friend named Reza to the village of Ahar, in the Elburz mountains to the north of the city. They arrived at around 6pm and stayed in a mountainside cabin near the shrine. After dinner they went to bed early, but then Taylor awoke, gasping for breath, in the middle of the night. He saw Reza fiddling with the paraffin lamp, thinking that it must be faulty, when the cabin was rocked by a loud pounding. Fearing that it was about to collapse around them, Taylor and Reza ran outside.

Then, in the darkness, they saw three men dressed in black; they had large, Mongolian eyes. Taylor feared that members of the shah's notorious secret police had come to arrest Reza but that thought passed from his mind when they started communicating with him and Reza telepathically. They were told not to be afraid and they were given time to dress before their abductors marched them down a mountain path. After a while they felt lush, Persian rugs beneath their feet. Before they knew it they were in a round room and were being invited to sit cross-legged on the carpeted floor and to look out of a huge window that seemed to take up the whole of one wall. They could see the distant lights of the city through it.

The room begun to shake before shooting upwards in the air; Taylor could feel the pressure building up in his eardrums. He could see Tehran, and then other great cities of the world – along with his native Birmingham. They were also privileged to enjoy the sight of aerial panoramas of some of the planet's greatest landscapes, such as mountain ranges, deserts and icecaps. The creatures gave them a telepathic commentary on this whistle-stop, guided tour of the world and also revealed that they lived on Earth, but in places in which humans hardly encountered them. Suddenly it was all over and Taylor and Reza found themselves back on the ground not far from the place from which they had been abducted. It was 2pm.

Three days later was a particularly detailed UFO sighting in the area. Lights were seen in the sky and two F4 Phantom jets were scrambled from Shahrokhi Air Force Base. The first suffered electrical failure as it approached the UFO, but when it turned away its equipment started working again. The second locked its radar onto the UFO and ascertained that the object was the size of a Boeing 707. The pilot thought that he saw the UFO launching a missile but when he tried to respond with a missile of his own the plane's electrical system failed; as with the first jet, however, when he turned away, everything returned to full working order.

Even more spookily, immediately after the abduction Reza told Taylor that he knew when and how he would die; he also revealed these details to his family. In early 1987 Reza was hospitalised with cancer of the liver. A few days before he died he checked out of the hospital and went to a shrine dedicated to Imam Reza, a descendent of Mohammed, where he handed out money. The next morning he was found dead in the room in which he was staying. There were no indications of suicide or foul play.

Military Abduction

Some have suggested that UFO witnesses have been abducted by military personnel in an effort to keep a lid on the UFO question, the reasoning being that if the public knew of the real extent of alien visits to this world then there would be widespread panic and civil disorder. Dr. Helmut Lammer, of the Austrian Space Research Institute, has studied the object, which, in ufology circles, is called military abductions of UFO witnesses (MILABs).

In his preliminary report on MILABs, which was published in 1996, Lammer found that the unmarked helicopters that are used for these abductions have been photographed on several occasions. There are usually other unmarked vehicles around when the victims are snatched, too. Military abductees are typically disoriented at the outset, either by drugs or a powerful, electromagnetic field. Then they are taken by helicopter or van to an underground facility, where they are subjected to medical examinations similar to those carried out by aliens. (However, one way in which to tell weather you have been snatched by the military or by aliens is by looking at the shape of the room: aliens examine their abductees in a circular, the military uses square ones.) Microchips are then implanted into abductees by means of a syringe; they are usually placed behind the ear and may transmit psychological data to a central data bank. (Interestingly, some abductees, such as Christa Tilton, say that they have seen aliens working alongside humans in underground facilities.)

According to Dr. Lammer, 'MILABs may be evidence that a secret military/intelligence task force has been in operation in North America since the early 1980s, and it's involved in the monitoring and kidnapping of alleged UFO abductees'. The purpose of such abductions seems to be to take out any alien implants and insert implants of their own, as well as to examine women to see if they have been impregnated with hybrid foetuses.

Only Half Abducted

In May 1973 Judy Doraty was driving with four other members of her family near Houston, Texas, when all five of them saw a bright light in the sky which appeared to be following them. Doraty recalled pulling over and getting out of the car in order to take a better look, but could not remember what had happened next.

This memory lapse annoyed her so much that seven years later she underwent regressive hypnosis. It was then she recalled that she had had what ufologists call a 'bi-locational experience' (in which the victim seems to be in two places at once): Doraty had simultaneously been standing by her car and aboard the alien craft. The aliens had used a strange beam of yellow light, which had particles swirling within it, with which to capture a calf, which they cruelly mutilated. They then apparently used various parts of its body to check the spread of a dangerous toxin.

The aliens appeared to be conventional greys – short and thin, with large heads, pale skins and tiny noses and mouths – but although they had large eyes they were not the black, featureless orbs that other greys have: this race had vertical, cat-like pupils and yellow irises. The aliens apologised to Doraty, explaining that she had been brought on board – or half brought on board – by mistake. Yet she caught a glimpse of her daughter being operated on by the aliens in the spacecraft. They assured her, however, that they did not intend to harm either of them.

Getting Our Own Back

The question remains: if aliens are always abducting humans, have humans ever abducted aliens? The answer is: they just might have. The aliens who were supposed to have been captured in the so-called 'Roswell Incident' were all dead, so this cannot really be called an abduction. However, there was an incident in the city of Varginha, Brazil, in 1996 that might just fit the bill. Varginha is a city of 150,000 souls, situated north-east of São Paulo and around 160 miles (257km) from the coast. During the mid-1990s there were numerous UFO sighting in the area.

On 20 January 1996 Varginha's Fire Department received a number of calls from people reporting a strange creature that was around the area. Four men were accordingly sent to apprehend it and when they arrived at the site of one sighting a crowd of people directed them to some woods. They were told that they had better hurry, as children were throwing rocks at the creature, so the firemen rushed down the hill towards the woods. There they saw what ufologists call a classic grey. It was short with long spindly arms, a big head, large eyes and no genitals; it also wore no clothes. But the odd thing about this grey was that it was red and shiny, as if it had been rubbed with baby oil; it furthermore smelt of ammonia. The creature put up no resistance and, with the help of net, the firemen caught it easily. It appeared to be angry, however, because it made a buzzing noise like a swarm of hornets. They dragged it up the hill, put it in a covered box and loaded it on to a lorry. Then they called the army, which took it away; military vehicles were later seen in the area.

About six hours later, three local girls – the 22-year-old Andrade Xavier, the 16-year-old Liliane Fatima and the 14-year-old Valquira Fatima – were taking a short cut home from work when they bumped into another strange creature. This one had three lumps on his head and big, red eyes and it was standing with one hand between its legs. They thought that it was the devil and ran home screaming. When the Fatima girls got home they told their mother what they had seen. She went back to the spot, but the creature had gone. However there was a strong smell of ammonia in the air and some strange footprints could be seen in the mud.

Two days later a dead creature answering this general description was seen in a box that was loaded on to a military vehicle. The story soon spread that seven aliens in all had been captured, including the dead one; two of them were badly injured. It was also said that five more were in the area and had gone into hiding. There were reports that creatures had been taken to hospital for medical treatment, as well as to the University of Campinas, where they would be studied. There were furthermore rumours that two USAF officers had been seen entering a local hospital, presumably to inspect the aliens. Two aliens – one male, one female – were said to have been taken from the hospital to the University of São Paulo under the escort of the USAF officers.

A young man who claimed to be a relative of a Brazilian soldier tried to sell a film purporting to show a captive alien to a Brazilian TV network for a mere $68,000, but his offer was turned down. However UFO investigators say that they have spoken to people who say that they have seen the footage.

But this time alien fever had reached such a pitch of excitement that a city official suggested changing the name of a local park to Seite Extreterrestes – the 'Seven Extraterrestrials'. The *US Wall Street Journal* then picked up on the story; although all the officials to whom the paper's journalists spoke denied any knowledge of the captured aliens. The army could not explain what its lorries had been doing in Varginha at the time in question. By then 60 witnesses had claimed to have seen a cigar-shaped UFO crash in the area. Furthermore, videotaped interviews of men in civilian dress who claimed to be part of the army detail which had rounded up the aliens were in circulation, although the paper said that it was impossible to determine the tapes' authenticity. Some of the original witnesses were now demanding $200 an interview, while a documentary about the incident that was aired on Brazilian television was so popular that it was repeated the following weekend. But then Brazil – even more so than the US – has long been home to UFO sightings, as well as to doctored photographs and other palpable UFO hoaxes. (Not that one would wish to dent the veracity of Antônio Villas-Boas' story – there are too many men out there, hoping.)

Encounters
of the
Worldwide Kind

Year of the UFOs

Gary Flatter could not believe his eyes as he jammed on the brakes of his truck. Crossing the road in front of him was a curious menagerie of animals – seven rabbits, a raccoon, a possum and several cats. At the same time he was aware of a weird, high-pitched noise. The animals had emerged from a nearby field, and when Flatter glanced over the hedge, he soon found out why. Two silver-suited figures were staring at him.

It was October 22, 1973, and America was in the middle of a wave of UFO sightings. Flatter had accompanied his friend, Deputy Sheriff Ed Townsend, to State Road 26 near Hartford City, Indiana, where a car driver had reported two strange creatures standing on the highway. When they arrived, the highway was empty, and the sheriff decided to get back to town. But Flatter opted to stay and look around.

When he turned a spotlight on the creatures, he saw that they were about four feet tall with egg-like heads that seemed to be covered by gas masks. Far from seeming upset by the light, they began putting on a show, jumping high into the air and floating down again. Then they flew off into the darkness, leaving a faint red trail.

Five days earlier, Paul Brown had seen two beings matching Flatter's description as he drove near Danielsville, Georgia. He told police that a bright light passed over his car with a swishing sound, and he saw a coneshaped object land in the road some 300 feet ahead of him. He skidded to a halt, and as tiny figures emerged from the object, he grabbed his gun from the glove compartment, crouching behind the open driver's door ready to challenge them. They advanced no further, returned to their craft and quickly took off. Brown claimed he fired some shots at the disappearing UFO, but did not hit it.

Rex Snow ordered his dog to attack two silver-suited figures cavorting in his brightly-lit backyard at Goffstown, New Hampshire shortly after midnight on November 4 of that same year. The dog, a German shepherd trained to obey commands, bounded towards the intruders. But when she got within 30 feet of them, they curtailed their antics and just stared at her. She stopped in her tracks, growled at them, then slunk back past her surprised owner and lay down inside the house, whining and clearly frightened. Snow said later that he too was overcome by a sudden sense of fear. He had his 38-calibre pistol with him, but was shaking too much to hold it properly.

He quickly followed his dog inside the house, and watched the weird beings through the window. They seemed to be luminous, with over-sized pointed cars, dark egg-shaped recesses for eyes, and large pointed noses. Their heads were covered in Klu Klux Klan type hoods.

The figures seemed to be picking things from the ground and putting them in a silver bag with slow, deliberate movements. After a time, they walked off towards some nearby woods. Snow rang the police, but before they arrived he saw the woods light up eerily, then become dark again.

Police were convinced that Snow had seen something odd and frightening. They said he was shaking like a leaf and still pale when he told them his story.

Three months earlier, on August 30, reliable witnesses in 22 towns in Georgia claimed they had seen strange craft in the sky. And on October 3, a deputy sheriff and four park rangers watched a saucer-shaped object 'the size of a two-bedroom house' manoeuvre over Tupelo, Mississippi, the birthplace of Elvis Presley. They described it as having red, green and yellow lights.

Dr J. Allen Hynek, who established a Centre for UFO Studies at Granston, Illinois, collected 1,474 authenticated reports of UFOs during that year. Major-General John Samford, a former director of intelligence at the Pentagon, admitted: 'Reports have come in from credible observers of relatively incredible things.'

And Senator Barry Goldwater, formerly a major-general in the US Air Force Reserve, said: 'I've never seen a UFO, but when air force pilots, navy pilots and airline pilots tell me they see something come up on their wing that wasn't a plane, I have to believe them.' .

A Gallup Poll at the end of 1973 showed that 15 million Americans believed they had seen a UFO – and 51 per cent of the adult population believed that UFOs were 'real'.

Water Diviners from Space

Farmer Pat McGuire claims that 5,000 acres of his ranch near Laramic, Wyoming, were turned from dry sagebrush desert into fertile grassland because aliens on a UFO gave him a piece of good advice.

McGuire, his wife, eight children and a couple who live on his farm all say that unidentified flying objects of various shapes and sizes have hovered over their land almost every night for seven years.

'They are mostly about 300 feet wide and 60 feet high, and there seems no limit to how fast they can go,' McGuire told reporters. 'At first we were frightened by them, especially after we found mutilated cattle on the spread. One evening in 1976, my brother-in-law and I saw a craft hovering over a young calf. We heard the beast bawling for quite a while, then, when the UFO flew away, it took the calf with it.'

The visits continued, but the herds were left untouched. McGuire and his family gradually lost their fear. Then, one night, the aliens made contact with the farmer, and took him aboard their craft.

He recalled the event under hypnosis administered by Dr Leo Sprinkle, a parapsychologist with the University of

Wyoming, and watched by an assistant psychiatrist.

In his trance, McGuire described the aliens as around six feet tall, with large eyes, thin lips and bald heads. They told him to drill a well in high plains country, near his ranch.

McGuire consulted geologists and drilling experts a few days later. They told him the land was 7,000 feet above sea level, and he had no hope of finding water there. But McGuire went ahead regardless, even though neighbours called him crazy. He bought the land, bored his way through the upper crust – and struck a massive underground stream just 350 feet down. Soon 8,000 gallons of pure soft water were gushing from the desert every minute.

In 1980, after studying McGuire's claims both while he was conscious and under hypnosis, Dr Sprinkle said: 'I believe the craft appearing over his farm could be goodwill ambassadors of an alien civilization. I believe people like Pat McGuire are being chosen to spread the word that they are among us. And I believe we will see full-scale contact over the next decade or so.'

Visitors to the Archery Club

When two UFOs arrowed in on the Augusta Country Archery Club, Virginia, William Blackburn got the shock of his life. Blackburn, who lived in nearby Waynesboro, was working alone at the club when he spotted two objects in the sky. He watched as the smaller of the two circled down to the ground, landing only 18 yards from where he was standing, open-mouthed with amazement.

Three extraordinary beings emerged from it, each three feet high, and wearing shiny suits the same colour as their craft. One had an extremely long finger, and all possessed piercing eyes which 'seemed to look through you'.

The aliens advanced a few paces towards Blackburn. Although he had a double-edged axe in his hands, he was frozen with fright, unable to move. 'They uttered some unintelligible sounds, then turned and went back to the ship, going in through a door which seemed to mould itself into the ship's shape,' Blackburn said. 'Then the ship flew up and disappeared.'

Blackburn told UFO investigators that he had reported the incident to a government agency – he refused to say which one – and had been interrogated, then warned not to mention what he had seen to anyone.

Not in the Curriculum

Brenda Maria's encounter with a UFO was almost too close for comfort. It happened in the grounds of the local high school in Beverly, Massachusetts.

At 9 pm on Friday, April 22, 1966, her neighbour's daughter, 11-year-old Nancy Modugno, burst into the room where her father was watching television, claiming that an oval, football-shaped object, the size of a big car, had just flown past her bedroom window, flashing green, blue, white and red lights.

When Brenda and another neighbour, Barbara Smith, arrived, Nancy was almost hysterical. Lights could still be seen flashing from the direction of the nearby school fields, but the two women, now joined by the girls' mother, Claire, suggested walking down there to put the child's mind at rest and prove that what she had seen was an aircraft.

On reaching the fields, however, they saw three brilliantly lit, oval, plate like objects flying in circles around the sky. One hovered over the school building, the others were farther away. Suddenly, the nearest UFO began to move towards them. Claire and Barbara turned tail and ran, failing at first to realize that Brenda was not with them. Eventually, at the top of a rise, they turned to witness a terrifying sight. Brenda was screaming and covering her head with her hands, and the object was hovering only 20 feet above her.

'All I could see was a blurry atmosphere and bright lights flashing slowly round above my head,' she recalled later. 'I was very, very excited – not scared very curious. But I was afraid it might crash on my head.'

When the UFO soared back towards the school, the three women raced back to their homes to alert neighbours, several of whom also saw the objects. One called the police. Two officers arrived in a patrol car, and drove into the school grounds while HQ alerted the Air Force.

At the sound of two approaching planes and a helicopter, the three UFOs flew away. The neighbours were certain, from the lights and engine noises of the investigating planes, that the objects they had seen were no ordinary aircraft. For they had been completely silent – even when hovering right above Brenda Maria's head.

Communicating by Numbers

UFO enthusiast Gary Storey claims he exchanged messages with an unidentified object that flew past his brother-in-law's house in Newton, New Hampshire early on Thursday, July 27, 1967.

Storey had set up a telescope to observe the Moon when a bright glow had attracted his attention. Changing lenses, he focussed on it, and saw a series of lights flashing in sequence along the side of what seemed to be a large disc.

On impulse, Storey's brother-in-law flashed a torch three times. The object suddenly went into reverse without turning round. Then it dimmed its lights three times in answer to the torch.

The two men, both former radar operators in the armed

services, could not believe their eyes. They flashed again, four times, then five. Each time the number was repeated by the disc, which was moving back and forth.

Suddenly they heard a jet approaching. 'The object extinguished all its lights until the jet passed,' Storey said. 'We thought it had left. Then it reappeared, an oval-shaped white object, at least ten times brighter than it had been before.'

The two-way flashlight conversation began again. The strange craft responded to one long and one short signal, repeating exactly the number and duration of every flashed message during nine or ten passes. Then it flashed all its lights once more, and vanished behind a line of trees.

Several scientists and UFO organizations investigated the claims of the two men, but none could find a conventional answer to the sighting. The local minister vouched for Storey's brother-in-law and his sister, saying both were God-fearing folk who would not lie or seek publicity. Their evidence was added to the dossier of sightings that suggest UFOs may be manned by intelligent beings keen to make contact with humans.

'Boomerang' Swoops on Factory

Workers on a repair gang at the Morenci copper-smelting plant in Arizona claim a massive spacecraft swooped on the factory in January 1981. It seemed to be examining one of the two 650-foot smokestacks by beaming a light ray down it.

The four men who were repairing the other stack said the UFO was shaped like a boomerang, and as big as four football fields. It had 12 small red lights on its surface, in addition to a large white searchlight beam underneath.

'It just sort of stopped in mid-air above the smokestack and shone the big light right into it,' said workman Randel Rogers, 20. His colleague Larry Mortensen added: 'I have never seen an airplane hover like that. I got the feeling that it wasn't aggressive. Certainly it did nothing to frighten us. Whoever was in it was friendly.'

A third member of the gang, Kent Davis, said that during examination of the stack, one of the red lights at the edge of the boomerang, suddenly darted away from the craft at fantastic speed, returning after a few moments. The whole object then turned, without warning, and shot off like a rocket.

The UFO was also seen by 100 members of a high school marching band holding a practice session on the football field of Morenci High School, just over a mile away. Director Bruce Smith said: 'I looked up and saw all these lights in the shape of a V. These was no sound. It hovered for a few minutes, then disappeared high into the sky.'

Alarm on the Farm

A shiny silver saucer paid three visits to a farm at Cherry Creek, New York, on August 19, 1965. The four young farmhands who watched it turn clouds green and leave a red and yellow trail as it descended were not too perturbed, but the farm animals were clearly alarmed. A bull bent the iron bar to which he was tethered – and the milk yield of one cow slumped from two-and-a-half cans to just one.

Halos Over the Mission

Anglican priest Father William Melchior Gill finished dinner at his Boianai mission in Papua, New Guinea, and decided to take a stroll in the compound. He looked up to see Venus shining brightly. But what was that new light just above the planet?

As he stared upwards, he noticed more brilliantly-lit objects rising and falling through the increasing cloud cover, casting halos of light on the clouds in passing. Then he spotted something even more fascinating. Figures resembling humans emerged from one object and began moving about on it. There were two of them, then three, then four. They were doing something on the deck. Teachers, medical assistants and children came out to watch the strange activities several hundred feet above the ground. A total of 38 people spotted the figures over a three-hour spell before darkness fell.

Father Gill was a calm, painstaking and methodical man. He took careful notes of what had happened, and obtained signatures from 25 adult witnesses for his report. He dated it June 26, 1959.

The following night, the strange shapes returned. A native girl alerted Father Gill at 6.02, just as the sun was setting. There were still 15 minutes of good light left to observe four creatures moving around the deck of what seemed to be a 'mother ship', while two smaller UFOs - hovered, one overhead, the other in the distance beyond some hills.

'Two of the figures seemed to be doing something,' Father Gill noted. 'They were occasionally bending over and raising their arms as though adjusting or setting up something.'

When one of the figures looked down, the priest stretched his arm to wave. He was astonished when it waved back. Another of the watchers waved both arms over his head. The two figures still on deck did likewise. Soon all four were on top of their craft, waving energetically.

One of the mission boys ran to collect a torch as darkness fell, and directed a series of Morse dashes towards the object. The figures could he seen waving back, 'making motions like a pendulum, in a sideways direction'. The

UFO advanced for perhaps half a minute, and the group of witnesses – now about 12 strong – began shouting and beckoning for their visitors to land. There was no response. 'After a further two or three minutes, the figures apparently lost interest in us for they disappeared below deck,' Father Gill said later.

The UFO stayed hovering over the mission for at least an hour, but later in the evening visibility became poor as low cloud moved in. Then, at 10.40, a tremendous explosion woke those at Boianai who had gone to bed. They rushed outside, but could see nothing in the sky.

Father Gill reported what he had seen to the Australian Air Attaché, who later contacted the US Air Force. The priest admitted that at one stage he thought the shapes 'might just be a new device of the Americans'. But the air force had no craft that could hover close enough for men to be seen on it, or could hover in silence. They had their own explanation – the sighting, they said, was 'stars and planets'. But as astronomer Dr J. Allen Hynek wrote 15 years later after visiting the mission site, 'I have yet to observe stars or planets appearing to descend through clouds to a height of 2,000 feet, illuminating the clouds as they did so.'

Father Gill himself wrote to a friend: 'Last night we at Boianai experienced about four hours of UFO activity. There is no doubt whatever that they are handled by beings of some kind. At times it was absolutely breathtaking ...'

There were nearly 60 separate sightings of UFOs over Papua that June. Trader Ernie Evernett gave what was probably the most vivid description. He saw a greenish object with a trail of white flame. 'It hovered about 500 feet above me,' he reported. 'The light faded apart from four or five portholes below a band or ring round the middle of the craft, which were brightly illuminated. The object had the silhouette of a rugby ball.'

A Hoax in the House

There was laughter in the British House of Commons when a Conservative MP quizzed the Air Ministry over a 'flying saucer' which alarmed villagers in Lancashire in March, 1957. Mr J. A. Leavey, who represented Heywood and Royton, demanded to know whether the Minister knew about 'The Thing'.

Parliamentary Under-Secretary Charles Orr-Ewing rose from the Government Front Bench and turned towards the Speaker. 'This object did not emanate from outer space,' he assured the House, 'but from a laundry in Rochdale . . .'

When the guffaws had died down, he added: 'It consisted of two small hydrogen balloons illuminated by a flashlight bulb and devised by a laundry mechanic.' The man responsible for starting the scare in Wardle, a village near Rochdale, was Neil Robinson, 35. He said later: 'I bought the balloons for fivepence each and sent them up as an experiment in tracing air currents, I never thought my little tests would be raised in Parliament.'

Panic in the Jungle

Battle-hardened American GIs in Vietnam grew used to the unexpected during the long years of jungle war. But on June 19, 1966, men at the 40,000-strong Nha Trang camp got the shock of their lives – and it came from the skies.

Hundreds of them were out of doors, watching movies on a newly-arrived projector, when suddenly a bright light appeared from nowhere. Sergeant Wayne Dalrymple described what happened next in a letter to his parents.

'At first we thought it was a flare, which are going off all the time, and then we found out that it wasn't. It was moving from real slow to real fast speeds. Some of the jet fighter pilots here said it looked to be about 25,000 feet up.

'Then the panic broke loose. It dropped right towards us and stopped dead still about 300 to 500 feet up. It made this little valley and the mountains around look like it was the middle of the day. It lit everything up.

'Then it went up – and I mean up. It went straight up and completely out of sight in two to three seconds. What really shook everyone was that it stopped, or maybe it didn't, but anyway our generator stopped and everything was black, and at the air base about half a mile from here all generators stopped and two planes that were ready to take off, their engines stopped.

'There wasn't a car, truck, plane or anything that ran for about four minutes. There are eight big bulldozers cutting roads over the mountain and they stopped and their lights went out, too. A whole plane load of big shots from Washington got here next afternoon to investigate.'

Dalrymple checked out all six of the affected diesel-powered and independently operated generators for faults, but found none. Later it was discovered that a Shell oil tanker anchored offshore had also lost power at around the same time, for no apparent reason.

The China Syndrome

Chinese airmen and scientists have also spotted strange craft in the sky. Though the government imposes a news blackout on such incidents, reports of two sightings in recent years have reached the West.

In July 1977, astronomer Zhang Zhousheng and several colleagues at the Yunnan Observatory watched a glowing object pass overhead, from north to west. 'It was yellow at the core with a giant spiral extension,' Zhang was reported as saying. 'It was very bright even in the moonlight, and its colour was greenish-blue.'

Ten months later, Air Force pilot Zhou Quington and other pilots were watching films outside their barracks in north-west China when a huge glowing object crossed the sky.

'It passed over our heads at 21,000 feet and disappeared behind some houses,' Zhou said. 'It seemed to have two large searching lights at the front and a bright tail-light. The lengths of the columns of light kept changing, creating a misty haze around the object.'

Chinese history also records curious sightings. Shen Kua was a famous scientist and scholar who lived at Yangzhou, beside the Yangtze river, more than 900 years ago. He wrote of 'a big pearl' that rose from marshes near the town, and hovered over a nearby lake. It had a round double shell and several people had seen it open. Inside was a bright, silvery light, the size of a fist, which dazzled anyone who looked at it. 'All the trees around had their shadows cast to the ground,' the scientist wrote. 'The shell would leave suddenly, as though flying through the waves. It seemed to be surrounded by flame.'

A Cure for Paralysis

A French doctor claims a UFO light beam cured a paralysis that human specialists had been unable to treat. The doctor had been partly immobilized by a wound received in Algeria in 1958, and was also nursing a leg injured while gardening.

On November 2, 1968, he was wakened by the crying of his baby son, and hobbled to the kitchen to get him a drink of water. He saw lights flashing outside, and walked out on to the terrace to investigate. He saw two objects which merged into one before descending towards the house. A beam of light shone on him, then the UFO vanished as abruptly as the image on a television set that has just been turned off.

As he rushed to tell his wife what he had seen, he realized he was running – his wound, which had shown no improvement during months in hospital, had suddenly healed.

Death of the Lavender Plants

A French farmer said he was immobilized by two figures he disturbed beside a UFO in his lavender fields. It was July 1, 1965, when Monsieur M. Masse spotted a craft shaped like an egg and about the size of a car on his farm at Valensole, in the Basses Alpes region. As he approached stealthily through his vineyard, he saw two 'boys' bending over lavender plants, and stepped forward to reprimand them.

He looked into the startled faces of two creatures unlike any he had ever seen. Both of the small 'men' had a large head, long, slanting eyes, high puffy cheeks, a slit-like mouth and a long, jutting chin. One of the creatures pointed a stick at him, and he was unable to move. They watched him for a while, then floated up in a beam of light into their craft. Its six legs turned, a central pivot began to throb, and the object floated upwards before vanishing.

It left a muddy hole in the bone-dry earth, and within days all the lavender plants close to the site had withered and died. New plants would not grow there for years.

Diplomatic Incident

Portugal was plagued by a mysterious invasion of inexplicable flying objects during August and September, 1977. A senior British diplomat was one of many who reported sightings.

The alarms started when dozens of residents in the town of Viano do Castelo claimed they had seen a strange craft in the night sky. Then fishermen at the port of Portimao, familiar with the layout of stars, began to notice a curious intense light in a spot where no star usually shone. Finally, 12 firemen in the city of, Guarda, returning from a call-out, reported a mysterious shining object circling in the sky.

In September, the British official for the Algarve region, Mr D. M. Armstrong, was alerted. An Englishwoman from the town of Alvor phoned to say that both she and her husband had heard a humming sound and seen an object hovering over their home.

The pro-consul fetched his binoculars and scanned the sky over Alvor, four miles away. He clearly saw an object flashing red, white and green lights, and estimated it was 25-30 degrees above the horizon.

In January, Mr Armstrong wrote to Lord Clancarty, a UFO enthusiast in London, who had tried to instigate an official investigation into flying objects.

He reported his initial contact with the UFO, and added: 'Subsequently I was able to see at least one – and some-times four – every night until mid November, when we had bad weather, and have had almost continual cloud cover since.

'On one occasion I was watching two over the sea. I could see both in my field of vision when one moved sud-denly and rapidly higher in the sky. I followed it and was interested to see, a few minutes later, that the second one had followed it.

'The effect is always the same, a rapid red-green-white flashing around what would appear to be a circular base. I have seen nothing by day, only in the evening.'

The diplomat added: 'As you may imagine, I received derision from my acquaintances, but when people came to dinner I would mention the fact and then take them into the garden with my glasses. Everyone had to admit there was "something unusual" there. The derision ceased at once.'

The Portuguese Embassy in London was unable to throw

any light on the sightings. But it then admitted to an 'incident' involving a Portuguese airliner. The pilot had radioed that he was being buzzed by a circling object that was nothing like any aircraft he had ever seen.

'Cigar' Over the Seine

Unidentified Flying Objects of different sizes operating together were reported from the small French town of Vernon in the early hours of August 23, 1954. Businessman Bernard Miserey had just parked his car in his garage at 1 am when he spotted a huge, long, silent, luminous cigar-shaped object suspended over the river Seine, 300 yards from where he stood. It cast an eerie glow over the dark houses of the town.

M. Miserey watched for several minutes. 'Suddenly from the bottom of the cigar came an object like a horizontal disc,' he recalled. 'It dropped in free fall, then slowed, swayed, and dived horizontally across the river towards me, becoming much brighter, surrounded by a halo. A few minutes after it disappeared behind me, going southwest at a prodigious speed, a similar object came from the cigar and went through the same manoeuvres.'

Five discs in all fell from the cigar, and shot off in different directions. After the last emerged, the cigar faded into darkness. M. Miserey went the next morning to tell the police what he had seen – and was informed that two policemen and an army engineer had seen exactly the same thing, at the same time.

Out Among the Berries

Two Norwegian sisters told police they spoke to a man from a flying saucer in August 1954. The women, aged 32 and 24, said they were picking berries in the hills near Mosjöen, in central Norway, when a dark, long-haired man in a buttonless khaki outfit motioned them into a hollow. In it they saw a saucer-shaped craft about 16 feet wide.

The man tried to communicate with words, gestures and drawings. But the sisters could not understand him, and he showed no signs of comprehension when they spoke to him in French, German and English.

Finally the stranger climbed back into the saucer, which rose fast into the air with a humming sound, like a swarm of bees.

Army Manoeuvres

Two British soldiers say they saw a UFO while on exercises with the Royal Armoured Corps in 1978. Mike Perrin and Titch Carvell were driving their Land-Rover on the Yorkshire moors when they saw a dome-shaped silvery object hovering 50 yards away, making a strange buzzing sound.

'It was about the size of five Land-Rovers and had portholes,' said Perrin, 27. 'Lights inside were flashing red and white. I tried to start our vehicle, but the engine was totally dead. We watched the UFO for five minutes, then it shot off and all the power returned to our engine.'

He added: 'It's army policy to dismiss UFO reports, but when we went back to the area next morning with a sergeant, we found a large circle of burnt grass where the object had hovered.'

'Draw What You Saw...'

When netball teacher Bronwen Williams spotted a strange object in the sky during a game she was supervising in February 1977, she knew exactly what to do. She ushered her nine pupils inside Rhosy-Bol county primary school at Anglesey, gave them pencils and paper, and told them to draw, without conferring, what they had seen. The pictures tallied remarkably – a cigar-shape with a black dome.

That same night policeman's wife Hilda Owen was looking out of her kitchen window when she too noticed a shape gliding silently in and out of the clouds. She drew it in lipstick on the window glass, and her husband made a copy on paper when he returned from duty. It might have been drawn by one of the schoolgirls.

Mrs Owen said the UFO appeared from a 'tongue of flame' over Aberffraw Common. 'At first I thought an aircraft was on fire, but within seconds the flame appeared to form a circle and a domed figure appeared,' she said. 'There was no mistaking the shape. I could see portholes quite clearly.'

The shape was still in the sky when her husband arrived home just after midnight. 'It was the colour of the setting sun and about twice the size of the sun as we see it,' he said. 'By the time I got my binoculars out, it had vanished.'

The Hole in the Hedge

Nine-year-old Gaynor Sunderland dashed home breathless and too scared to speak one day in July 1976. Her mother Marion calmed and comforted her, then listened to her description of what was later described as 'probably the best UFO encounter ever documented in Britain'.

Gaynor had seen a strange, silver, saucer-shaped object land in a field a mile from her home in Oakenholt, near

Flint, North Wales. She lay quietly, terrified but fascinated, peering through a gap in a hedge. Two silver-suited people emerged from the craft and probed the ground with equipment. They were short and angular with large pink eyes, and seemed to be a man and a woman.

The craft they came from was around 35 feet long and nine feet high. It had a band of yellow windows along the side, and a flashing box on the top. There was a loud humming noise when it took off again after about half-an-hour.

Gaynor told her mum: 'I was cold and terrified – I was sure both of them had seen me.' But for 18 months, her story remained a family affair. Mrs Sunderland explained: 'Gaynor was frightened of ridicule.' At last the child plucked up the courage to tell others. She was twice questioned under hypnosis, and produced drawings of what she had seen.

UFO watcher Jenny Randles, of *Flying Saucer Review*, said: 'Gaynor's description is among the most detailed ever recorded.'

The Clicking 'Dolls'

Three women in northern England claim to have seen white, doll-like aliens emerge from UFOs. All reported their experience to researchers, but insisted on remaining anonymous.

The first encounter came in September 1976. Two women, aged 63 and 18, were walking near their homes at Fencehouses, Tyne and Wear, when they saw a small, oval object, and found themselves hypnotically attracted by it. As they approached, two beings, 'the size of large dolls', appeared. They had large round eyes and white hair. They seemed startled, and retreated quickly.

Exactly three years later, a 23-year-old woman was in her bedroom at Felling, Tyne and Wear, at 4 am when a glowing, glittering bell-like disc actually entered the room. 'There was a buzzing noise everywhere and I felt paralyzed,' the woman told investigators. 'Then 12 white creatures appeared, small, like dolls. They were making clicking noises and seemed to be watching me. One even touched me. Then they disappeared.'

UFO researcher and author Jenny Randles said: 'I am convinced the women are telling the truth about genuine experiences. In one way the reluctance of the witnesses to get involved lends credence to their stories. At least we know they are not after cheap publicity.'

At nearby Killingworth, a 21-year-old nurse who would agree only to be identified as Linda, reported a UFO flying between two houses in February 1978. There was a deafening build-up of sound, and her mother hid under the bedclothes, convinced a plane was about to crash. But Linda looked out of the window, and saw, only feet away, a silvery object with a string of coloured lights. She said it looked 'like a tin containing expensive cigars'.

Weird Happenings in the South West

A courting couple, a builder's wife and a deck-chair attendant all saw something strange in the sky on the night of Saturday, May 21, 1977. All the sightings were near Poole, Dorset.

Pretty 18-year-old tax officer Karen Iveson, and her boyfriend, apprentice technician Cliff Rowe, 19, had just parked their car on a lonely road near Parley Cross when a beam of light struck the back of it.

'We couldn't see what was causing it, but it scared us a little, so we decided to move on,' said Cliff. Back on the road, they saw what had disturbed them.

Karen said: 'A large, silvery disc-shaped object hovered over a field, and a silver-green, cone-like beam of light shone down from the centre of it. We stopped to watch, and it seemed to stay there for ages. Then it suddenly veered off fast and dropped behind some trees, much lower than any plane could go. We both panicked afterwards. It was not like anything I'd ever seen before.'

Builder's wife Pauline Fall, 31, saw the same thing only miles away as she drove down a dark country lane near the village of Longham. A beam of light fell across the car bonnet four or five times, 'as if something was tracking us', but at first Pauline could not see where it came from.

'One minute there was nothing in the sky, the next there it was, looking like the underside of a big dinner plate,' she recalled. 'Out of the centre came a silvery white light, narrow at the top and widening into a cone. It was solid light, as if a line had been drawn round it.

'I'm not normally one to panic, but the pit of my stomach went ice-cold. A friend with me was unnerved, too.'

Pauline kept driving for home, even though the beam seemed to be getting shorter as the object descended. 'Then it just disappeared as if it had been swallowed up by the ground.'

When Pauline got home to Wimborne, husband John thought from the look of terror on her face that she must have had an accident. Her hands were as cold as ice, and it was an hour before warmth began to creep back into them.

Odd things happened to Pauline's car after that night. Petrol consumption shot up, and the engine, perfectly all right when John was at the wheel, inexplicably cut out when Pauline was driving. She refused to take it out alone at night for four months.

'I've done a lot of soul-searching, but I haven't found a logical explanation for what I saw,' Pauline said. 'I just wish someone could tell me what it was, where it came from, and what it wants from us.'

The third person to see the craft was deck-chair attendant Richard Morse, 27, who spotted a flickering light behind clouds as he hurried to a bus in Poole. 'I thought it was the Moon, then I saw the Moon in another part of the sky.

'Just looking at it made me feel weird. It was a flying saucer shape with another shape on top and a beam of white light from its centre to the ground. Time seemed to stand still as I watched, then the thing started to move off, banking very fast, before it disappeared. It wasn't like anything from this planet ... I was really glad to hear others had had similar experiences that night, because my friends were starting to think I was mad!'

Along the coast at Parkstone, Dorset, Mrs Ethel Field had a strange encounter in March, 1978, when she left her husband and daughter watching television to bring in washing from her back garden.

'Suddenly I saw this object in the distance, rising from the sea,' she said. 'It ascended and came closer. It was circular with a dome on the top. Beneath it were several lights, shielded by hoods that looked like eyelids. When the lids slid back, there were spotlights lighting the ground.

'I was frightened and stunned. It hovered directly above me. The lights were so strong that I put my hands up to screen my eyes.

'Then I saw two figures standing in front of an oblong window. They had longish faces and were wearing silver suits and what looked like skull caps. They seemed to be standing at some controls.

'I felt that some power was holding me where I stood. I waited, shielding my eyes. Then one of the figures turned away from me to look at his companion. The minute he did that, I felt a release and ran petrified to call my husband and daughter. They laughed at me and wouldn't leave the TV.'

Mrs Field spent several sleepless nights worrying about

what she had seen. Then red blotches appeared on the hands that the light had hit. Soon her hands were raw, with skin flaking and scaling all over them. 'I went to several doctors about it, but only one listened seriously to what I told him,' she said.

A Scottish Saucer

Two ten-year-old girls at Elgin, north-east Scotland, described 'a silver-coloured saucer with a bump on the top' which they had seen hovering in a wood. The craft glowed with a red light and a silver-suited man stood beside it. Mrs Caroline McLennan, mother of one of the girls, said: 'When my daughter told me about it, I remembered having heard a strange whirring noise and saying to my neighbour, "Sounds like a flying saucer." The girls led us back to the wood and we found a big patch of flattened grass. The leaves on the trees nearby were scorched.'

Backwards in Time

Salesman Alan Cave, 45, of Taunton, Somerset, remembers the precise moment he became a 'time traveller'. As he was driving, one October morning in 1981, from Bath to Stroud, his car passed directly beneath a strange, orange, cloud-like object in the sky.

'I know it was exactly 11 o'clock,' said Alan, 'because a newsreader announced the headlines on the car radio. But then I glanced at my watch and it said 8 o'clock. My digital pen said 9. Both were right when I set out. 'Then the speedo started going back – it was weird. It lost 300 miles, though a mechanic has since told me it was impossible.'

Alan doesn't believe in flying saucers, 'but something very odd occurred in those few seconds and I wouldn't like it to happen again.'

The British Flying Saucer Research Bureau later said they investigated several reports of a UFO in the same area. They added that checks on aircraft movements had not provided an explanation.

Scary New Year

Weird sightings were reported throughout Britain on New Year's Eve, 1978. Schoolboy Andrew McDonald, 13, claimed he was buzzed by a UFO as he rode his bike home through Runcorn, Cheshire.

'I heard a hum like a high-pitched engine' said Andrew. 'I looked up and there was a big white light with a very bright trail above me. It stayed with me for about ten seconds, then soared up into the sky. I could feel it trying to lift me off the ground.' Andrew was so unnerved that he could not cycle any farther.

In London, night club waitress Patricia White saw a blazing white shape as a taxi drove her home through Wembley. 'It shone like a big, bright star, but it was following the cab,' said Mrs White, 34, of Harrow. 'I was petrified, and so was the cab-driver.'

Scared witnesses also reported seeing unexpected lights and shapes above Newcastle upon Tyne, Sheffield, Manchester, Norwich. and places in Scotland. But the Ministry of Defence said: 'We are not being invaded. We think it is just some space debris burning up.'

Marooned on the Moors

A good turn proved a nightmare for Lillian Middleton. She drove out onto the lonely Northumberland moors to rescue a friend who was stranded – and ended up being chased for miles by a terrifying UFO. The ordeal began at 2.30 am on August 21, 1980, when the bedside phone rang at Mrs Middleton's home in Seaton Delaval, Whitley Bay. The 33-year-old woman agreed to drive out to her friend, whose car had run out of petrol. But as she reached the moors, she saw a bright flash of light.

'I thought a plane had caught fire or exploded in mid-air,' she said later. 'I slowed down and peered out of the window. I was shocked to see a huge rugby ball shape giving off a brilliant light and hovering in the sky. It suddenly zoomed down towards me. I was terrified.

'It seemed about the size of two big cars. I put my foot down and was soon doing 70 mph in an effort to get away, but the thing kept with me, hovering just above the roof. It moved to the side from time to time, as if it was trying to see who was inside. After what seemed an eternity, I saw my friend beside his car. He too had seen the shape.'

The UFO followed them all the way to the petrol station, a few miles away. A taxi driver and a couple in another car had also watched it approaching. Armed with a can of petrol, Mrs Middleton set out again for the stranded car.

'This time the UFO zoomed right down low to car rooftop height,' she recalled. 'My friend became as scared as I was, and we turned round and went back to the service station. I wasn't going down that lonely road again. I cried with relief that someone else had seen what had happened. The other couple were still at the garage, and we all watched the thing for some time until it suddenly shot off at speed and disappeared.'

Mrs Middleton rang the police, who were sympathetic.

Indeed, their own chief inspector had also reported seeing the UFO. But the experience left its mark. 'I was in a state of shock for several weeks,' Mrs Middleton said. 'Now I won't go out driving after dark, and for a long while I couldn't bring myself to look into the night sky.'

Aliens over Russia

For years, the Kremlin authorities mirrored the attitude of air chiefs in America – UFOs did not exist. Sightings were dismissed as cloud formations, planets and the like, or fobbed off with some other logical explanation; and people who reported strange craft were considered idiots who also believed in goblins and fairies.

But in the late 1970s, *Pravda* began publishing accounts of mysterious visitors all over the country. And they were every bit as astonishing as those reported in the West.

Dr V. G. Paltsev, a veterinary surgeon, was making his country rounds, 500 miles from Moscow, when he came across a grounded craft. Three small humanoids with egg-shaped heads and long fingers were standing beside it, but as he approached he was knocked out by a strange force.

He came round to find his watch had stopped. Above him, a glowing saucer shape was disappearing. Dr Paltsev went home, and carried on working as if nothing had happened. But at night, he repeatedly dreamed that he had been carried onto the craft while unconscious.

A doctor interrogated him under hypnosis – and decided that the vet probably had been taken for a ride on the saucer.

Dr A. I. Nikolaev, a respected historical sciences professor, spent three months in hospital recovering from his close encounter. He and three academic colleagues were on a camping holiday in southern Russia when they came across a metallic, saucer-shaped craft partly hidden in long grass. One of them threw some stones, which seemed to disappear inside the object.

All four men then felt a strange force. Dr Nikolaev was knocked out. The others, though drowsy, dragged him away. Two stayed with him while the third went for help: but both sentries soon fell asleep.

When they woke, two three-foot figures in space suits and helmets were staring at them. At the first signs of life, the small humanoids scurried back to their craft, vanishing from sight through the hull. The object glowed, then disappeared.

Professor F. Zigel, who led the official team of investigators into the case, said: 'There is no doubt that a spaceship landed – possibly because of illness among the crew.'

Only days later, three other scientists saw an alien craft just 67 miles from Moscow. They too were camping, and that night in their tents they heard a babble of loud voices. None of them recognized the language, but all felt a sense of unaccountable fear.

It was half-an-hour before they dared look outside – and there stood a shining violet-coloured object, about 80 feet high, looking something like a giant electric light bulb'. It rose, swayed slightly, then soared upwards into a fluorescent cloud.

Next morning, the campers found a circle of flattened grass 500 feet from their tents ... and called in the investigators.

The Russians showed unexpected interest in an English UFO sighting. Hope and Ruby Alexander spotted a bright, triangular light hovering over Hayes Road, Bromley, Kent, as they drove home one evening in 1978 after a concert. Their sighting was reported in the local newspaper, which noted that there seemed no explanation for it. The two women preferred to let the subject drop.

But two years later, the paper received a postcard from the Soviet science city of Novosibirsk. Someone signing himself V. I. Sanarov asked for a copy of the article, and any further information available. Hope said: 'We were astonished at the interest after all this time.'

Charles Bowen, editor of the British magazine *Flying Saucer Review*, said: 'Soviet scientists have a great interest in UFOs. For several years the Soviet Academy of Science has been ordering three copies of every issue of the Review, and last year I had half a dozen letters from people in the Soviet Union asking for information on the subject.'

The Bermuda Riddle

Dozens of ships have disappeared in the Bermuda Triangle since the US Navy craft *Cyclops* sank with 300 men in 1918. Aircraft, too, have vanished without trace, including five Avenger bombers from Fort Lauderdale that reported having lost their bearings on a routine flight in 1945, and were never seen again. But in 1978 it was a mystery arrival which baffled the experts who keep watch on the area of sea between Florida, Puerto Rico and Bermuda.

Radar crews at the Pinecastle Electronic Warfare Range near Astor, Florida, suddenly found a zig-zagging shape on their screens at a time when no military or civilian planes were expected. And it quickly became clear that this was no ordinary plane. The object was moving in a very erratic way, changing direction at incredible speed, suddenly stopping, then accelerating, within seconds, to 500 mph. Officers scanned the skyline with binoculars, and saw a circular craft emitting curious red, green and white lights. Nobody knew what it could be.

'It manoeuvred in such a way and at such speeds that it could not have been an airplane or helicopter,' said one technician. 'I've never seen anything like it and I don't want to see anything like it again.'

At 5.30 am on September 27, 1979, two children on the island of Bermuda claimed they were immobilized by strange noises from a UFO. Laquita Dyer, 13, and her brother Melvin, 11 were sleeping in separate rooms when both heard a loud, rasping, buzzing sound coming from above their roof.

'When I tried to get up, I couldn't move at all, I was paralyzed,' said Melvin. His sister said: 'I tried to get to the window, but couldn't.' After about ten minutes, the noise switched to a softer tone, then stopped. Only when it died away could the children move their limbs again.

The children's ordeal came only hours after many other people reported a UFO streaking across the sky to the south of the island. Jeffry Schutz, a consultant with the US Department of Energy, was one of them. He was with his mother and sister on the patio of his home. 'At about 9.45 pm we saw an object travelling from west to east, climbing at an angle of 45 degrees,' he said. His sister Betsy, 23, said: 'It was a yellowish, whitish ball, faster than a satellite but slower than a shooting star. It was climbing in a clear sky, with a white vapour trail. We watched it for 20 seconds, then it vanished with a greenish glow.'

English teacher Nigel Kermode and his wife Julie also saw the UFO from their porch. He said: 'It was much too bright and big to be an airplane.' She said: 'It seemed to lose momentum and then speed up again. Then it just disappeared.'

Local tracking stations had no explanation for the sightings.

The Green-suited Superman

A strange flying humanoid dropped in on a family in Puerto Rico on July 12, 1977, according to a man who lived in the town of Quebradillas. He said he and his daughter were at home that day when a small figure ducked under their fence and approached the house. He presumed it was a child, and asked his daughter to switch on the lights.

This seemed to alarm the visitor, who immediately

doubled back. The couple saw that he was about 3 feet tall, in a green suit with padded feet, and a green helmet with a transparent face-plate. Affixed to this was an antenna, and on his back was a box, attached to his belt. The figure also had a tail.

To the amazement of the watchers, the alien ducked under the fence again, pressed the front of his belt, and took off, zooming Superman-fashion, towards some flashing lights in the distance.

Celebrity Sightings

One film star who has seen a UFO is German-born actress Elke Sommer, who in 1978 was in the garden of her Los Angeles home when a shiny orange ball, about 20 feet in diameter, appeared out of the blue. 'It was glowing and floating about like a big moon,' she said. 'It came towards me and I fled into the house. When I reappeared, it had vanished.'

Boxer Mohammed Ali was on a training session in New York's Central Park in 1972 when he encountered a UFO. He said: 'I was out jogging just before sunrise when this bright light hovered over me. It just seemed to be watching me. It was like a huge electric light bulb in the sky.'

Statesmen and politicians who have seen UFOs include John Gilligan, Governor of Ohio, who in 1973 reported seeing a UFO near Ann Arbor, Michigan. He described it as 'a vertical shaft of light which glowed amber.'

Sir Eric Gairey, Prime Minister of the Caribbean island of Grenada, tried unsuccessfully in 1978 to have the United Nations officially investigate UFOs. He said he himself had seen one – 'a brilliant golden light travelling at tremendous speed.'

But the most famous UFO spotter of all is Jimmy Carter, who, while still Governor of Georgia, in 1973, was sitting out on a verandah with 20 other people after an official dinner at Thomastown when, in his words, they witnessed a UFO 'which looked as big as the moon and changed colour several times from red to green.' He launched a $20million study into UFOs after becoming US President.

Pot Shots

For six years defence chiefs were baffled by the UFOs that periodically rained down on the Arizona plains. They showed up on radar at a Colorado tracking station, but when the apparent landing sites were checked, there was no sign of them.

Then, in July 1979, federal narcotics agents were told by an informer that Mexican drug smugglers were using home-made rockets to shoot marijuana across the border into America. When dates were checked, at least one drug consignment coincided with the UFO sightings.

Major Jerry Hix said at the tracking station: 'It wasn't strong enough to spark off a nuclear alert, but it did have us a bit puzzled.' A narcotics agent said: 'It sure as hells adds a new meaning to the old saying about taking pot shots...'

Lord Dowding's admission

The Royal Air Force's late Air Chief Marshal, Lord Dowding, was a staunch believer in UFOs. As early as 1954, he said: 'I have never seen a flying saucer, yet I believe they exist. Cumulative evidence has been assembled in such quantity that, for me at any rate, it brings complete conviction.

'There is no alternative to accepting the theory that they come from an extraterrestrial source. For the first time in recorded history, intelligible communication may become possible between the Earth and other planets.'

Although Dowding admitted never having seen a flying saucer, he would have received numerous reports of UFO sightings by the fliers under his command during World War Two.

UFOs Down Under

Something strange was happening in the skies above Australia and New Zealand at the end of 1978. During one ten-day period, six pilots separately reported curious objects flying alongside their planes.

Radar stations recorded inexplicable bleeps on their screens. Wellington air traffic controllers watched for three hours as objects darted erratically around at remarkable speeds. Above Cook Strait, ten shapes blipped across the screen, 'radically different in behaviour from normal aircraft'. Then, at midnight on December 30, a television crew aimed their camera at a blazing light approaching a plane. And experts who analyzed their amazing pictures said: 'It may be a spaceship.'

The TV crew were from Channel O in Melbourne, Australia. Anxious to check out the spate of strange sightings, they boarded an Argosy turbo-jet used to make newspaper deliveries between Wellington, Christchurch and Blenheim, New Zealand. Captain Bill Startup, an airline pilot for 23 years, had seen gleaming oval objects over the Cook Strait during his regular run a few days earlier. Now, as he flew over the same area with the TV team, they were there again.

Reporter Quentin Fogarty, 32, said: 'We saw a blazing white fireball about 50 miles ahead. It was brilliantly lit at the bottom and seemed to have orange rings round it.' Cameraman David Crockett began shooting film as his

294

wife Ngaire switched on the sound equipment. As the plane got closer, Crockett became convinced that the shape in his sights was not a natural one. Then he noticed smaller objects around it. They were moving in 'intelligent' fashion. They seemed 'in control of the situation, not of this world'.

Captain Startup said: 'One object resembled a large ball of light. No aircraft would have the kind of acceleration that this thing did. It came within 18 miles of us and we decided to go in closer. It went above us and then below, and shot away at tremendous speed.' Co-pilot Bob Guard added: 'We watched the objects for almost 20 minutes. It was almost like watching strobe lights.'

Next morning, the TV team examined the evidence. Leonard Lee, a 32-year-old documentary film producer and a senior member of the Channel O news staff, said: 'The film sent a shiver down my spine. Every time I looked at it, my whole body tingled. We realised we had obtained something absolutely phenomenal, but we decided to make no claims about it other than what our film crew had seen.'

The film was sold to countries all round the world, and screened on news bulletins. Interest in it was astonishing. For the first time, professional cameramen had captured evidence of what seemed to be a mechanical craft from somewhere other than Earth.

But there were plenty of doubters, even in Melbourne. Professor Ronald Brown, head of the city's Monash University chemistry department, said: 'All my training as a scientist tells me the spacecraft theory is extremely unlikely. Looking at the film, I think it is quite feasible that an unusual shower of meteors could have had a similar appearance.'

The professor, one of the world's leading galacto-chemistry specialists, added: 'It is possible that life forms exist elsewhere in the universe, but I do not believe other creatures would be able to shift a solid object such as a spacecraft at such enormous speed. An incredible amount of energy would be required to propel such a craft, and science already knows that the universe contains only a limited amount.'

But TV man Lee brushed aside such sceptics. 'It seems a natural reaction from certain people,' he said. 'They dismiss something simply because they cannot explain it.' He decided to take the film to America for appraisal by UFO experts.

Navy physicist Dr Bruce Maccabee, also a senior official with the National Investigations Committee on Aerial Phenomena agreed to study the film frame by frame. Lee arrived in America in January 1979 with his evidence in a suitcase, sealed with a secret combination, and handcuffed to his right wrist. 'The very existence of this film makes it extremely important,' Dr Maccabee explained. 'It has a whole lot of organisations jumping, wanting to get a look at it. So much of our work ends up in 'Shooting down turkeys. This is one turkey that deserves the closest possible research.'

He spent weeks poring over the film, examining some frames with digital computer enhancement processes. He saw a perfectly-formed glowing triangle, which he estimated to be the size of a house.

Another frame showed an oval with a slight dome protruding. A third section of the film captured a circular object travelling at immense speed. Dr Maccabee said: 'The computer study unarguably shows that the images could not have come from stars or planets, or from the ground or sea surface.'

The physicist also flew secretly to New Zealand to interview the eye witnesses. 'I didn't want anyone to know about the project,' he explained. 'It had been given a lot of publicity earlier in the year, and I wanted to do my inquiries with a minimum of fuss.' He taped statements from Captain Startup and his co-pilot, from cameraman

Crockett and his wife, and from reporter Fogarty, who had been admitted to hospital after being emotionally drained the entire episode.

Dr Maccabee also listened to tapes of conversations between Captain Startup and air traffic controllers, who had spotted inexplicable objects on their radar screens on the evening in question. 'All the witnesses agreed to submit to lie detector tests if their veracity was questioned,' he said.

Finally, the Navy expert concluded that the film and the interviews were a significant advance in UFO research. Stanton Friedman, a nuclear physicist and another of America's foremost aerial phenomena experts, added: 'We arc definitely dealing with a genuine unidentified flying object. What makes this sighting so important is not just the film, but the wealth of additional evidence. Few reported sightings have ever had so much attention focussed on them, and the quality and quantity of the research has been impressive.'

The Channel O film was not the only pictorial evidence of UFOs during the spate of sightings early in 1979. A New Zealand camera team took to the air, and filmed 'an illuminated ping-pong ball, rotating, pulsating and darting around' over South Island. And private detective José Duran filmed what he described as 'a man from outer space' from the garden of his home near Adelaide.

UFO experts who examined Duran's cine-pictures agreed that they seemed to show a 'human embryo-like' object disembarking from a flying saucer and hovering between two 'spacecraft'. Duran said he first saw a red and amber light, moving slowly from the north-west towards the south-east.

'I watched through binoculars for a little while, then the light seemed to approach me,' he added. 'I filmed it from the garden. There was a strange kind of flashing, and although it was travelling very slowly with no sound, I thought at first it might be a plane.

'To my surprise, when I developed the film, I saw something I hadn't actually noticed when I was taking it. There was a white object travelling from an angle. It stopped for a couple of seconds above what I thought was a plane. It made a jerking movement above the flashing light, then moved off in a different direction. The whole movement on film looks like a large V-sign. Moving between the spacecraft was a humanoid, flesh-coloured at one end, but the rest of its body covered in a blue shroud. Microscopic examination of the film has shown two more humanoids in and around the spacecraft.'

Experts from Contact International, the British UFO research body, spent months analysing the film, and decided that the balls of light were probably alien craft. Research officer Derek Mansell said: 'The lights cannot be those of aircraft, and space agencies have confirmed that there was no debris entering the Earth's atmosphere at that time and place.'

More Sightings in 1979

UFO sightings in early 1979 were not confined to Australasia. From Israel came reports of a rash of red balls and flashing lights. In northern Italy, dozens of villages on the slopes of the Gran Sasso mountains were plunged into darkness after a UFO was seen hovering over a hydro-electric plant. Technicians said their equipment suddenly went haywire.

In America, TV reporter Jim Voutrot was amazed when he read about the New Zealand pictures. For he too had captured UFOs on film, at around the same time. And his shots were remarkably similar.

It happened as Voutrot cruised near Pease Air Force Base, a Strategic Air Command bomber post in New Hampshire, with Betty Hill, the woman who claimed she was abducted by aliens in 1961. 'She was telling local meetings on her lecture circuit that UFOs were being sighted over the air base,' Voutrot explained. 'Several reporters did stories with Betty, but I'm a sceptic, and wanted to do one when I was sure there was no chance of being set up.

'So I called Betty one day and we were out looking about five minutes later. Suddenly we saw a big, round white object in the sky. I was surprised and bolted out of the car to start filming. Then, all of a sudden, it was gone. I haven't the faintest idea what it was – quite honestly, I've never seen anything like it, before or since. But I know it wasn't the Moon or Venus, it wasn't aircraft landing lights, and it wasn't a balloon.'

Mrs Hill said: 'We were driving up a hill, just a hop and a skip from Pease, when Jim yelled, "There's one." He was out of the car before I could stop it, and was filming.' When the 15-second film was magnified and examined, investigators discovered an undetected second object in the sky, with a tail of light behind it, like a comet. Further magnification disclosed yet more similar tailed lights, just as on the New Zealand film. And the fast, erratic movements of the larger light defied explanation.

Voutrot checked with sources at the Pease base, and found that nothing had been reported, visually or from radar watches. Tower staff told him their air space often played host to unidentified pieces of 'junk'. Air Force spokesmen were also unable to help. But CIA documents have revealed unexplained objects sighted over air bases in Maine, Montana and Michigan in the past.

The most incredible report of a sighting from that period came from South Africa. In January 1979 Mrs Meagan Quezet, a former nurse, said a pink unidentified object landed near her home in Krugersdorp, west of Johannesburg and a squad of dark-skinned little men emerged from it.

Mrs Quezet said she saw them just after midnight, as she

took her son André, 12, for a walk because he could not get to sleep. 'As we walked down the road, we both saw a pink light come over the rise,' she said. 'Suddenly we came across this thing standing in the road about 20 yards away. In front of it were five or six small beings. These people were dark-skinned, as far as I could tell. One of the men had a beard and seemed to be the leader.

'I said hello to one of them, but I couldn't understand what he was saying. I told André to run off and bring his father, and as he did so the creatures jumped about 5 feet into the air and vanished through a door into their craft. The door slid closed and the long, steel-type legs began to stretch out. Then it disappeared into the sky with a humming noise.'

Both Mrs Quezet and her son said the craft had bright pink lights on either side of the door. The humanoids appeared to be wearing white or light pink suits and white helmets.

Many scientists and astronomers around the world poured scorn on this flurry of UFO sightings. They dismissed them as an unusual concentration of meteors, or space debris burning up on entering, the Earth's atmosphere. Sir Bernard Lovell, director of Britain's Jodrell Bank radio astronomy station, said the reports were 'pure science fiction'. But, as we shall see, the facts about UFOs were often even stranger than fiction.

Clear For Landing

The town of Arès, in the French Bordeaux region, has made UFOs an offer it hopes they will not refuse – a safe spot to land. Engineer Robert Cotton, who works at Bordeaux Airport, came up with the idea for the Ovniport. He believed UFO pilots were reluctant to land because ordinary airports get too crowded. He persuaded officials at Arès to set aside land on the town borders for a landing strip. We have installed a number of landing lights and markings so we believe it can easily be spotted by UFO pilots,' said the mayor of Arès, Christian Raymond. So far, none have arrived to to entertain the tourists who regularly turn up to watch and hope.

Encounters
of the
Closest Kind

Perils of Washing Day

A terrified Devon housewife claimed she was grabbed by aliens, and beamed onto a spaceship as she pegged washing out in her back garden in February 1978.

The woman, who asked to remain anonymous when quizzed by UFO researchers, said she first saw a blue shining shape approach her ' home in Ermington, near Plymouth, from the north.

'The light hovered over the garden,' she said. 'I was petrified. I dropped the washing. Suddenly I was completely enveloped in bubbles of light. I saw three beings who looked like men. They did not speak. They were each about five feet tall, wearing bluish metallic-like clothing.

'They grasped me by the arms and we were lifted up in a beam of light into a kind of room. There were more of the men there. I was given the impression – I don't know how – that I would come to no harm.

'A little later, I found myself back on my lawn. I felt a sharp blow on the back of my neck. I was stunned but not hurt. When I looked round, the thing set off at great speed and disappeared.'

The woman, identified only as Mrs G, told her strange story to Contact UK, one of the largest British UFO investigating organisations. Bernard Delair, one of its senior members, said: 'We take this report very seriously. Her story is very graphic and fits many others.'

A Whirlwind Ride

Missile engineer Daniel Fry claimed in 1950 that a smooth, oval shaped capsule landed near him in New Mexico, and voices invited him to take a trip in it. He said he was whisked to New York and back – a round trip of 8,000 miles – in less than an hour. The voice told him that expeditions from the UFO's planet had been visiting Earth for centuries to try to help human development, but had yet to meet people who were sufficiently intelligent. Fry later claimed that the CIA suppressed his story for 12 years.

No Sign of the Dog

Angler Alan Morris claims that a UFO crew kidnapped his dog. Morris, of Bethesda, Wales, told police he was fishing in a river near his home when a ball of pulsating light approached.

'It hovered for a while over where I was sitting, then landed in a nearby field,' he said. 'I moved closer to get a better look.'

Morris said he saw a hatch open in the side of a saucer-shaped craft. A metallic-looking ladder dropped down to the ground, and three beings climbed down it. 'They were about seven feet tall, with antennae on their heads,' he recalled. 'They each seemed to be carrying spades and containers.'

When the figures started digging, Morris's dog suddenly ran towards them. The fisherman stood up to call him back, then blacked out.

By the time he opened his eyes again, the saucer had vanished, leaving only burn marks on the spot where it had been. And there no sign of the dog.

Strangers From the Sea

In November 1980, a concert pianist and her friend claimed that they had been whisked aboard a UFO and given a gruelling examination. Again the full story came out only after hypnosis sessions.

Luli Oswald, 55, said she and Fauze Mehlen, 25, were driving along the coast road near Rio de Janeiro, Brazil, when they saw a fleet of strange craft emerge from the ocean. 'When they came out of the water, it was like a mushroom with water spilling over it,' she told police. 'Then we noticed a big black one ahead of us. It seemed to be about 300 feet across with a small dome on top.'

Mehlen, who was driving, lost control of the car. It began weaving crazily across the road, with the doors opening and shutting by themselves. Then, suddenly, the nightmare ended. Shaken, they stopped at a restaurant for coffee ... and discovered that it was two hours later than they thought.

'The man was panicked and she was trying to calm him down,' a restaurant spokesman said. 'They told me what had happened, and I told them that others have had trouble on that curve of road. One of my friends was chased by a UFO there.'

Miss Oswald went to a top hypnotist, Dr Silvio Lago, in an attempt to fill the two-hour gap in her memory. 'I can see two small UFOs above us, ' she said when under his influence. 'I'm feeling sick, nauseated. Our car is being grabbed by the top. A light from the small ones is holding us, the light clasps the car. We're being held prisoners by this light. It's horrible . ..'

The pianist began sobbing with terror as she continued to relive the experience.

'We have entered the black disc from the bottom, she said. 'The car is inside the UFO, but we are outside the car.

They are putting a tube in my ear. There are tubes everywhere ... they are pulling my hair.

'They look like rats ... oh, how horrible! They have huge, horrible rat ears and their mouths are like slits. They are touching me all over with their thin arms. There are five of them, their skin is grey and sticky ..."

Miss Oswald said she saw Mehlen lying unconscious on a table as the aliens examined him with a strange ray of light that smelt of sulphur. The rat figures communicated without speaking, but she said one of them did talk to her. 'He said they came from Antarctica, she recalled. 'There is a tunnel that goes under the South Pole, that's why they come out of the water. Others are extraterrestrials.'

After two hours, the examinations were over, and the couple found themselves mysteriously back in the car and on Earth.

'You Will Forget Everything ...'

A 16-year-old American high school boy reported seeing a space creature in his backyard. He described the creature as being very tall, with large green eyes and no nose. The boy said he stepped from his back door and walked towards the visitor ... and remembered no more until he awoke the next day.

The boy agreed to undergo tests at the Southwest Montana Mental Health Centre, Annaconda. The youth had forgotten what happened after he encountered the creature, but under hypnosis he revealed that three aliens had dragged him into a spacecraft. They had examined him under a bright light and then told him he would forget the entire incident.

Dr Kent Newman, who conducted the experiment, said: 'I believe that boy honestly reported what he had experienced.'

A similar view is taken by Dr Leo Sprinkle of the University of Wyoming, Laramie, whose tests revealed that most space encounters took place against the subjects' wills and that they were generally terrified and highly emotional. They often experienced physical effects and amnesia.

Dr Sprinkle said: 'I don't know whether these people experienced physical or out-of-the-body encounters, but my personal and professional bias is to accept their claims as real.'

Dr Alfin Lawson, of California State University, Long Beach, is more cautious. After placing several abduction witnesses under hypnosis, he said: 'Their stories are at least partially true. But that does not mean that their experiences are necessarily "real" physical events – any more than hallucinations are.'

The Price of Passion

Aliens from other planets have hijacked men and women for sex, during visits to Earth. That is the incredible claim of British UFO Research Association investigator Barry King.

He reported the case of a lady from Devon in England. She alleged that one night, she was driving her car quite close her home when her car engine cut out. Thinking that this was quite unexpected, she got out of the car to look under the bonnet. As she stepped onto the road, the lady was seized from behind. In the sudden attack, the lady fainted.

She came round to find herself naked and bound to a table, a blue blanket covering her. Three men in pale blue tunics, all around 5 ft 6 in tall, with fair skin and round, emotionless eyes, conducted a medical examination. Then two left.

Mrs X. claimed the man who stayed with her placed a small pin-like device on her thigh, which made her feel numb and semi-paralysed. He then raped her, calmly and without emotion. Mrs X. passed out again, and next found herself lying beside her car. The engine worked perfectly.

'She drove home and told her husband the whole story,' Mr King said. 'She is a level-headed woman, and I am satisfied she is telling the truth. I believe such cases happen more frequently than we know. Many victims would be very unwilling to talk about it.'

A Brazilian farmer was certainly reluctant to talk about his close encounter of the intimate kind – until he was forced to go to the authorities with radiation sickness.

The man had an extraordinary story to tell. He claimed a shining egg-shaped capsule landed in one of his fields on the cold starlit night of October 15, 1957.

A group of humanoids in close-fitting grey overalls and grey helmets bundled him aboard the craft, stripped him, sponged him with a freezing cold liquid, then took a blood sample from his chin with a suction cup.

He lay naked and frightened on a couch after they left. Then a naked woman appeared, a woman unlike any he had seen before. Soft, blonde hair framed a triangular face with large blue, almond-shaped eyes and a pointed chin.

She had a well-formed figure, a narrow waist, broad hips and long legs. The prisoner thought her the most beautiful creature he had ever seen. She smiled down at him, then put her arms round him and began to rub her face and body against him.

'Alone with this woman, who clearly gave me to understand what she wanted, I became very excited,' the man later told the authorities. 'I forgot everything, seized the woman and responded to her caresses. 'It was a normal act and she behaved like any other woman, even after repeated embraces. But she didn't know how to kiss, unless her playful bites on my chin meant the same thing.

'Finally she became tired and was breathing heavily. I was still excited but now she withheld herself from me. Before she left, she turned and pointed first to her belly, then with a kind of smile at me, and finally at the sky in a southerly direction.'

At first, the young farmer told only his mother about the curious experience. He thought no one else would believe the strange tale. He could hardly believe it had happened himself.

But in the days that followed, his health deteriorated quickly. Nausea and headaches kept him indoors. His eyes burned and he could not sleep. Then small red circles appeared all over his body.

Local doctors who examined him called in a specialist from Rio de Janeiro. He confirmed their tentative diagnosis – the farmer had been exposed to radiation. And the dark scars and thin, tender skin on his chin were evidence of blood-letting.

The doctor was convinced that his patient was not inventing his wild story. He told sceptical police officers: 'There is a complete lack of any direct or indirect signs which might indicate mental illness.'

Over the Seas and Far Away

Some people abducted by UFOs have found themselves landed hundreds miles from where they were picked up.

In 1968, Dr Geraldo Vidal and his wife were driving near Bahia Blanca when something peculiar happened. Hours seemed to shoot by in seconds. When the spell was broken, they were still driving, but the road and scenery had changed. They stopped to check their whereabouts and eventually found they were in Mexico.

The couple could not account for being 3,000 miles from home. They had no recollection of bright lights, no feeling of being lifted into the air. And they were as mystified as Mexican motor mechanics by the strange scorch marks on the bodywork of their car.

Jose Antonia da Silva remembered more about what happened to him after he vanished from Vitoria, Brazil, on May 9, 1969. He had an astonishing story to tell those who found him wandering, dishevelled and trance-like, in Bebedouro, 500 miles away, four days later.

He said creatures about 4 feet tall had plucked him from the ground and carried him off to another planet. Incredulous Brazilians had to admit that something odd had happened to him. He was clearly frightened, constantly darting his eyes skywards, and terrified of bright lights.

All three of these mystery journeys happened during the UFO age after World War Two, when most people had read of eerie experiences in their newspapers. But what can have been the thoughts of the soldier who was found stumbling in bewilderment around Mexico City's main square in 1593 kitted out with the uniform and weapons of a regiment stationed more than 9,000 miles away?

Questioned by the Inquisition, he said he had no idea how he got to Mexico, no sensation of travelling, and only a dim memory of a few blank seconds. His last recollection was of standing outside the Presidential Palace of Manila in the Philippines on sentry duty, and being told that the governor had just been assassinated. Three months later a ship from the Far East arrived in Mexico with the news that the governor had been murdered on October 25 – the very day the soldier had been found in the square.

Attacked by Aliens

Forestry worker Bob Taylor was shocked and bleeding as he told police, of his strange encounter in a Scottish wood. He said two bizarre-looking creatures had tried to drag him into their craft, but he had fought to stop them. He had no way of knowing whether he succeeded.

Taylor, the 50-year-old father of seven grown-up children, worked for the Development Corporation at Livingston, West Lothian. Early on Friday morning, November 9, 1979, he and his placid red setter were alone on Dechmont Hill when he became aware of a strong and curious smell of chemicals. As he approached a clearing to investigate, the dog started barking furiously. Taylor emerged from the trees to see an object that looked like a spacecraft.

Suddenly two aliens appeared from it. They were, he said, like landmines or wheels with arms. They approached Taylor slowly, then grabbed at his trousers, ripping them at the seams, and leaving scratch marks on his legs. The frightened forester tried to fight them off, then passed out. When he woke, both they and the craft had vanished, but he had dim memories of being pulled unconscious towards it. Eventually, Taylor struggled the few hundred yards to his van; but the wheels stuck fast in mud as he reversed, and he abandoned the vehicle to stagger the mile and a half to his home.

Police sealed off the hill while careful checks of the area were made. They found Taylor's van with the engine still running – in his panic he had forgotten to switch off the ignition. In the clearing, they found several deep triangular marks in the ground, and two parallel caterpillar tracks. Nearby were two small ruts which could have been made by the heels of someone being dragged.

Taylor's boss Malcolm Drummond also checked the secluded clearing, and was puzzled by the absence of any marks leading to the triangular indentations. He said: 'Bob Taylor is not a man to make something up. If he says he was attacked by some creatures, then there must have been something there. Bob was shocked and upset by the incident.' West Lothian police said: 'We're taking this seriously. The marks in the ground look as if they were made by the legs of some machine.'

Later the same day, a 72-year-old Glasgow woman reported seeing a pale white ball in the sky. Mrs Mary Hunter, of Easterhouse, said: 'I called a neighbour and we watched it for some time. I am sure I saw it split up in half and come together again, then suddenly it just vanished.' She added: 'I don't drink so I wasn't seeing things.'

Encounters
of the
Historical Kind

Puzzles of the Past

Though reports of UFOs have increased dramatically in the last 40 years, they are by no means unique to the 20th century. Researchers have documented more than 300 sightings before 1900. Monks at St Albans in Hertfordshire saw 'a kind of ship, large, elegantly shaped and well equipped, and of a marvellous colour' on the night of January 1, 1254. And in 1290, the abbot and monks of Byland Abbey, Yorkshire, noted 'a large round silver disc' flying over them.

Author W. Raymond Drake, of Sunderland, Tyne and Wear, who has written many books on UFOs, says: 'The belief in beings from the skies who surveyed our Earth persisted in human consciousness throughout the Middle Ages.' The most spectacular display from that time was probably the one recorded at Basle, Switzerland, on August 7, 1566. Giant glowing discs covered the sky, to the consternation and amazement of the locals.

In March 1716, Sir Edmund Halley, the British astronomer who gave his name to the world's most famous comet, reported seeing a brightly lit object over London for two hours.

On December 11, 1741, Lord Beauchamp claimed he watched a small oval ball of fire falling over London. About 750 yards up, it suddenly levelled off and zoomed eastwards, its long fiery tail trailing smoke as it rapidly disappeared.

And on March 19, 1748, Sir Hans Sloane, later president of the Royal Society, observed a dazzling blue-white light with a reddish-yellow tail dropping through the evening sky. It was, he said, 'moving more slowly than a falling star in a direct line.'

A stream of saucer-shaped objects were seen flying over

the French town of Embrun on September 7, 1820. Witnesses reported that they too changed direction, performing a perfect 90 degree turn without breaking their strict formation. And in 1882, astronomer William Maunday saw a huge disc moving quickly as he studied the north-east horizon from London's Greenwich Royal Observatory. It passed the Moon, he said, then changed into a cigar shape.

In America, too, strange things happened in the sky during the 19th century. In 1878, Texas rancher John Martin was out hunting south of Denison on January 22 when he saw an object coming down from the sun, 'about the size of a large saucer'.

Nine years later, in April 1897, more than 10,000 people were said to have seen an airship over Kansas City, Missouri. Charles Fort, who had also reported a 'large, luminous craft' over Niagara Falls back in 1833, wrote of the Kansas sighting: 'Object appeared very swiftly then appeared to stop and hover over the city for ten minutes at a time. Then, after flashing green-blue and white lights, it shot upwards into space.' The same craft was reported over Iowa, Michigan, Nebraska, Wisconsin and Illinois. The *Chicago Record* newspaper reported that it actually landed in fields near Carlinville, Illinois, but took off when curious townsfolk approached.

Alexander Hamilton, a member of the House of Representatives, had an even more incredible story to tell. He made a sworn statement to the effect that on April 21, 1897, he was awakened by a strange noise outside his home in Le Roy, Kansas, and watched a 300-foot cigar-shaped craft, with a carriage underneath, land near his farm. 'The carriage,' he said, 'was made of glass or some other transparent substance alternating with a narrow strip of material. It was brilliantly lighted and everything within was clearly visible. It was occupied by six of the strangest beings I ever saw. They were jabbering together, but I could not understand a word they said.' Hamilton said he and two of his men tried to move even closer to the craft, but the beings turned on some strange power, and the UFO soared up into the sky.

Both Britain and New Zealand seemed besieged by UFOs in 1909. People in more than 40 towns across Britain reported strange shapes and lights in the sky, most of them during the third week in May. At Caerphilly in Wales, a man said he met two curious figures in fur coats as he walked near his home at 11 pm on May 18. 'They spoke in excited voices when they saw me, then rushed back to a large cylindrical object which lifted off the ground and disappeared.'

The New Zealand sightings were almost all of cigar-shaped objects. Hundreds of people reported them over both the North and South Islands, by day and at night, during the six weeks from the end of July to the start of September. In February 1913 it was Canada's turn with groups of UFOs appearing over Ontario on six separate days.

In those days, with aeroplanes still in their infancy, space and space travel were mere dreams, fantasies to be indulged in the pages of novels by H. G. Wells. It would take two world wars to produce the hotbed of technological invention that began to make exploration of the universe a possibility. During the 1960s and 1970s, science learned about space at first hand. And what was learned cast new light on some puzzles of the past.

The Siberian Space Catastrophe

It was the greatest space disaster of all time. A stricken interstellar craft changed course towards the nearest planet, its nuclear engines overheating uncontrollably. The crew were racing against time, and they lost. Just a mile from the surface, there was a blinding flash, and both they and their ship were blasted to oblivion. And it happened on Earth ... on June 30, 1908.

That is the latest startling theory from scientists trying to explain one of the most baffling mysteries of the 20th century, the Great Siberian Fireball. For years, investigating teams returned from the desolate and devastated explosion site around the Tunguska River, unable to attribute the amazing damage they found to anything but a gigantic meteorite plunging from the heavens. Then human achievements in the arms and space races threw new light on the affair.

It was just after dawn when the fireball was first spotted. Caravans winding their way across China's Gobi Desert stopped to watch it scorch across the skies. Soon people in southern Russia picked it up, a cylindrical tube shape, glowing bluish-white, leaving a multi-coloured vapour trail. It was getting lower all the time. Then at 7.17 am came the explosion. To the peasants of the sparsely-populated area of swamps and forests, it seemed like the end of the world.

'There appeared a great flash of light,' said farmer Sergei Semenov, who was sitting on the porch of his home at Vanarva, 40 miles south of the centre of the blast. 'There was so much heat that I got up, unable to remain where I was. My shirt was almost burned off my back. A huge ball of fire covered an enormous part of the sky. Afterwards it became very dark.' At a nearby trading post, customers shielded their faces against the intense heat. Seconds later they were flung into the air as shock waves of enormous force reached the village. Farmer Semenov was also bowled over, and knocked unconscious. Ceilings cracked and crumbled, windows rattled and shattered. Soil was gouged out and flew through the air.

Closer to the Tunguska, the devastation was even worse. Tungus guide Ilya Potapovich had relatives who owned a herd of 1600 reindeer. 'The fire came by and destroyed the forest, the reindeer and the storehouses,' he told investigators later. 'Afterwards, when the Tungus went in search of the herd, they found only charred reindeer carcasses. Nothing remained of the storehouses. Clothes, household goods, harnesses ... all had burned up and melted.'

The pillar of fire that followed the explosion was seen from the towry of Kirensk, 250 miles away. So were the thick black clouds that rose 12 miles above the Tunguska as dirt and debris were sucked up by the blast. The accompanying thunderclaps were heard 50 miles away. A seismographic centre at Irkutsk, 550 miles south of the Tunguska, registered tremors of earthquake proportions. Hurricane-force gusts shook windows 375 miles from the explosion. Five hours later, British meteorological stations monitored violent air waves across the North Sea. When scientists all over the world later compared notes, they discovered that shock waves from the Siberian blast had twice circled the globe. And when exploration teams arrived at the spot where it had happened, they understood why.

Virtually all the trees in an area 40 miles wide had been blown over and scorched. Giant stands of larch had been uprooted and snapped as if they were twigs. The earth, too, looked unreal. Leonid Kulik, who led the early investigations for the Soviet Academy of Sciences, reported: 'The peat marshes of the region are deformed and the whole place bears evidence of an immense catastrophe. Miles of swamp have been blasted ... the solid ground heaved outwards from the spot in giant waves, like waves in water.' Kulik's researches revealed that the explosion had been seen or heard by people in an area four times the size of Britain. He revised his initial theory that the blast was caused by a single meteorite, concluding that an entire shower of meteorites was responsible.

Yet that hypothesis posed problems. Whenever meteorites had hit the Earth before, they had left craters. In Arizona, a hole 570 feet deep and nearly three quarters-of-a-mile wide had been gouged by the largest one thus far known. There were other inconsistencies, too. Though trees for miles around had been blown over, some at what appeared to be the centre of the explosion were still standing, gaunt and eerie after losing their foliage and branches. In addition, some Tungus had reported finding unusual pieces of shiny metal, 'brighter than the blade of a knife and resembling the colour of a silver coin'. Others claimed that, since the blast, their reindeer had contracted a strange new disease which produced scabs on their skins.

For years scientists argued about the fireball. Was it a gaseous comet, which would not leave a crater on impact? Was it a meteorite that had exploded in mid-air? Then, in August 1945, America exploded an atomic bomb 1,800 feet above the Japanese city of Hiroshima. And when Soviet scientist Aleksander Kazantsev saw the blitzed area, he realized he had seen scenes of identical devastation – in Siberia.

At Hiroshima, trees directly under the blast still stood, while those at an angle to it were flattened, along with buildings. The mushroom cloud, the blinding flash, the shock waves, the black rain of debris – all had been noted in 1908, nearly 40 years before the nuclear age. Kazantsev was convinced that he had the answer to the Tunguska riddle. But it was far from being proved scientifically. So he alerted his colleagues to the possibilities in a novel way. He wrote a science fiction story in a magazine that mingled fact and fiction, surmising that a nuclear-powered spaceship from Mars had exploded over Siberia.

Other scientists took up the nuclear theory, though keeping an open mind about the space suggestion. They compared the Tunguska evidence with what happened when both Russia and the United States held H-bomb tests. And in 1966, Soviet investigators V. K. Zhuraviev, D. V. Demin and L. N. Demina issued a definitive paper which declared that the Siberian fireball had been, without doubt, a nuclear explosion. Further studies, both in Russia and

The World's Greatest Alien Abduction Mysteries

America, revealed that the energy yield of the blast was 30 megatons – 1,500 times greater than at Hiroshima.

Soviet experts examined and dismissed suggestions that the blast was caused by anti-matter or a black hole from space. In both cases, they argued, a crater would have been caused on impact. Professor Felix Zigel, an aerodynamics teacher at Moscow's Institute of Aviation, and geophysicist A. V. Zolotov both re-examined the evidence and the site, and discovered that the area of destruction was not oval in shape, as had been thought, but roughly triangular. To Zolotov it seemed that the explosive material had been in 'a container' when it detonated, a shell of non-explosive material.

Professor Zigel went through eye-witness statements about the cylindrical shape, the trail of fire behind it, and the trajectory of its flight, and came to the conclusion that the object had 'carried out a manoeuvre' in the sky, changing direction through an arc of 375 miles, before it blew up. Soil samples from the blast site revealed tiny spherical globules of silicate and magnetite, a magnetized iron.

Dr Kazantsev, whose science fiction story had prompted the new direction in Soviet investigations, commented: 'We have to admit that the thing long known as the Tunguska Meteorite was in reality some very large artificial construction, weighing in excess of 50,000 tons. We believe it was being directed toward a landing when it exploded.' The Russians claim that no UFOs were sighted for decades after the crash. When they were again reported, the craft were smaller and seemingly more manoeuvrable.

The greatest space disaster of all time? If there was a crew on the UFO, they were not the only victims of the blast. Soviet doctors believe thousands of Siberian peasants died as a result. Residents of the scattered villages around the Tunguska river were renowned for their good health and long life. Many survived long past their 100th birthdays. But after 1908, local medical men reported a big increase in 'premature' deaths from 'strange maladies'. By the time teams investigating the nuclear explosion theory exhumed some of the long-dead bodies, science had found a name for such maladies. It was radiation sickness.

The Undertaker's Secret

For nearly 100 years, the secret of what carpenter William Robert Loosley saw in an English wood remained locked away in his desk drawer. But when his great, great granddaughter, clearing out her attic, discovered his report, experts were forced to the startling conclusion that a flying saucer may have visited Buckinghamshire on an autumn night in 1871.

Loosely was a highly respected member of the community of High Wycombe, now a thriving town, but in those days a small village. The carpenter woke, hot and uncomfortable, at 3.15 on the morning of October 4, and decided to take a walk in his garden to cool off. What happened next was detailed in the manuscript he locked away.

A fight like a star moved across the sky, 'brighter than the full moon'. Then came a clap of thunder – 'odd because the sky was clear'. The object flew lower, stopped, then carried on descending, moving from side to side. It seemed to touch down in nearby woods.

Next morning, Loosely walked to the landing site, and after a long search, struck something metallic as he poked his walking stick into a pile of leaves. Scrabbling with his hands, he uncovered a strange metal container, 18 inches high and covered with curious knobs.

'Almost at once the thing moved a trifle,' Loosely noted. 'With the sound of a well-oiled lock it opened what looked like an eye, covered with a glass lens and about an inch across. Seconds later another eye opened and sent out a beam of dazzling purple light.'

Then a third eye appeared, and shot out a thin rod, a little thicker than a pencil. Loosely decided to leave, but as he moved away, the machine started to follow, leaving a trail of three small ruts. The undertaker came to a clearing, and noticed that the whole surface was criss-crossed with similar ruts.

The metal box stopped briefly, and a claw shot out into the undergrowth. The purple light shone on the corpse of what seemed to be a dead rat. Then the rod sprayed liquid on the body and the rat was pushed inside a panel that. opened on the side of the machine.

Loosely dropped his walking stick in his hurry to get away, and the object picked that up, too. Then it followed him into another clearing, and started herding him, 'like an errant sheep', towards another, bigger metal box.

The undertaker was now close to panic. He looked up and saw a strange moon, like globe in the sky, which seemed to be signalling with lights. But before he could work out the sequence, it vanished. He fled back to his home.

As he lay in bed that night, unable to sleep, Loosely saw, through the window, a light falling into the clearing he had visited during the day. Then it rose again, and disappeared into the clouds. Baffled as to the meaning of all this, the bemused man jotted his experience down on paper, and locked the manuscript in his desk.

After it was discovered, almost a century later, science fiction expert David Langford studied the document, and later wrote a book about it. He said: 'The manuscript has withstood every test of authenticity. It is clearly not a fabrication, because the man's death in 1893 absolutely rules out the possibility that he could describe the scientific concepts apparent in his tale'.

Spring-heeled Jack-man or Space Monster?

Was Spring-heeled Jack, the mysterious monster who terrorized Victorian England, really a misunderstood alien left behind by a UFO? The idea is now being taken seriously by some of those still seeking answers to one of the world's most bewildering riddles.

Spring-heeled Jack was the nickname given to an awesome giant spotted at places from London to Liverpool during a 68-year reign of terror. Early reports of a frightening figure bounding across Barnes Common in London were dismissed as hysterical nonsense. Then, in February 1837, 25-year-old Jane Alsop answered a loud knocking on the door of her home in Bearhind Lane, Bow. She found a shadowy figure on the step, so tall that she had to raise her candle to see the face. With a bellow of agony, the visitor crashed headlong into her, loping away when her screams brought her father and sisters racing to help.

Miss Alsop later told the police: 'His face was hideous. His eyes were like balls of fire, and he vomited blue and white flames. His hands were like claws, but icy cold.' She said he was wearing what looked like an oilskin garment under a black cloak, and had a fishbowl over his head.

The description tallied with similar reports from women who claimed they had been attacked on Blackheath, Barnes Common and beside Clapham churchyard. Then there was another frightening account. Lucy Scales and her sister were leaving their brother's home when they were confronted by a strange creature in Green Dragon Alley, Limehouse.

Lucy said later that a cloaked figure had sprung from the darkness, spitting flames which temporarily blinded her. Her screams summoned her brother, who found the girls sprawled half senseless on the cobblestones, then looked up to see a giant towering over him. Incredibly, the figure bounded out of sight over a 14-foot brick wall.

In January 1838, the Lord Mayor of London, Sir John Cowan, gave the monster official recognition. During a meeting at the Mansion House, he read out a letter from a panic-stricken Peckham resident who described a terrifying creature performing phenomenal feats of jumping. Immediately a flood of similar stories poured into the police from people who had stayed silent for fear of ridicule. The newspapers labelled Jack Public Enemy Number One. As sightings spread from London to the Home Counties, vigilante squads were organised and rewards offered to anyone capturing the beast. Even the Duke of Wellington, then nearly 70, went out on horseback in an attempt to hunt it down.

In February 1855, people in five South Devon towns woke to find that there had been a heavy snowfall – and that mysterious footprints had appeared in it, running for miles, over fields, along the top of walls, on rooftops and across enclosed courtyards. Some said it was the Devil. Some attributed them to a ghostly, unknown animal. Some blamed them on Spring-heeled Jack.

In the summer of 1877, a figure which fitted his description appeared outside an army post at Aldershot. Two sentries, both crack shots, fired at almost pointblank range when he refused to halt. But he just bounded away leaving no blood on the ground where he had been hit. According to the London *Morning Post*, the intruder was 'no ordinary mortal – if in fact he is mortal at all.'

Four months later, residents of Newport opened fire when they cornered Jack on rooftops. Again he escaped unhurt. In Lincoln he bounded out of range when vigilantes chased him. Finally, he astounded hundreds of watchers with a display of prodigious athleticism in the Everton district of Liverpool on September 10, 1904, leaping from building to building, sometimes covering 30 feet at one jump. Then, after 15 minutes, he vaulted effortlessly over a row of terraced houses ... and was never seen again.

For years, experts argued about Jack. Was he, as some claimed, a rich eccentric playing diabolical games? Was he an unknown animal? A phantom? Then, in July 1969, Neil Armstrong took man's first step on the Moon, watched by millions on television. A few of them saw the tremendous bounding steps of the astronaut, and remembered the story of Jack. He leapt like that. According to Jane Alsop, he had worn similar clothes – a jumpsuit and helmet. Maybe, they said, he was no ordinary mortal. Maybe he was an alien from another planet, and unaffected by Earth's gravity. And maybe he could have taught humans a lot – if only they had not greeted him with panic.

The Village that Disappeared

Police are still trying to discover why an entire village of 1,200 people and even the dead from their graves vanished without trace into the dark of a northern winter. The mystery began in 1930, when trapper Armand Laurent and his two sons saw a strange gleam crossing Canada's northern sky. Laurent said the huge light changed shape from moment to moment so that it was now cylindrical, now like an enormous bullet.

A few days later a couple of Mounties stopped at Laurent's cabin to seek shelter on their way to Lake Anjikuni – where, one of them explained, there was 'a kind of problem'. The Mountie asked a puzzled Laurent if the light he'd sighted had been heading toward the lake. Laurent said it had.

The Mountie – nodded without further comment, and in the years that followed the Laurents were not questioned again. It was an understandable oversight. The Royal Canadian Mounted Police were already busy at that time with the strangest case in their history ...

Snowshoeing into the village of the Lake Anjikuni people, another trapper named Joe Labelle had been oppressed by an odd sense of dread. Normally it was a noisy settlement of 1,200 people and today he'd expected to hear the sled dogs baying their usual welcome.

But the snowbound shanties were locked in silence, and no smoke drifted from a single chimney.

Passing the banks of Lake Anjikuni, he found boats and kayaks still tied up at the shore. Yet when he went from one door to another, there was only the unearthly quiet. And still leaning in the doorways were the men's cherished rifles. No Eskimo traveller would ever leave his rifle at home.

Inside the huts, pots of stewed caribou had grown mouldy over long-dead fires. A half-mended parka lay on a bunk with two bone needles beside it.

But Labelle found no bodies living or dead, and no signs of violence.

At some point in a normal day – close to mealtime, it appeared – there had been a sudden interruption in the day's work, so that life and time seemed to have stopped dead.

Joe Labelle went to the telegraph office and his report chattered into the headquarters of the Royal Canadian Mounted Police. Every available officer was despatched to the Anjikuni area. After a few hours search, the Mounties located the missing sled dogs. They were tethered to trees near the village, their bodies under a massive snowdrift. They had died of cold and hunger.

And in what had been the Anjikuni burial ground, there was another chilling discovery. It was now a place of yawning open graves from which, in sub-zero temperatures, even the bodes of the dead had been removed.

There were no trails out of the village, and no possible

means of transportation by which the people could have fled. Unable to believe that 1,200 people could vanish off the face of the earth, the RCMP widened its search. Eventually it would cover the whole of Canada and would continue for years. But more than half a century later, the case remains unsolved.

Could UFOs also be responsible for other vanishing acts over the years? In 1924 two experienced RAF pilots called Stewart and Day crash-landed in the Iraqi desert during a routine short flight. When they failed to arrive, rescue parties were sent out. They soon found the plane, and footsteps leading away from it showed that the two men had set off on foot in the direction of their destination. But after a short distance the footsteps stopped. There were no signs of a skirmish, no other footprints in the sand, no other marks at all. The men's track just stopped suddenly, one foot in front of the other, indicating that they had been walking normally when something happened. The two were never seen again.

In 1900 three tough fishermen set out from Lewis in the Outer Hebrides to relieve three lighthouse keepers at the Flannan Isles beacon. They found nothing wrong at the lighthouse. There were no hints of damage or accident, no disorder, no signs of panic, no missing boats, no loss of fuel, no messages ... and no men. The three keepers had simply vanished of the face of the Earth.

In 1909, Oliver Thomas, an 11-year-old boy, walked out of a Christmas Eve party at his home in Rhayader – and disappeared for ever. Merrymakers dashed outside when they heard a sudden cry that seemed to come from the sky above the house, but they saw nothing.

Life on Other Planets

Many scientists believe that human life itself came from space developing from viruses and bacteria brought to Earth by giant comets. Sir Fred Hoyle, for 20 years professor of astronomy at Cambridge, was scoffed at when he first put forward the theory in 1940. But now scientists all over the world believe he was right.

Hoyle was one of the first to identify giant dust clouds that float silently through space, swarming with the ingredients of life. He claimed a comet plunged through one of these clouds 4,000 million years ago, picking up viruses and bacteria that became locked in the frozen water in its tail.

When the comet – our first UFO – crashed into Earth's atmosphere, friction melted the globules, and the life-forming cells were showered into the mists of the cooling planet to produce plants, animals and humans.

Dr Chandra Wickramasinghe, of University College, Cardiff, believes that millions of comets, 'dirty snowballs' of frozen gases and dust, bombarded Earth, carrying randomly constructed genetic molecules that took root here.

He pointed out to an international conference in Maryland that the Greek philosopher Anaxagoras had similar ideas in 500 BC, arguing that the seeds of plants and animals swarmed in the universe, ready to sprout wherever they found a proper environment.

New scientific techniques have proved that the dust clouds of space contain such chemicals as methane, formic acid, formaldehyde and other substances crucial to forming simple life cells. One cloud showed traces of cellulose – the vital glue of molecular chains.

Could comets have created life on other planets, and in other forms? Dr Sherwood Chang, of the Ames Research Centre in Mountain View, California, says that the millions of impact craters on Mars and Venus were formed mainly by comets. And in the words of Dr Wolfram Thiemann, of the University of Bremen, West Germany: 'Chemical evolution is definitely growing on other planets and in interstellar material. There is more and more evidence that there are other planets like Earth in outer space.'

Sir Bernard Lovell, one of the world's leading radio astronomers, believes there are about 100 million stars in our galaxy, the Milky Way, that have the right chemistry and temperature to support organic evolution; and there are billions more galaxies in the observable universe. The odds against Earth being the only planet with life are therefore ... astronomical.

Space Collisions and Explosions

Other planets have played a crucial part in the development of life on Earth – and are even responsible for the shape of Earth as we know it. That was the controversial theory put forward in 1950 by Immanuel Velikovsky, a Russian born doctor and psychoanalyst who settled in America, in his book entitled *Worlds in Collision*.

Velikovsky claimed that cataclysmic disasters recorded in the Bible and echoed in the ancient writings of the Mayas, Chinese, Mexicans and Egyptians were all due to convulsions in the universe, which sent Venus and then Mars into orbits too close to Earth.

Venus, according to Velikovsky, was part of Jupiter until an explosion sent it crashing into space more than 4,000 years ago. It hurtled towards the sun, blazing brightly, and trailing a slip stream of dust and gases. Earth moved into the outer edges of this slip stream in the middle of the 15th century BC, and a fine red dust coloured our rain. 'All the water that was in the Nile turned to blood,' stated the Biblical Book of Exodus. Then came showers of meteorites, and according to the Mexican Annals of Cuauhtitlan, the sky rained 'not water but fire and red-hot stones'.

When gases coalesced to form petroleum, 'people were drowned in a sticky substance raining from the sky', in the words of the Mayas' sacred book, Popol-Vuh. Elsewhere, the petroleum was ignited by oxygen in the Earth's atmosphere, and a terrible deluge of fire was recorded from Siberia to South America.

Finally, said Velikovsky, Earth was subjected to the full gravitational pull of the new planet and was tugged off its axis. Hurricanes and floods destroyed islands, levelled cities and altered the face of continents. 'Heaven and earth changed places,' wrote the Cashinatia of western Brazil. The Persians watched in awe as three days of light were followed by three days of darkness.

It was then, Velikovsky argued, that Moses led the Israelites across the Sea of Passage. Freak gravitational and electromagnetic forces, as well as the convulsions of the Earth's crust, piled up the waters on either side of the seabed. As the Egyptians pursued their former slaves, a powerful electric bolt passed between Earth and Venus, and the waters flooded back into place, drowning them.

The few survivors of the worldwide catastrophes faced starvation. But suddenly food fell from the skies – manna from heaven to the Israelites, ambrosia to the Greeks, honey-like madhu to the Hindus. Velikovsky believed it was created either by bacterial action or by electrical discharges in the Earth's atmosphere working on hydrocarbons in the trail of Venus.

Just as Earth was getting accustomed to its new seasonal timings, Venus swung past again, in about 1400 BC, with equally disastrous effects. Then it settled into an orbit that left our ancestors in peace. But in the 8th century BC, it passed too close to Mars, dislodging the smaller planet, and pushing it into an orbit which clashed with that of Earth. Again there were geophysical upheavals, recorded in the Bible by the prophets Isaiah, Hosea, Joel and Amos, and in the Iliad by Homer. Once more the calendar had to be revised, because a year of twelve 30-day months was no longer accurate.

Velikovsky said that Mars returned every 15 years until 687 BC, the last time it caused great disturbances, when, according to the Chinese Bamboo Books, Stars fell like rain and the Earth shook. In some parts of the world, the rising sun dipped back below the horizon as Earth again tilted on its axis. Then both Venus and Mars settled into orbits that no longer influenced us.

The controversial theory explained many aspects, of ancient myth, legend and history, not least why Mars replaced Venus as predominant god among the Greeks and Romans. But it outraged scientists in 1950. One curator of a planetarium who backed Velikovsky was sacked.

Velikovsky had flown in the face not only of accepted scientific principles but of Darwin's theory of an ordered evolution. Yet in the next 30 years, as Space travel revealed many more facts about Venus and Mars, his theories were proved right time and time again. He was ridiculed for saying that Venus had a comet-like tail, that it was much hotter than Earth, and that its atmosphere was far heavier than that of Earth. American and Russian probes proved the truth of his claims. He was derided for saying that Mars had a surface of craters, and that its atmosphere contained the rare gases argon and neon. Again, space explorations found he was correct.

Neither Venus nor Mars were exactly unidentified flying objects, but the powers they unleashed terrified and puzzled our ancestors. And even today there are flying objects we can identify, but which are every bit as baffling as UFOs.

The Day It Rained Animals

By the known laws of nature, frogs, fish, mice and peri-winkles do not fly. Yet all have fallen from the skies for no apparent reason, and without explanation.

At Sutton Park, Birmingham, in June 1954, shoppers in a crowded street were astonished by a deluge of tiny, pale frogs. They bounced off umbrellas and hats, fell into shopping baskets, and hopped so profusely about the road and pavements that screaming women dashed into the stores to escape them. By the time the downpour stopped, as suddenly as it had begun, hundreds of the small creatures had been crushed or killed, and hundreds more had hopped away into sewers, alleys and gardens.

But that shower was nothing compared to what had happened centuries beforehand in Sardinia. According to ancient Egyptian books in the library at Alexandria, a frog-fall on the island lasted three days. Frogs clogged the roads and ponds, blocked doors and poured into houses. The people could do nothing to stop the invasion. A Greek scribe wrote: 'All vessels were filled with the frogs. They were found boiled and roasted with everything the Sardinians tried to eat. The people could make no use of water because it was all filled with frogs, and they could not put their feet on the ground for the heaps of frogs that were there. Those that died left a smell that drove the people out of the country.'

Flakes of meat up to three inches square showered down on the American state of Kentucky from a clear blue sky in March 1876. One astonished fieldworker boldly ate some, and said they tasted like mutton. In May 1890, a shower of bright red rain drenched Messignadi, Calabria, southern Italy. The Italian Meteorological Society identified it as birds' blood.

Fish up to five inches long fell on Aberdare, South Wales,

in a dense downpour in February 1859. They covered the roofs of houses and children scooped them up in the streets. Specimens sent to the British Museum were identified as minnows, and put on show at the zoo in Regent's Park, London.

A terrible thunderstorm swept the English city of Worcester in May, 1881. A donkey pulling a cart was struck dead by lightning in Whitehall, and hailstones tore leaves from trees and battered crops to the ground. In Cromer Lane, gardener John Greenhall raced to shelter in a shed, and watched astonished as the hailstorm suddenly turned into a deluge of periwinkles. They bounced off the ground and shredded the leaves of his plants, covering some parts of the ground to a depth of several inches. When the storm had passed townsfolk flocked to the area and collected the molluscs for hours. One man filled two buckets. Another picked up a huge shell and found it occupied by a hermit crab.

A rain of sprats, smelts and whiting fell on the county of Kent at Easter 1666. Some traders cheekily picked them up and sold them in Maidstone and Dartford. Hordes of yellow mice tumbled from the sky over Bergen, Norway, in 1578. Thousands fell into the sea and swam ashore like a tide. Norwegian legend has it that such showers are nature's way of replacing lemmings lost in periodic mass suicides when they rush over cliff-tops into the ocean.

What can be the real reason for these amazing falls of live creatures? The most common explanation is that they have been sucked up by whirlwinds and waterspouts elsewhere on the Earth's surface, and carried by the wind to be dumped unexpectedly where least expected. But if that were so, why are frogs not accompanied by some evidence of their environment, such as pondweed, mud or tadpoles? How can the wind select only sprats or whiting from an ocean full of different species of fish?

Charles Fort, a 19th-century American writer, believed

that such living showers originated in some kind of immense Sargasso Sea, somewhere in the atmosphere. These periodic showers replenished stocks or spread species to new parts of the globe. Sadly, nobody has yet located Fort's aerial sea.

If comets were the vehicles that brought humans to Earth, could they still be raining life down on the planet?

First UFO on Film

Two Swiss astronomers from Basle observed a spindle-like object, surrounded by a glowing outer ring, pass in front of the sun on August 9, 1762. The sighting corresponded with a shape seen over Mexico by hundreds of people in the 1880s. The photograph which Professor Bonilla took there through a telescope at Zacate observatory on August 12, 1883 is believed to be the first photograph ever taken of a UFO.

Deities from Space

Man has found no use for the eerie, empty spaces of Peru's southern coastal plain. Nothing lives in the dry-as-dust flatland which stretches from the Pacific Ocean to the snow-capped Andes. But in 1939, two men in a plane looked down on it, and discovered complex lines and geometric patterns of astonishing precision, stretching for miles across the arid wastes. And ever since, people have been asking: was this once a landing place for aliens? Could it have been an intergalactic spaceship terminal for giant UFOs, which may have brought to Earth, in prehistoric times, the real ancestors of man?

Archaeologists and scientists have never been able to explain the sudden, dramatic evolutionary and technological leap made by *Homo sapiens* 10,000 to 15,000 years ago. There are no genetic clues to the sudden doubling in size of the human brain. In the trail pursued by the experts, there seem to be more missing links than clues.

But some say the clues are there ... in the deserts of the world and the legends of early civilized man. They point to god-like visitors from space who passed on skills and technological knowledge to primitive man, and may even have interbred with him.

The amazing patterns in the Peruvian desert near the city of Nazca cover an area 37 miles long and one mile wide. The plain consists of yellow soil covered by a thin layer of stones. Each line was made by removing the surface layer of stones. The task was comparatively simple, but the undeviating accuracy of the lines is stunning, stretching dead straight for mile after mile, passing over the horizon, crossing gullies and climbing slopes. Their precision matches anything modern engineering can achieve. And they show up clearly only in very high-level aerial photographs.

Scientists who tracked them, with difficulty, on foot, could find no reason for them. They led nowhere and matched no astronomical pattern. But from far out in space, deserts would seem the most obvious place to land on a planet with as much surface water as ours. When America aimed astronauts at the Moon, it chose the lunar equivalent of our deserts. There are reasons for believing that UFOnauts may have done the same thing.

Drawn on the desert floor beside the Nazca lines and patterns are birds, spiders and fish. These too are virtually invisible at ground level. Scientists dismiss them as ancient worship objects. But that may be exactly what they were – invocations to the aerial gods to visit Earth again.

A bird was almost certainly the only other thing known to early Peruvians that could defy gravity by flying; and their desert bird has a tail that fans out like the blast-off trail of a rocket. The spider looks like the object we now recognize as America's spindle-legged Moon-landing craft. And the fish? They might represent the gods themselves.

On the fringes of another desert, the Sahara, lives a primitive tribe discovered just over a century ago by explorers from the West. The Dogons of Mali still worship intelligent, fish-like amphibians who, they insist, came from the sky. They called themselves the Nommos, who landed in a whirling, spinning ark, and had to live in water.

They told the Dogons that they came from a tiny but heavy star called Sirius, which followed an elliptical orbit round the brightest constellation in the sky. Early explorers who listened to the tribe's story nodded their heads with amused condescension.

Then, amazingly, in the 1950s, astronomers using the most modern radio telescopes discovered the tiny, heavy, elliptical orbiting star, exactly as the Dogons had described it. So faint was it that no optical telescope had previously detected it. So how did this African tribe know of its existence?

The Dogons were not the only people visited from above during prehistory. The Sumerians called their gods the Oannes, and they, too, were amphibians. They brought the secrets of mathematics, writing and astronomy to the people of the Tigris and Euphrates valleys of Mesopotamia, long acknowledged as the birthplace of human civilization.

Berossus, a Babylonian priest, described the Oannes god as part man and part fish. He plunged into the sea 'to abide all night in the deep, for he was amphibious'. In Philistine legend, the God was born from an egg which dropped from heaven into the Euphrates. Like the deities of the Dogons', the Oannes had a connection with Sirius. Their worshippers venerated the figure 50 – the exact orbital period of the star, as mentioned by the Dogons.

The theory that spacemen visited our ancestors is reinforced by the art of many ancient peoples. In 1950, archaeologists uncovered the tomb of an ancient Mayan priest at

Palanque, Mexico. Clearly visible on the drawings was the figure of a man in a capsule. He was surrounded by levers and machinery, and there was a fiery trail, like exhaust fumes, at the back of the craft.

In caves below the Sahara mountain range of Tassili N'Ajjer, on the present-day borders of Algeria and Libya, there is a lasting pictorial record of the daily life of a tribe forced to move on when their green oasis was drowned by the shifting sands. Drawings show water buffalo, birds and parties of armed hunters. Among the groups are the clear figures of what we can now recognize as space travellers. They are no bigger than the hunters, but they wear space suits and helmets – round headgear with antennae.

Mysterious markings on the floor of the barren Gobi Desert in Outer Mongolia puzzled explorers and archaeologists for centuries. They were not made by any known form of fire or gunpowder. Then scientists found identical marks in the sands of the Nevada Desert in America ... after the United States triggered its first atomic bomb test in 1944. Had a nuclear-powered UFO visited the, Gobi in the long-distant past?

Across the Himalayas, Indian priests still chant the Ramayana in praise of gods who arrived on Earth in 'vimanas', strange flying machines propelled by quicksilver and fierce winds. The words of the hymn relate that, 'at the gods' bequest, the magnificent chariot rose up to a mountain of cloud on an enormous ray as brilliant as the sun and with a noise like a thunderstorm...'

Such discoveries have suggested to scientists that other marvels of ancient man may be associated with the knowledge and skills of extra-terrestrial visitors. How did the ancient Egyptians build the pyramids, and how did they discover the seemingly magical properties of the Pyramid shape? Who built the giant stone figures on lonely Easter Island, and why? What was the secret wisdom of the ancient Greek oracles?

Even some devout Christians are beginning to wonder whether their religion is based on the visits of a space race. In the Old Testament Book of Ezekiel, the Hebrew prophet records how, in the 6th century BC, he watched a weird cloud descend in the desert beside the River Chebar in Babylon. It was amber, the colour of glowing metal, and surrounded by 'fire infolding itself'. Four objects came out of it, each a wheel within a wheel with a ring of eyes; and out of them came man-like creatures in suits of burnished brass, with 'crystal firmaments' on their heads. He might have been describing a 20th-century astronaut.

American UFO researcher Raymond E. Fowler is not alone in using such descriptive passages to determine the realities of Biblical legends. A pillar of fire guided Moses through the wilderness, and the prophet Elijah was taken to heaven in a fiery chariot. Both were clearly seen later with Jesus on the Mount of Transfiguration, glowing in contact with the 'cloud' on which they stood.

In the New Testament, a pillar of fire and voices from 'the host of heaven' told the shepherds of Christ's birth at Bethlehem; and a bright 'star in the East' guided the wise men to the crib where lay the child born to a virgin. Jesus had magic, mystical powers, and ascended to heaven on a cloud. 'If I have told you of earthly things and ye believe not,' He told disciples, 'how shall ye believe if I tell you of heavenly things?'

The 'angels' of the Lord were messengers of God who came from the skies. Daniel called them 'watchers' – and were allowed to intermarry and eat human food. Were they really UFO aliens who came to educate or 'save' a primitive people? Was the blinding light that converted Paul on the road to Damascus a UFO bringing Jesus back, to repeat His message? And will the Second Coming, with fearful sights and signs in the sky, a host of clouds and angels, really be an invasion fleet of spacemen in UFOs?

Space Shrine

British housewife Phyllis Henderson believes Jesus Christ was a flying saucer pilot from Saturn. She turned the garage of her home at Warrington, Cheshire, into a church after joining the Aetherius Society, a movement begun by George King, who claimed he met Jesus when a UFO landed at Holdstone Down, Devon, during the early 1950s.

Phyllis, 59, and her husband Steuart, 62, received temporary planning permission to use their asbestos and brick garage as a shrine. Then neighbours complained that their services were too noisy. 'The complaints are nonsense,' Phyllis said. 'Our church has only got seven members and they do not make a lot of noise.'

Encounters
of the
Grisly Kind

Mysterious Wreckage in Space

The wreck of a spaceship from another planet is in orbit round the Earth – and could contain the bodies of alien beings. That was the astonishing claim of Russian scientists that made front-page news in 1979.

Top Soviet astrophysicist Professor Sergei Boshich revealed that scientists first spotted wreckage floating 1,240 miles above Earth in the 1960s. They identified ten pieces of debris, two of them measuring 100 feet across, in slightly different orbits, and fed their findings into a sophisticated computer, to trace the age of the wreckage.

'We found they all originated in the same spot on the same day – December 18, 1955. Obviously there had been a powerful explosion.' Man's first space rocket went up in 1957.

Another top Russian astrophysics researcher, Professor Aleksandr Kazantsev, said the two large pieces of debris gave clues about the shape and size of the craft. 'We believe it was at least 200 feet long and 100 feet wide. It had small domes housing telescopes, saucer antennae – for communications, and portholes.

'Its size would suggest several floors, possibly five. We believe alien bodies will still be on board.'

Moscow physicist Dr Vladimir Azhazha ruled out suggestions that the debris could be fragments of a meteor. 'Meteors do not have orbits,' he said.

'They plummet aimlessly, hurtling erratically through space. And they do not explode spontaneously.

'All the evidence we have gathered over the past decade points to one thing – a crippled alien craft. It must hold secrets we have not even dreamed of.'

Russian geologist Professor Aleksei Zolotov, a specialist in explosions, added: 'The wreckage cannot be from an Earth spaceship – the explosion happened two years before we launched the world's first satellite, Sputnik 1.

'A rescue mission should be launched. The vessel, or what is left of it, should be reassembled here on Earth. The benefits to mankind could be stupendous.'

Leading American scientists were at first stunned, then excited by the revelations. Dr Henry Monteith, a physicist working on top-secret nuclear research at the Sancia Laboratories in Albuquerque, New Mexico, said the evidence warranted further investigation.

'It certainly sounds like a solid study by the Russians,' he added. 'It's very exciting – we could even send up a space shuttle. If it is an alien spacecraft, it would be the find of the century. It would conclusively prove the existence of intelligent life elsewhere in the universe.'

Dr Myran Malkin, director of the NASA Space Shuttle office of space technology said: 'We would consider a joint salvage attempt if the Russians approached us.'

And nuclear physicist Stanton Friedman said: 'If we retrieved the fragments, there's a chance we could put the pieces back together.'

The British reaction was more cautious. Dr Desmond King-Hele, a space researcher at the Royal Aircraft Establishment in Farnborough, Hants, said: 'There are more than 4,000 pieces of wreckage orbiting the Earth. Each has a catalogue number to identify it. We would like to know the catalogue number of this wreck. It is possible to date wreckage after a considerable number of observations.

'Like the Americans, we would be interested to look at this if the Russians make the information available.'

American physicist William Corliss recalled an article written by astronomer John tagby in the US magazine *Icarus* in 1969 – a time when government agencies had just decreed that UFOs did not exist.

He wrote that ten moonlets were orbiting the Earth after breaking off from a larger parent body. And he traced the date of the disintegration ... December 18, 1955.

'Bagby could not explain the explosion,' Corliss said. 'He was only interested in proving that the objects were out there, and dismissed them as natural phenomena. It seemed the safest thing to do at that time ...'

Other UFOs have successfully negotiated Earth's atmosphere, only to crash on to the surface of the planet, according to several American researchers. But proving their claims is virtually impossible, they say, because governments have kept all the incidents secret.

The Secret of the Dead Aliens

It was the worst storm New Mexico had seen in years. The wind and rain raged all night, and in the middle of it all, rancher Bill Brazel heard a strange explosion. At first light he saddled his horse and rode out to make sure his sheep were all right, What he found that morning, July 3, 1947, was to make his farm world famous – and spark off a UFO controversy that continues to this day.

His fields were covered by small beams of wood and thin sheets of metal. The wood looked like balsa and felt as light, but it was actually very hard, did not burn, and would not break. Some pieces carried strange hieroglyphics. The metal looked like tin foil, but could not be dented or bent. Then Brazel noticed a huge battered disc. As he rode closer, he saw something even more strange. There were beings, who were not human, lying beside the object. Some were alive, but they could not speak. Brazel raced back to the house and called the sheriff. He alerted nearby Roswell Army Air Field.

Intelligence chief Major Jesse Marcel led the investigating team. As ambulances carried off the burned bodies, and army trucks arrived to collect the wreckage, he threw an immediate security cordon round the fields, and told rancher Brazel not to talk about what he had seen.

New Mexico was then the hub of America's atomic, rocket, aircraft and radar research. Roswell was the home of
the 509th US Air Force Bomb Group, the only combat-
trained atom bombers in the world. Marcel had no idea
what the crashed craft was, but he knew it should not have
been over such a sensitive defence area.

On July 8, his attempts to keep the affair under wraps
were jolted when the base's public information officer,
Walter Haut, issued a press release without the authority
of his commander, Colonel William Blanchard. It read:
'The many rumours of the flying disc became reality yes-
terday when the intelligence office of the 509th Bomb
Group gained possession of a disc through the cooperation
of some local ranchers.

'The object landed at a ranch near Roswell some time last
week. It was picked up at the rancher's home, inspected at
Roswell Army Air Field, and loaned by Major Marcel to
higher headquarters.'

Wire services quickly passed the news on to papers all
over the world, and the Army came under pressure to
release more details. But reporters now found the story
had changed. A rash of denials poured out of Roswell and
Washington. A senior Air Force officer assured the public,
via a Texas radio station, that the wreckage was the
remains of a Rawin balloon. Newspapers were issued with
a picture of him and another officer examining a balloon.

The official line soon cooled curiosity about what had
happened at Roswell. But some UFO researchers were
unsatisfied. Eventually Charles Berfitz, author of books on
the riddle of the Bermuda Triangle, took up the trail. And
in 1980 he published a book, co-written by William Moore,
which accused the government of covering up the real
facts – that the Roswell craft was a spaceship containing
six aliens.

He quoted Grady 'Barney' Barnett, a government engineer, who told friends he was one of the first to reach the

site on the morning of July 3. 'I was out on assignment,' said Barnett, 'when light reflecting off some sort of large metallic object caught my eye. It was a disc-shaped object about 25 or 30 feet across.

'While I was looking at it, some other people came up from the other direction. They told me later they were part of an archaeological team. They were looking at some dead bodies that had fallen to the ground. I think there were others in the machine that had been split open by explosion or impact. I tried to get close to see what the bodies were like. They were like humans but they were not humans.

'The heads were round, the eyes were small and they had no hair. They were quite small by our standards, and their heads were larger in proportion to their bodies.

'Their clothing seemed to be one-piece and grey in colour. You couldn't see any zippers, belts or buttons. They seemed to be all males and there were a number of them. I was close enough to touch them. While we were looking at them, a military officer drove up in a truck and took control. He told everybody the Army was taking over and to get out of the way.

'Other military personnel came up and cordoned off the area. We were told to leave and not talk to anyone about what we had seen – that it was our patriotic duty to remain silent.'

Berlitz and Moore could not get the story from Barnett himself. He died in 1969. His version of the events was related by friends to whom he had talked in 1950. Rancher Brazel was also long dead, but his son Billy told how his father had found the debris.

'Father was very reluctant to talk about it,' Billy said. 'The military swore him to secrecy and he took that very seriously. I don't know what the craft was, but Dad once said the Army told him they had definitely established it was not anything made by us.

'He told me the occupants of the ship were still alive, but

their throats had been badly burned from inhaling gases and they could not speak. They were taken to California and kept alive in respirators, but they died before anyone had worked out how to communicate with them.'

Berlitz and Moore also quoted a California university physics professor, Dr Weisberg, who said he examined the disc. 'It was shaped like a turtle's back, with a cabin space inside about 15 feet wide. The interior was badly damaged. There were six occupants, and an autopsy on one revealed they resembled humans except in size.

'One body was seated at what appeared to be a control desk on which hieroglyphics were written. They were peculiar symbols. It was definitely not a known language. There was no evidence of a propellor or a motor. No one could understand how it was driven.'

A Los Angeles photographer, Baron Nicholas Von Poppen, claimed he had taken pictures of the crash ship after being approached by two men from military intelligence. He said they offered him a top-secret assignment at an exceptionally high fee – but warned that if he revealed anything he saw or photographed, he would be deported.

Von Poppen, who had developed a system of photographic metallurgic analysis, said he was escorted to the Roswell air base and took hundreds of pictures, which he had to hand over at the end of each day. He described the craft as about 30 feet wide, and the cabin 20 feet across. Its floor was covered with plastic sheets on which there were symbols. There were four seats in front of a control board covered with push buttons and levers, 'and in each seat, still strapped in, was a thin body, varying in height from two to four feet'.

The Baron added: 'The faces of all four were very white. They wore shiny black attire without pockets, closely gathered at their feet and necks. Their shoes were made of the same material and appeared very soft. Their hands were human-like though soft, like those of children. They had five digits, normal-looking joints and neatly-trimmed nails.'

Berlitz and Moore said Von Poppen smuggled one negative from the craft, and locked it away in a safe place, to be
opened only after his death. When he died, in 1974, aged
90, no trace of the negative was found.

The authors claim that Major Marcel was interviewed
again about the Roswell incident in 1978, after he had
retired to Houma, Louisiana. Asked whether the wreckage
he had collected from the ranch was really a weather balloon, he said: 'It was not.'

He went on: 'I was pretty well acquainted with everything in the air at that time, both ours and foreign. I was
also acquainted with virtually every type of weather-
observation or radar-tracking device being used by the
military and civilians. It was something I had not seen
before, and it certainly wasn't anything built by us. It most
certainly wasn't any weather balloon.'

In that case, why say it was? Marcel said Brigadier-
General Ramey ordered the cover story 'to get the Press off
the Army's back'. Berlitz alleged that the bodies and
wreckage were secretly shipped around the country by
truck and train for analysis at various scientific centres.

'We have been able to track down people who have a
clear recollection of the crash, technicians who examined
the alien machinery and clerks who checked the bodies
into various establishments,' he said. 'Their stories tally
too well for the whole story to be just a legend.'

Berlitz believed the facts were covered up to avoid causing
public panic, and for military reasons. Any nation that could
work out how the disc was powered would have a massive
advantage over its rivals in the missiles and space races.

Only successive incoming presidents were allowed to
share the military secret. 'Eisenhower, Kennedy and
Johnson carried it to their graves, Nixon, Ford, Carter and
Reagan have to live with it.' Berlitz recalled that Jimmy
Carter had promised to make government information on

UFOs available to the public if elected. When the author rang the White House, he was told that reopening of UFO investigations was not warranted.

Berlitz commented: 'His silence was undoubtedly prompted by the fact that he had learned something which convinced him to keep quiet about the whole issue.'

UFO Crashes

Have other UFO crashes been covered up? The secretive attitude of the authorities makes it impossible to confirm reports that spacecraft have fallen into the hands of earth-bound investigators.

There were strong rumours in the late 1940s that a flying saucer had come down just outside Mexico City, and that the wreckage – and the bodies of three silver-suited occupants, all only three feet tall – had been loaded onto trucks and taken to the United States for study.

Raymond E. Fowler, an authority on UFOs, particularly in New England, received what could be confirmation of the rumours when he gave a lecture on UFOs at Boston.

In his book, *UFOs: Interplanetary Visitors*, published in 1979, he said that an assistant minister at a Boston church told him he was working for the Pentagon in naval intelligence at the time.

A colleague in Mexico was assigned to help investigate an air crash. 'When he arrived, the area had been roped off and personnel were loading remains of an oval object and its occupants into trucks ... He was quickly ordered out of the area by a superior, who told him not to mention what he had seen.

Fowler traced the minister's former colleague to Belfast, Maine, where he was living in retirement. He denied all knowledge of the incident, and said his friend must have made some mistake. But the minister stood by his story.

Fowler concluded that the man might be afraid to reveal his secret because he was living on a pension from the Navy.

In 1957, fishermen at Ubatuba Beach, Brazil, claimed they had seen a flying object explode and fall into the sea. They also produced fragments of ultrapure magnesium which, they said, came from the debris. The authorities were sceptical, even though they could not explain how simple fishermen could have come by magnesium which, as later tests showed, was forged by a directional casting method not even invented by 1957.

Ten years later, Raymond Fowler came across what may have been another UFO crash. He met Mr and Mrs Bill Marsden, who recalled driving towards Mattydale, a suburb of Syracuse, New York, during the winter of 1953-4. It was 3 am on a Sunday when they came across the flashing lights of four or five police cars, and slowed, thinking there had been an accident. The road was clear, but something in a nearby field caught Mr Marsden's eye. He told Fowler:

'I saw an object which appeared to be 20 feet in diameter and possibly 15 feet high in the centre. It had phosphorescent lights of several colours spaced over the surface. These lights were strong enough to make clearly visible quite a few men walking around the object and examining it. Some were uniformed and some were not. One had what appeared to be a large press camera with a strap and was taking pictures.'

On the Monday morning, Mr Marsden called his local newspaper to ask why it had no story of the event, then phoned the sheriff's office. He claims he was told: 'Yes, we know about that, but it is a military secret and we can't discuss it.' But when the newspaper talked to the sheriffs office, and to the Air Force, reporters were told that no such incident had taken place. The sheriff also denied that anyone had told Mr Marsden about 'a military secret'. Mr Marsden let the matter drop, even though he had checked the field next morning, and found indentations and tyre tracks. He knew that UFO supporters faced ridicule.

When Fowler made inquiries with the sheriffs office in 1967, he was told that the only objects which had fallen to earth during the winter of 1953-4 were a weather balloon, a wing tank from a plane, an aircraft, and a sand filled imitation bomb dropped by mistake from another plane. None of those objects tallied with what Mr Marsden had seen – and he stuck to his story.

Bizarre Autopsies

America has recovered a total of more than 30 bodies from crashed alien craft, according to researcher Leonard Stringfield. Many have been given autopsy examinations, and all are preserved either at the Wright-Patterson Air Force base in Ohio, or at the underground Air Force complex near Colorado Springs.

Stringfield, who says that the aliens are between 3½ feet and 5 feet tall and slender, with oversized hairless heads, made his astonishing claims after talking to two doctors and six Forces personnel involved in the recovery and analysis of bodies over the last 30 years. He added that a specially trained force called the Blue Berets is on constant standby, ready to move instantly should a UFO crash.

All Stringfield's sources asked to remain anonymous, and he refused to identify them, even when questioned about his book. This is what he says they told him:

A doctor who observed an autopsy in the early 1950s described the alien corpse as just over 4 feet tall. It had a large, pear-shaped head with Mongoloid eyes recessed in the face. There were no eyelids, ear lobes, teeth or hair.

A former major and pilot in the Air Force observed strange bodies in an underground preservation chamber at Wright-Patterson during 1952 after secret Air Force instructions were sent out ordering pilots on UFO missions to shoot down strange craft.

Another USAF pilot watched three crates being delivered to Wright-Patterson in 1953. He was told they contained bodies from a flying saucer crash in Arizona. An officer said the humanoids were still alive when rescuers arrived, but died despite receiving oxygen.

An army intelligence officer viewed nine alien bodies which had been frozen at Wright-Patterson in 1966, and was told that there were 30 in various government establishments. The same man learned later that five UFOs had crashed in the Ohio, Indiana and Kentucky region between 1966 and 1968.

An Air Force sergeant and air policeman, identified only as Carl, said he was blindfolded and driven to a secret location to guard a room. When he peeped inside, he saw three small bodies, around 3 feet tall, with abnormally large heads.

A doctor who was present at an autopsy said the bodies had no digestive tracts or sex organs. And their blood was colourless.

Stringfield, whose claims were published by Mutual UFO Network, of Seguin, Texas, also tells of some aliens who got away. A colonel told him that, in 1968, he confronted strange beings who emerged from a saucer at the Nellis Air Force Base in Nevada. A beam of light paralysed him, and he could only watch as the figures returned to their ship and took off.

Some aliens are not so expert in controlling the humans they meet. And even when their crafts land safely, they face new hazards on Earth.

Bullets that Bounced Off

A group of farmworkers and their wives astonished police at Hopkinsville, Kentucky, when they burst into the station at midnight. They said they had just fired shotguns and rifles at goblin-type aliens from a UFO – but that their bullets had bounced off the creatures.

The Sunday evening of August 21, 1965, turned into a nightmare for the family at Kelly, a sprawling cluster of houses seven miles from Hopkinsville. The Langfords of Sutton Farm, eight adults and three children, had returned after church services when one of the youngsters saw a brightly glowing object descend behind a barn. People on nearby farms saw it too, but the family dismissed it as a shooting star.

Then, at around 8 pm, the dogs in the yard began barking. Two of the men went to the door to investigate, and saw, 50 yards away, a creature in a glowing silver suit, about 3 feet tall, coming towards them. It had a huge head, long arms that nearly reached the ground, and large webbed hands with talons. The men grabbed a 12-gauge shotgun and a 22-calibre pistol, and fired at close range. The being was knocked over – but to the amazement of the watchers, it then jumped up again and scurried away.

The stunned family locked themselves inside, turned off all the inside lights, and put on the porch lamps. Then one of the women screamed. She looked out of the dining room window and saw a face peering in at her, with wide slit eyes behind a helmet visor. The men rushed into the room and fired, but again the creature, although hit, ran away.

A total of almost 50 rounds were blasted at the five aliens over the next 20 minutes, but none of the bullets stopped them. Radio newsman Bud Ledwith, who interviewed the family next morning, said: 'Whenever one of the creatures was hit, it would float or fall over or run for cover. All the

shots that struck them sounded as though they were
hitting a bucket.

'The objects made no sound. The undergrowth would
rustle as they went through it, but there was no sound of
walking. The objects were seemingly weightless as they
would float down from trees more than fall from them.'

When caught by bullets or flashlights, the aliens, who
seemed to approach with their hands in the air, would
drop their arms and run. But they kept coming back,
apparently making no attempt to enter the house, but just
standing and staring at it. After 20 minutes, the creatures
melted away into the night. But the scared family stayed
alert for another two hours before daring to venture outside, and drive to the police. Officers who visited the farm
could find no trace of the visitors.

It was an amazing story, but only Bud Ledwith seemed
interested in investigating it seriously. An officer looked into
it for the United States Air Force Bluebook file on UFOs,
since he happened to be in the area and heard news of it on
the radio. Several points of his report were later found to be
erroneous. After interviewing Mrs Lenny Langford, one of
the women involved, he declared that she, her sons, their
wives and some friends had attended a service of the Holy
Roller Church that evening, and were, emotionally unbal-
anced 'after working themselves into a frenzy. ' In fact, Mrs
Langford belonged to the Trinity Pentecostal Church, whose
services are perfectly traditional.

Other investigators tried to find out whether there were
any travelling circuses in the area, apparently believing
that the farmers had seen escaped monkeys. Monkeys that
floated? Monkeys in bullet-proof vests?

Bud Ledwith firmly believed that his witnesses were
telling the truth, that they were simple folk who had no
motive for trying to perpetrate a hoax. And as Dr J. Allen
Hynek, whose Centre for UFO Studies later probed the
case, pointed out, they had nothing to gain from publicity,

and later 'suffered horribly from curiosity seekers, reporters and sensation mongers.'

The case was later used as an example in the secret air force training manual on UFOs – to show that humans can be dangerous for aliens! The textbook added: 'At no time in the story did the supposed aliens shoot back, although one is left with the impression that the described creatures were having fun scaring humans.'

The family decided they could no longer live at the farm and sold it.

Demons and Demon-ships

Aggression by man against UFOs is nothing new. According to the US Air Force Academy textbook, supposed airships were treated as demon-ships in Ireland in about 1000 AD, 'and in Lyons, France, "admitted" space travellers were killed in around 840 AD'.

Vigilantes on the Alert

UFOs fever in Virginia led to the formation of local vigilante groups in January 1965. The *Richmond Times Despatch* quoted Sheriff John Kent of Augusta County as saying UFO reports had got 'completely out of hand' and had become 'dangerous to country residents'.

A posse from the Brands Flat area of the Shenandoah Valley armed itself to go looking for creatures said to have landed in a UFO. But the Sheriff said that even if little green men had arrived, residents 'had no right to mow them down'.

That was not a view shared by Attorney General Robert Button. When a Fredericksburg justice of the peace consulted him, he replied, somewhat tongue in cheek: 'There is apparently no state law making it unlawful to shoot little green men who might land in the state from outer space.'

In March 1966, a man driving near Bangor, Maine, did shoot at a UFO. He spotted the metallic, oval shape hovering over a field, and got out of his car to investigate, taking with him the 22 pistol from his glove compartment. When the mysterious flying object swooped towards him, scraping the tops of the bushes, the man began firing and heard bullets ricochet off metal as the craft passed overhead; before climbing out of sight at tremendous speed.

Not everyone was so unwelcoming. After a spate of local sightings, the mayor of Brewer, Maine, had a giant billboard put up inviting UFO travellers to settle down in the town.

Bullet Proof

Police at Fort Beaufort, South Africa, fired shots from only eight yards when a glowing metallic object landed on June 26, 1972. But the bullets had no effect. The machine merely took off with a humming noise.

Encounters
of the
Arresting Kind

Grilling the Police

Patrolman Gene Bertrand did what any good cop would in an emergency when faced with a hostile intruder – he dropped to one knee and drew his revolver. But he was faced with no ordinary intruder. The object hurtling towards him really was out of this world.

Bertrand had been called into headquarters at Exeter, New Hampshire, to investigate the story of a kid who had come in 'all shook up about some object that had chased him'. Norman Muscarello had been hitch-hiking home from Amesbury along Route 150 in the early hours of September 3, 1965, when a glowing red object had appeared in a field beside the road, and moved towards him.

Bertrand knew the boy. He said: 'He's real tough, but something must have really scared him. He could hardly hold his cigarette and was as pale as a sheet.' They drove out in the squad car to the field. They parked and sat in the car for several minutes. Nothing happened.

'I radioed the station and told them there was nothing out here,' Bertrand recalled. 'They asked me to take a quick walk in the field before coming back in. I must admit I felt kind of foolish walking out on private property after midnight, looking for a flying saucer.

'We walked out, me waving my flashlight back and forth, then Norman shouted, "Look out, here it comes!" I swung round and could hardly believe what I was seeing. There was this huge dark object, as big as a barn, with red flashing lights on it. It barely cleared the trees, and it was swaying from side to side.

'Then it seemed to tilt and come right at us. I automatically dropped to one knee and drew my service revolver, but I didn't shoot. I remember suddenly thinking that that would be unwise, so I yelled at Norman to run for the cruiser. He just froze in his tracks. I had to almost drag him back.

'The thing seemed to be about 100 feet up. It was bright red with a sort of halo effect. I thought we'd be burned alive, but it gave off no heat and I didn't hear any noise from it. I did hear the horses in a nearby barn neighing and kicking in their stalls, though. Even the dogs around the area started to howl. My brain kept telling me that this doesn't happen – but it was right in front of my eyes.'

Bertrand's partner, patrolman Dave Hunt, arrived while the UFO was still in sight. The three stood watching in amazement for ten more minutes. 'It floated, wobbled, and did things that no plane could do,' Bertrand said. 'Then it just darted away over the trees towards Hampton.'

As the policemen went back to their office to write their reports, Bertrand's mind went back to the woman he had met an hour earlier on Route 101. She was sitting in her parked car, 'real upset' about a red glowing object that had chased her. He had sent her home without thinking much about it. Now he knew what she had seen.

Others had seen it, too. The men had not long been back in the station when a telephone operator from Hampton called. A man from a public call box claimed he had been chased by a flying saucer ... and it was still out there. The line went dead before he could say more, and though the officers tried to locate him, they could not do so.

Air Force investigators who interrogated Bertrand and Hunt told them to keep quiet about what they had seen, so that it would not get into the newspapers. But a local newspaper reporter had already got the story.

Unable to keep it quiet, the authorities began issuing a string of curious denials. The Pentagon at first blamed the sighting on a temperature inversion that had caused 'stars and planets to dance and twinkle'. Officers Bertrand and Hunt protested that such a statement put their reputations as responsible policemen at risk.

Then the Pentagon claimed that Big Blast Coco, a high-altitude Strategic Air Command exercise, was responsible. The town of Exeter was within the traffic pattern used, said the war chiefs, adding: 'During their approach the aircraft would have been displaying standard position lights, anti-collision lights and possibly over-wing and landing lights.'

But Bertrand had an answer to that one, too. He wrote another protest letter, saying: 'Since I was in the Air Force for four years engaged in refuelling operations with all kinds of military aircraft, it was impossible to mistake what we saw for any kind of military operation ... Immediately after the object disappeared we did see what probably was a B-47 at high altitude, but it bore no relation at all to the object we saw.'

The two officers also pointed out that they saw the UFO at 3 am, nearly an hour after the exercise ended.

Grudgingly, the Air Force gave way – but only a little. 'The early sightings . . . are attributed to aircraft from Operation Big Blast Coco,' their final statement said. 'The subsequent observations by officers Bertrand and Hunt occurring after 2 am are regarded as unidentified.'

Even such a small admission was a huge advance for UFO believers frustrated by years of officialdom's stubborn refusal to acknowledge that there could be such things.

The following March, Exeter was again visited by a UFO. One Sunday night, a police sergeant checking doors in the town around 10 pm saw a fast-moving white light falling to the west. He climbed a hill to get a better view, and saw what looked like a lighted egg with rotating red, white, blue and green lights underneath it, moving slowly back and forward. Then it plunged quickly down to hover over power lines.

The sergeant radioed headquarters, and a lieutenant arrived, carrying binoculars. He had always been sceptical about UFOs, despite the sightings the previous September by his own men. Now, as he peered at the egg-shaped object with a bright white dome on top, he was converted.

Officer Bertrand and a newspaperman also saw the UFO. But this time nobody made a fuss about it. The town was clearly determined to live down the notoriety aroused by the earlier sightings.

No Pictures Please

Police chief Jeff Greenhaw lost both his wife and his job because of what he claims he saw on the night of October 17, 1963. But he stuck to his story.

It was just after 10 pm when he took the call at his home in Falkville, Alabama. A woman said she had seen a UFO with flashing lights land in a field west of town. Greenhaw, 26, was off duty at the time, but decided to investigate anyway. He took his camera.

As he drove up a gravel road towards the remote landing site, he saw a figure in the middle of the track. It was about the size of a large human, but was clad in a silvery suit that looked like tin foil. Antennae seemed to sprout from its head. As it moved towards him, he shot four flash pictures, then turned on the revolving light on top of his car. The figure turned and ran, 'faster than any human I ever saw'.

Greenhaw agreed to publish his pictures, which showed the blurred shape of an astronaut-type figure. But within four weeks, he was to regret it. His wife left, unable to cope with the publicity and 'side effects'. Greenhaw's car engine blew up, then a caravan he owned went up in flames. Finally, on November 15, he was asked to quit his job.

Whether he saw an alien or a hoax invented by someone with a grudge was never established. Many other people reported odd lights that night, but despite their evidence, Greenhaw's superiors felt that his credibility had diminished.

In Pursuit of the Unknown

The black Chevrolet shot past the courthouse at Socorro, New Mexico, far faster than it should have done. Patrolman Lonnie Zamora gunned the engine of his patrol car into action, and swung out into Old Rodeo Street in pursuit. He noted the time for his report – it was 5.45 pm on April 24, 1964. Zamora would never catch the speeder, but he would remember that day for the rest of his life.

As he accelerated out of town, he noticed a flame in the sky, a mile or so to the south-west. He also heard a roar. The noise came from the direction of a dynamite storage shack. Had it blown up? He decided to abandon the chase and investigate.

He swerved off the road and on to a rough gravel track. The tapered blue and orange flame seemed to be descending against the setting sun. He lost sight of it as he struggled to steer the car up a small hill. Three times he had to reverse and try another route as gravel and rock spun the wheels.

At the top of the hill, Zamora looked round for the shack. Then a shiny object 150 yards away caught his eye. 'It looked at first like a car turned upside down,' he recalled. 'I thought some kids might have turned it over. I saw two people in white coveralls very close to the object. One seemed to turn and look straight at my car. He seemed startled, to quickly jump somewhat.'

The officer began manoeuvring his car closer, with the idea of giving help. When he next looked at the object, the two figures – small adults or children – had vanished. The oval shape was whitish, like aluminium. He stopped the car, and radioed to HQ that he was leaving to investigate a possible accident.

As he put down his microphone, Zamora heard two or three loud thumps, 'like someone hammering or shutting a door hard'. Then the roar began growing louder and increasing in frequency. 'It was nothing like a jet,' the policeman told investigators. 'I know what jets sound like.'

Now he saw the blue and orange flame again, and the object was going straight up into the air. He noted that it was oval and smooth, and saw no doors or windows, only a red insignia drawing, about 30 inches wide. As the roar increased, Zamora turned and ran – 'I thought the thing was going to blow up.'

He ran past his own car, stumbling as his leg struck the back bumper, and kept going, glancing over his shoulder a couple of times to see what was happening. The craft was still rising slowly from the deserted gully where it had landed. The officer dived over the top of a ridge and spread himself on the ground, covering his head with his arms.

As the roar stopped, he gingerly peeped over the hilltop. The object was speeding away towards the south-west, about 10 to 15 feet above the ground. Then it suddenly lifted higher into the sky and flew off rapidly, without sound or smoke, finally disappearing behind nearby mountains. Zamora radioed his story in to the duty desk sergeant, and a second squad car sped to the scene. The reinforcements noted 'landing marks' about 2 to 3 inches deep in the hard-packed, sandy surface. Greasewood bushes and grass around them were scorched and smouldering.

Air Force investigators arrived a few days later, intent on finding some natural explanation for what patrolman Zamora had seen. They tried hard to establish that some man-made craft had been in the area, but without success. Colleagues described Zamora as a solid, well-liked citizen, a down-to-earth character of integrity. Cynics said that residents living near the site had seen and heard nothing; that the scorch marks could have been caused by a cigarette lighter; that the 'landing marks' could have been created with a small shovel, or by moving boulders; that the land was owned by the town's mayor, who would welcome the publicity and tourists attracted by a UFO report.

Other investigators, however, were forced to admit that Zamora had probably seen some real phenomenon of undetermined origin. One of them was Dr J. Allen Hynek, who, talking later of the scorn some critics poured on UFO reports, said: 'It is paradoxical that the testimony of policemen, which in some cases might be sufficient to send a man to the electric chair, is in instances like this often totally disregarded.'

Similar sightings, of a white, aluminium-like oval shape, were reported right across the United States that spring. It was seen at La Madera, New Mexico, Helena, Montana, and Newark, New jersey. The last witness also reported seeing curious child-size creatures beside the grounded craft.

The Acrobatic Disc

Detective Sergeant Norman Collinson watched a disc-like object perform 90-degree turns at incredible speed above the town of Bury, Lancashire, in April 1976. The officer, who later became an inspector, said: 'After a while it streaked away at an even higher speed, reaching the horizon in around two seconds.'

Cops in Confusion

A woman police officer and a male colleague spotted a long, cigar-shaped object hovering 500 feet above the select residential district of Rickmansworth, Hertfordshire, at 3.25 am on November 29, 1979. It was brilliantly lit along its entire length, and had red lights above and below it. It made no sound. WPC Anne Louise Brown, 21, admitted later: 'I was scared stiff when it was above our car. I don't know what it was, but it was definitely too big and too bright to be a plane or a star. I told my colleague he must be crackers to report it back. I was sure people would think we were potty.'

Minutes later, two other officers, both men, saw the same shape above nearby Chorley Wood, and gave chase in their Panda car after alerting HQ. It flew quietly out of sight, but two hours later they spotted it again.

Hertfordshire police checked with West Drayton air traffic control, and confirmed that no planes were in the area. Inspector George Freakes said: 'This is being treated seriously. We are convinced the officers saw something – they are very genuine types – but as yet no one can explain exactly what it was.'

Patrolmen from several forces gave chase when a UFO was spotted over Will County, Illinois, south of Chicago, in the pre-dawn darkness of November 25, 1980. And the mysterious shape led them a merry dance.

Sheriff's deputies Lieutenant Karl Sicinski and Sergeant Jay Mau were first to see the UFO, about 1600 feet up and two miles away. It drifted south, shot off east, then turned north, and finally ended up to the south-east of them.

It was faster than any plane I ever saw,' said Sicinski, who flew fighter jets during his days in the US Navy. 'I've never seen any aircraft that can manoeuvre as tightly as this object did. It was huge and very bright. It was shaped like a tear drop lying on its side and had a pinkish-whitish cast to it.'

Policemen in neighbouring towns Frankfort, New Lenox and Mokena overheard Sicinski radio his report to HQ, and saw the shape he mentioned. Frankfort patrolman Sam Cucci was driving west towards the UFO when he spotted it rising, getting brighter, then dimming its lights.

'Suddenly I lost sight of it,' he recalled. 'I asked two other

squad cars where it was and they said, "It's behind you."
So I whipped the squad car round to the east and, with the
two other units, gave chase at about 60 mph. I put on my
spotlight, but the UFO veered away and then just dissipat-
ed, like it was a light and someone shut it off.'

In New Lenox, officers Carl Bachman and Charles Proper
watched the UFO zigzag across the sky for 20 minutes. 'I
won't forget that night,' said Bachman. 'There is something out there we don't know about.' Proper said: 'It was
a bright light, and all of a sudden it just went straight up
and disappeared. In a matter of just one or two seconds it
was out of sight.'

Mokena patrolman Tom Donegan, who also saw the UFO,
said: 'It makes you wonder who's out there watching us.'

In March 1981 came news of a bizarre encounter involving a chief of police. Miguel Costa, in charge of the force at
Melo, Uruguay, was driving with his wife Carmen and
friends Armando and Maria Pena along a gravel road near
Tacuarembo when a huge shape, gleaming with orange
and yellow lights, loomed out of the early morning darkness in front of them.

Costa stopped the car, and, on impulse, flashed the headlights. 'All of a sudden the UFO hesitated, then zigzagged
up and back as if answering our call,' the police chief
reported.

'As soon as we started out again, it was there following
us. I again stopped the car and flashed my lights. Again
the UFO wavered in reply. We drove on once more on the
twisting road and the UFO stayed with us, always about
half-a-mile away. This went on for almost 30 miles. That's
when the strangest thing of all occurred.

'We were all glued to the windows watching as the disc
suddenly shot towards the ground as if it was going to
crash. It stopped 50 to 100 yards from the earth, and we
could clearly see its round, dome-like shape with a large
flat plate underneath. There was a slight ring of cloud

around the dome. The top was reddish but the bottom was a brilliant glowing white.'

Feeling somehow menaced by the craft's new, lower flight path, Costa turned the car round and headed back towards Tacuarembo, the nearest town. The blazing light of the UFO remained constant in the rear window. Costa pulled over and parked under some trees.

'We walked over to a little clearing and looked up,' the policeman said. 'A second disc was moving some distance behind the first. They never touched, but they seemed to be travelling together. They moved up and down and clouds started to form.

'They passed over the top of the clouds and lit them up like a halo. Then they faded, getting smaller and smaller until finally they had gone. It was dawn. They had been over us for 90 minutes. We looked at each other without speaking. We still couldn't believe what we had seen.'

Chief Costa paused, then added: 'I never believed in UFOs before, but I realise now that I have seen something special and unreal.'

Follow that UFO!

Five policemen saw a multi-coloured flying object hovering above the town of Dumfries, Scotland, late in 1979. Two of them later described the sighting at a press conference.

The officers were called in after a flood of calls from people going home after the pubs shut. They saw the huge shape for about 20 minutes before it streaked away over nearby hills.

Sergeant Bill McDavid, 39, said he drove to within a mile of the thing. It was larger than any aircraft and seemed to be 500 feet up. Its shape was like that of an airship, with five or six white lights shining from separate compartments.

PC James Smith said: 'I never believed in UFOs up to now. It was raining at the time and the cloud base was very low. The shape remained stationary for 20 minutes then vanished over the hills to the west.'

Mary Blyth, 22, and her sister Vicky, 19, were just two of the people who rang the police after they spotted the UFO. 'The lights appeared from nowhere,' said Mary. 'We just stood there and stared at it in amazement.'

Glasgow weather centre said it was not unknown for low clouds to reflect bright lights from Earth, but a spokesman added: 'If light from the ground is reflected, it is usually just a yellowish glare. I have never heard of a cluster of coloured lights in the way that has been described. I have no explanation as to what these people really saw.'

Two patrolmen in Minnesota spotted a glowing white ball after being called out by Farmington housewife and computer programmer Karen Anondson in September, 1979. 'It was definitely a UFO,' said patrolman Dan Siebenaler, of the Farmington Force. 'I am familiar with what is in the night sky, and that thing did not belong there.' Steve Kurtz, an officer from neighbouring Apple Valley police, said: 'It was something unexplainable, I've never seen anything like it before.' Mrs Anondson, 32, said she had seen the ball at least nine times as she drove home from work. 'It's become a normal thing,' she said. 'I look for it when I come out of the office.'

A few months earlier, a Minnesota deputy sheriff reported a frightening encounter with a UFO. Val Johnson was driving his patrol car on a lonely road near Warren when he saw a bright light about 21 miles away. 'I drove towards it to find out what it was,' he recalled. 'After I'd gone about a mile, the light rushed towards me. It was a brilliant light, so brilliant it was almost painful.

'I remember the brakes locking when I applied them, and I remember the sound of breaking glass. Then I lost consciousness for about 30 minutes. When I came too, I radioed for help.'

Officers who examined the car found that both the windshield and a headlight had been broken, and the top of the hood was dented. Even more curious was the fact that the two spring-loaded whip antennae on the roof had been bent at an angle of 90 degrees. 'The damage to the hood, windshield and headlight might have been caused by stones or rocks,' said UFO researcher Allen Hendry. 'But there's no explaining how the antennae, which are extremely flexible, got to be bent that way.'

Doctors who checked the deputy sheriff after his ordeal had to treat burns round his eyes. They were of the kind welders suffer when they fail to use protective masks.

A dozen policemen in Tennessee watched a UFO for two hours in February 1980. It amazed the people of three towns in two counties with its aerial antics, hovering, then shooting off at impossible speeds and incredible angles.

Deputy sheriff Franklin Morris, from Winchester, first heard about the strange sight over his radio, and raced to a hill to get a good view of it. 'At first I thought it might be a plane, but there was no noise at all, no engine, no rocket. It hovered a while, three or four minutes. Then it decided to take off, and moved so fast you could hardly watch it. I've seen some pretty fast jets in my time, but never anything like this

Winchester patrolmen Milton Yates and Gerald Glasner saw bright red and white lights coming towards them as they drove on the east side of town. 'It was coming towards us and it stopped, sat two or three minutes, then shot off at 500 or 600 mph,' said Sergeant Yates. 'The way it took off it couldn't have been an aircraft. It had no moving lights, no noise, just those flashing lights, and it went round in circles. I feel real sure it was a UFO.'

Officer Glasner added: 'It was not like anything we've got here on Earth. The speed, manoeuverability, those

374

flashing lights, the silence.' Officers from nearby Monteagle and Cowan also watched the UFO in amazement. When they checked with the National Weather Service station at Nashville, officials could offer no alternative natural explanation.

Two Michigan policemen chased a multi-coloured, shapeless craft for more than 26 miles in March 1980 after picking it up in the sky over Gladstone. 'It was glowing orange, with a green light in the rear, red lights top and bottom, and a blinking white light at the front,' said patrolman David Mariin, 26.

The men radioed for assistance as they followed the lights for nearly an hour through winding roads and dense forest. Two more police units joined the chase, and the four officers in the other cars all saw the object above the trees as it darted from side to side, leading them on, then vanishing at astonishing speed.

'I was rather sceptical about UFOs before this,' said Mariin's partner, Mark Hager, 22. 'But this made a believer out of me.' The men checked with nearby K. I. Sawyer Air Force base, but were told nothing unusual had been spotted on the radar. 'No one there seemed very interested,' Mariin said. 'It was almost as if they didn't want the public to know.'

Three policemen were among hundreds who saw a gigantic bullet-shaped object which cruised through the night sky above Kansas and northern Missouri for four hours on November 18, 1980. Adair County deputy sheriff Charles Cooper and Missouri highway patrolman Bob Lober were amazed when it flew backwards without turning round. And patrolman Mike Leavene said: 'I've never seen anything like it before.'

People in at least 22 towns reported seeing the UFO as it crisscrossed the two states. Don Leslie, a 42-year-old welder from Milan, Missouri, said: 'It was at least as big as a football field.' Roger Bennett, 40, of Huntsville, Missouri, said: 'It was so big it would make a B-52 bomber look like a Piper Cub.'

He added.. 'It was like a big fat cigar, travelling very high from east to west. You could hear a faint rumbling when it was overhead. Just before it disappeared above some clouds it ejected about six smaller objects in a fanshaped burst. They sped off in different directions.'

Truck driver Randy Hayes, 26, also saw the UFO drop its 'satellites'. He said: 'They were round and had a bluish glow. The mother ship was so big, it blocked out a lot of stars.'

In Trenton, Missouri, photographic student Rick Hull, 19, took a picture of a triangle of lights, which looked like a boomerang. He said the object seemed to make a banking movement, thus revealing lights from 'the windows of a cockpit'. Music teacher Buddy Hannaford and his wife Karla both saw lights 'as if from the cabin of a plane'. Karla, who watched through binoculars, said: 'The thing was delta- or triangle-shaped, with two white lights and a red beacon on the bottom. It passed right over our house.'

The object was picked up on radar at the Federal Aviation Administration station north of Kirksville, Missouri. Technician Franklin West said: 'It went through the area four or five times. I estimated the speed at about 45 mph. I'm not saying it was a flying saucer. I am saying it was an unidentified flying object, because I couldn't identify it.'

Forest Phantom

Two policemen responded to a 999 call at Hainault Forest, Essex, early one May morning in 1977, and spotted a tent-like object glowing red through the trees. They watched it 'pulsing' for three minutes, then it dissolved into the darkness.

Arrest That Saucer!

PC Chris Bazire and WPC Vivienne White spotted a flying saucer 500 to 700 feet above Salisbury Plain, Wiltshire, in November 1977. 'It was oblong with a domed top and flat bottom,' they reported. 'It was travelling very slowly at first, then shot off at tremendous speed, leaving a vapour trail.'

The Burning Cross

Two Devon policemen hit the headlines in 1967 when they chased bright lights in the shape of a pulsating cross. Constables Roger Willey and Clifford Waycott spotted the glowing UFO over Hatherleigh at 4 am one night while on routine patrol in their car, and pursued it for some distance along a narrow lane before it shot off across fields. Critics said the shape could have been aircraft refuelling in mid-air from a tanker plane, which would explain the cross-like effect, and the British Defence Ministry confirmed that such exercises were going on in the area. But they had been completed by nightfall the previous evening.

Encounters
of the
Concentrated Kind

The Warminster Visitations

Warminster was for centuries just a quiet, unremarkable country town on the edge of Salisbury Plain. Little happened to disturb the day-to-day routine of its 14,000 inhabitants. Then, early on Christmas Day, 1964, a strange drone jolted postmaster Roger Rump from his sleep at his home in Hillwood Lane. He heard a violent rattle, like tiles being ripped off his roof. The Thing had arrived.

Two weeks later, his neighbours, Mr and Mrs Bill Marson, were woken three times in one night by the sound of 'coal being tipped down our outside wall'. Then Mrs Rachel Attwell, wife of an RAF pilot, was roused by a curious noise at 4 am. She looked out of her bedroom window in Beacon View and saw a cigar-shaped, glowing object hovering in the sky, bigger and brighter than any star. Another housewife who spotted it, Mrs Kathleen Penton, described the craft as 'something like a railway carriage flying upside down, with all the windows lit up'.

Soon more and more people were scanning the heavens above Wiltshire. On June 2, a total of 17 people – including Mrs Patricia Phillips wife of the vicar of Heytesbury, and three of her children – watched the cigar-shaped Thing for 20 minutes in the late evening. By the end of 1965, three people had even taken pictures of it. And strange things were starting to happen.

A flock of pigeons mysteriously fell from the sky. Naturalist David Holton examined the bodies, and declared that the birds had been killed by soundwaves not known on Earth. Then a farmer found that several acres he had left fallow were now a mass of weeds – silvery thistles of a type considered virtually extinct in England since 1918. And in Warminster itself, the East Street garden of Harold and Dora Horlock became another horticultural attraction when ordinary thistles that normally grew to only 5 ft 6 in soared to nearly 12 ft tall.

The curious freaks of nature brought newspaper reporters and television teams flocking to the town. And as news of the UFO sightings spread, observers from all over Britain turned Warminster into a Mecca. They were not disappointed.

Few months went by without a sighting. The months turned to years, and the area of activity was pinpointed as a triangle roughly bordered by Warminster, Winchester in Hampshire, and Glastonbury in Somerset. Local folk became used to their curious visitors. Mysterious lights in the sky, agitated animals, stalling cars and electrical equipment going haywire became almost commonplace. Then, in November 1976, the Thing made contact with the humans.

Tough German parachutist Willy Gehien had seen many years service with the French Foreign Legion. He liked to keep in practice, and in mid September 1976 he was on his way from home at Bishops Castle, Shropshire, to the Army Parachute Centre at Netheravon, near Salisbury, when he decided to stop for the night and sleep in his estate car. After searching in vain for a camping site, he pulled in beside a farm gate near Upton Scudamore, a village two miles from Warminster on the Westbury road.

He fell asleep after locking all the doors, but woke shivering in the early hours to find the hatch-door at the back of the vehicle wide open. He slammed it shut, turned the key, and curled up again in his blankets. The same thing happened.

'Normally I sleep very lightly and hear the slightest sound,' he said. 'But I had heard no one open that door. Feeling uneasy and a little unsettled, I decided against more sleep and started to prepare a cup of coffee on my camping stove. It was 3 am.'

Then, above the sound of a distant train, Gehlen heard a strange humming sound, 'like a swarm of bees in flight', and became aware of a figure standing behind the farm gate 10 yards away. 'The sheer size of this person made me wonder – he was almost 7 feet tall – but I was not frightened. I assumed it was the farmer guarding his animals against rustlers, and explained that I was only camping there for one night. There was no answer. Instead he shone a sort of square-shaped torch at me from his chest. The light was dark orange, and I thought he needed some new batteries.

'I got on with making my coffee, and when I looked up a minute or so later, he had gone. Then I heard the humming noise again, and saw a large shape lift off the ground. There was a pink, pulsating glow underneath it, and I watched it disappear across the field. It lifted to about 45 degrees above the ground, but I assumed the farmer was towing something up a hill. It was only after it became light that I realized there was no hill.'

The baffled ex-airman discussed the incident with pals in his local pub when he got home. When he took their advice, and consulted a UFOlogist, he realised that he might have met an alien.

Londoners Steve Evans and Roy Fisher have made frequent special trips to the Warminster area since 1971 to try to spot UFOs. They claim to have seen at least 30, and to have had two even closer encounters.

The first happened as they gazed at the sky from the top of Cradle Hill, one of several vantage points around the town. 'A forcefield seemed to move through the grass like a snake, crackling furiously like static electricity,' Evans said. 'It came straight for Roy's feet, then veered suddenly to the right. Sheep in the field were going frantic. When daylight came, we found flattened grass, as though something had landed.'

That same weekend, the two friends had an unnerving experience at the top of nearby Starr Hill. Evans said: 'We got the distinct feeling we were being watched. I glanced over my shoulder and saw a figure in a sort of white boiler suit, with a white hat, running towards a clump of bushes. I started to chase him. I was making a row myself, crashing through the bracken, but I swear he wasn't making any noise at all. After a while he slowed down, looked back for a second, then disappeared into the bushes.'

Fisher added: 'When Steve ran, I followed instinctively, even though I didn't know what he was after. I reached the bushes after he did, then someone brushed up against me and ran away. I didn't see or hear anything, but it wasn't my imagination. He felt as solid as a man of average size and weight.'

Sally Pike and her husband Neil also saw something strange on Starr Hill. They were among a group of eight UFO watchers, and had spotted two unidentified, high-flying objects when they all felt the air grow suddenly warmer.

Sally went on: 'These two figures appeared. They were about 7 feet tall, and it was as if they were made out of smoke. We could see their outlines down to their waists, then they gradually started to fade away.

'When Nell approached them, he just seemed to blend in with them. He couldn't see them when he got close, but we watched him walk straight through the figures and out the other side. The figures remained in the same place for about half-an-hour, then disappeared.'

Ken Rogers was so intrigued by the mystery of the Warminster Thing that he moved to the town from London to study the evidence at first hand. One night he came across an enormous white object blocking a track down Cradle Hill.

'It was a classic saucer design, perfectly outlined,' he recalled. 'As I got closer, I got very hot and my hands started perspiring heavily. I walked on right through it, whatever it was. It seemed like fog, only you could see every detail absolutely clearly.'

Rogers, a director of the British UFO Society, added: 'I think UFOs are extraterrestrial forces. It is the most likely explanation for them. Remember, 50 years ago people were laughing hysterically at the notion of man going to the Moon. I believe it would only take a race maybe 50 years more advanced than us to make visits of this kind feasible. I don't think the UFOs mean us any harm, so I'm not frightened at all. I believe they are studying our progress.'

Tapping the Energy

Why did the Thing choose Warminster? That is the question that has baffled UFO experts. The town is close to a large army base, and large areas of nearby Salisbury Plain are used by the armed forces for exercises. Many so-called UFO sightings over the years have been traced to flares or equipment used by the Services. But many remain inexplicable. And a Territorial Army commander is among those who have reported the enigmatic cigar-shaped flying saucer. It stalled his car in January 1979.

UFOs have, in fact, often been spotted near military bases all over the world. The US Defence Department admitted in secret papers that unexplained aerial activity over missile sites and nuclear silos gave cause for concern during 1975. But the number of separate sightings at Warminster, over so many years, is unique. And if UFOs, as many people believe, are manned by intelligent beings, they would know that there are far more important military targets on Earth than Salisbury Plain.

Two other theories for the sightings, which have averaged two a week since 1964, have been put forward. Just outside the Warminster Triangle lies Stonehenge, thought by many to be an ancient 'computer' for astronomers. And the town of Warminster itself lies at the crossroads of 13 ley lines, the mysterious straight lines formed by monuments, graves, burial grounds, stone crosses and other ancient holy places.

Several scientists and historians believe the ley line network had strange powers that tapped the energy of the Earth centuries ago, and that Stonehenge was a powerhouse for this energy. Man has lost the ability to use it, though some people claim to receive electric shocks from some of the ancient stones. Could it be that aliens can tap the energy, or are attracted to Earth because of it? Could it even be that they guided the early Britons to create the lines during earlier visits to Earth?

Other UFOlogists look to the West Country to account for the concentration of activity over southern England. The historic town of Glastonbury forms one outer limit of the Warminster Triangle. Legend says that the Holy Grail – the cup from which Jesus Christ drank at the Last Supper – was brought to Britain by Joseph of Arimethea, and buried at Glastonbury Tor in about AD 60. Many people who have spotted the bright lights in the sky over Warminster sincerely believe that they herald the Second Coming.

Brazilian Balls of Fire

For some unknown reason, Brazil seems to get more than its fair share of UFOs. They are seen frequently, coming from both the sea and the sky. On June 27,1970, Mrs Maria Machado looked from the window of her Rio de Janeiro home as she prepared lunch, and saw a metallic grey disc with a transparent dome apparently sailing on the ocean. Two figures in shiny clothes were moving around on the deck. Her husband, four daughters and a policeman also saw the strange craft. After 40 minutes, it skimmed along the surface and took off, leaving behind a white hoop-like object which floated out to sea.

On September 12, 1971, typewriter mechanic Paulo Silveira claimed that two figures in one-piece blue suits dragged him inside a shining disc. He told the authorities he was driving home at Itaperuna, north of Rio, when the disc blocked the road. A luminous beam shot out and his car engine died, then his own energy drained away. He heard an engine start as the two aliens, about the size of ten-year-old children, carried him aboard; then he went into a coma. He came round to find them carrying him out again. They laid him beside his car, returned to their craft, and took off. Another motorist found him dazed, blinded and disoriented, and drove him to hospital. He had lost three hours of his life.

In January 1981, farmer Domingos Monteiro Brito claimed that he met two strange beings when a grey, glowing flying saucer landed at dawn on his land at Camaracu Island. The aliens, who resembled humans, asked him a string of questions in his own language – how many people lived in his village, were there any large uninhabited areas nearby – but he was too paralysed with fear to remember if he answered. The craft took off again, but the beings told him they would be back.

Early in 1980, the 30,000 residents of Tres Coroas, south of Rio, experienced one of the strangest concentrations of UFOs ever recorded. Over 20 days, balls of fire chased cars, flames erupted without burning anything, and scores of strange shapes zoomed through or over the city.

Bicycle-shop owner Joao Jose de Nascimento was driving home late at night when a fire-like object appeared beside his car, apparently following him. He said: 'It was strange and I was afraid. I felt it was trying to capture me.'

When he got home, his son Vicente told him he had seen another UFO – an onion shape revolving in the sky, with lights which switched from green to orange to blue. Estate agent Roberto Francisco Santana said he saw the shadow of a saucer travelling very fast as he drove through the town with his wife and children. Then he spotted two more saucers flying over city buildings. 'While I was looking up, I smashed into a car in front of me,' he said. 'The things we saw were very frightening.'

Military police commander Antonio das Gacas Santos said he raced to a neighbour's home when he noted that the back garden was curiously illuminated. 'I saw very clearly a creature about the height of a human being with its arms extended. I couldn't see any physical details, but I heard a low whining noise, like a puppy. My neighbour touched the creature, then fell back shocked. I was afraid, but it wasn't a normal fear. I still get goose-bumps thinking about it.'

Rio psychiatrist Dr Gloria Machado was astonished by what she saw when she and her husband Mario, president of the Brazilian Association of Parapsychology, arrived in the city. 'There was a fire which didn't burn anything, and flashes of light which exploded in the tree tops,' she said. 'Indoors I saw a box of matches floating in mid-air, bottles breaking for no reason, chairs flying around ...'

Her husband persuaded people watching a brightly-lit UFO to try to communicate with it. He said: 'We began uttering the letters of the alphabet, and heard sounds from behind the lights. The letter D came back long and hard. Lights started flashing everywhere, and we heard something that sounded like a beating rhythm. Then suddenly everything went dark.'

Lawyer Josefino de Carvalho, who watched the experiment, said: 'I'm sure that we are dealing with intelligent beings.' And police chief Santos said: 'I now believe that on other planets in other solar systems there exist forces which can manifest themselves here.'

Through Earth's Windows

Aerial researchers in America think they have located two UFO windows on this world – the sleepy New England town of Winsted, Connecticut, and the Michigan Rectangle of the mid-West.

The people of Winsted have grown used to strange shapes in the sky over the last 20 years. In February 1967, a businessman was one of three witnesses who reported an object that hovered over the town for 15 minutes before disappearing with red and green flashes. Only a few nights later two girls heard lawnmower-like sounds from a barn, and saw three humanoid creatures approaching their house. A passing car frightened them of, and minutes later the girls – and their neighbour – saw a UFO rising from a nearby hill.

Later that year, a cone-shaped object with red lights was spotted on two consecutive nights, and a month later a shape flashing red and green lights was observed hovering noiselessly over trees before zooming off at high speed.

In 1968, the sightings included a very bright globe, a balloon near the Moon, and an orange moon-like shape on a night when the Moon itself could not be seen.

In 1976, 13 girl campers and their leader heard a high-pitched whine as they climbed Blueberry Mountain, just outside the town. They looked up to see a silver, flat-bottomed saucer, about 25 feet wide. It was surrounded by a purple mist and had a red dome on top. It hovered for 30 seconds before vanishing.

In 1977, a policeman and three other people saw a red-topped object hovering soundlessly near the town's sewer treatment plant, and examining the ground two yellowish-white beams of light. The same year, people saw UFOs apparently diving into the local reservoir, and splashing upwards again.

Connecticut UFO investigator Ted Thoben is one of those who believe Winsted is a window through which UFOs arrive on Earth. He says: 'Windows are a magnetic deviation in the terrain, where these things slip through, But I don't believe they come from another planet. I think they exist at a different vibratory or frequency rate so that we cannot see them most of the time. They inhabit the same space that we do, and places like Winsted are the exchange point between different dimensions.

'That theory is far more logical than saying that UFOs are from outer space. For one thing, the Earth is in the boon-docks of the Milky Way galaxy. I can't see that after 2,000 years, some distant planet still finds us so intriguing they could allot so much effort to come here when there are so many other planets out there.'

The term Michigan Rectangle was coined by David Fideler, head of the local Anomaly Research organization. After studying reports of strange happenings from north of Kalamazoo to the Indiana state-line in the south, he said.: 'The rectangle may be a centre of window phenome-na – in other words a gateway from the ordinary world to the supernatural, where unreality leaks into the reality of the everyday.'

Fideler has chronicled strange shapes and lights in the sky as far back as 1897. There have also been many reports in the area of phantom panther-like creatures and of Bigfoot – a human-shaped creature covered in hair, with brilliant, gleaming red eyes. And Fideler says that before the white man arrived at Lake Michigan, the Indians called it Magician Lake. He believes geophysical and electromag-netic disturbances could account for the region's bizarre events.

The 1897 sightings included a brilliant white light, a huge ball of fire and a mysterious airship. A woman also report-ed hearing voices from the sky. In April that year, at least a dozen people watched an unexplained light fly over the centre of Kalamazoo.

In 1950, a DC-4 crashed with 58 people aboard – and Fideler says a curious ball of light was seen in the sky at the time. In 1966, a policeman was among those who saw a UFO 'so bright you couldn't look straight at it'. Then a

UFO nearly 40 feet long was spotted cruising above a highway, blinking its lights at drivers. In 1970 there was a mysterious explosion, heard four miles away and a gaping 40-foot hole was ripped in the ice of Upper Scott Lake. Chunks of ice flew more than 100 feet from the lake.

In 1974, police cars chased a UFO for 45 minutes. It flashed white and coloured lights, moved at 35-40 mph, and kept a height of around 600 feet before vanishing. Two years later a misty, glowing figure was reported floating a few feet above the ground, and in 1978 an unusual shape shot beams of light down on to Cook nuclear power station.

Fideler says: 'There are too many bewildering reports from a small area over a long period of time for them to be simply dismissed as unrelated incidents, or the ravings of crackpots.'

UFOs – All in a Row

The year 1954 saw an unprecedented wave of UFO sightings over central Europe and French parapsychologist and science writer Aimé Michel found a fascinating link as he studied some of the most reliable reports. When he plotted them on a map of his country, they formed a straight line running between and beyond the towns of Bayonne and Vichy.

All They Need is Love

The people of a tiny town in the Arizona desert claim UFOs have been visiting them for more than 30 years. And they say they bring only one message for the human race: 'We love you.' The town is Childs, an isolated settlement on the East Verde River between Flagstaff and Phoenix. Clarence Hale, 64, said: 'We've seen hundreds of UFOs – I first saw one 1947. We see so many of them we don't pay them any attention any more.'

His wife Mamie Ruth, 62, added: 'We can tell when the starships are around. We don't even have to go outside and see them any more. It's a feeling we get, a really warm and kindly feeling. It's a sort of love-thy-neighbour feeling deep inside, a feeling of humanity.

'We truly believe that aliens from outer space are trying to talk with the people of Earth. The strong feeling of love and compassion we get is their way of contacting us. They are trying to make Earth and the universe a better place to five – there is absolutely no reason to fear them.'

The good neighbours claim to have provided the authorities with evidence of UFO landings – powder and strands of silvery 'angel hair'. Power-plant manager Cliff Johnson found five circles of the powder on his new-mown lawn one morning, each about 12 feet in diameter.

'The powder was greyish-white until I touched it,' he said. 'Then it turned black, like soot. There was no other evidence that anything had touched down, just those big circles of powder. Some of them had spots of ash in the middle of them.'

Clarence Hale also found powder circles after watching a 'starship' land outside his home. 'It was about 8 am and I saw it coming in over the ridge line at about 30 mph, a big saucer-shaped ship, about 200 yards long. I could see windows and portholes with lights shining through. The ship was a silver colour, like metal. It landed and took off. When it left I found the powdery ash.

'I also found the angel hair. It looks like fine cobweb, but it feels synthetic. It sits on the trees and bushes after take-off. I gathered up about 30 feet of it one night, but when I wadded it up in my hand, it just vanished.'

University and government laboratories which tested the powder were unable to pinpoint its chemical make-up or origin. A top research scientist for the US Geological Survey admitted: 'It has got us baffled. We could not match it with anything we know on Earth.'

The people of Childs believe the 'angel hair' may be a device to protect humans from damage during take-offs. Kathy Soulages said: 'I think it's a flame retardant. Whenever a starship comes anywhere near where it might harm someone, it ejects the material to protect us from the heat.'

Terror in the Outback

Eerie objects in the sky have worried the people of a small town in Australia for more than 12 years. Trucks and cars have been chased and threatened, and one man even took a shot at a UFO with his rifle.

The town is St George, 300 miles west of Brisbane. Max Pringle, editor of the local paper which serves the 2,500, residents and the outlying farms, said: 'There have been several hundred sightings, since 1967, most of them by upright citizens, not the sort to look for publicity. Nobody knows why these things are scaring the wits out of the people here. God only knows what's behind it.'

Pringle says he saw his first UFO in 1977. 'It was orange, shaped like a football, and soaring silently about 500 feet off the ground,' he recalled. 'It had green flashing lights on top, and red underneath. I was stunned – I'd never seen anything like it.' By 1980, he had seen at least two dozen more.

Lorry fleet boss Jack Dyball claims he was buzzed by a silver-grey craft in 1975 as he drove near the town in a truck. 'It headed straight for me, then suddenly pulled up and flew out of sight,' he said. 'I tell you, it really fright-ened me. It wasn't a plane, it had no wings. I really thought it was going to crash into me. When it lifted off, I saw big blue flames coming out of five burners in the back.'

Rancher's son Murray Beardmore took a pot shot at an orange UFO in September 1978. It flashed red and green fights as it flew in front of a truck he and two friends were driving. Beardmore says he stopped the truck, grabbed his rifle, and fired one shot. Then, scared, they drove very fast to his home with the UFO trailing them. At one stage, their engine inexplicably cut out. The boy's father, John Beardmore, said: 'He was really shaken when he got here. All three of them were pretty ashen-faced. My sister, my wife and I all saw the thing, and got in the car to chase it, but it disappeared.'

The Broadhaven Triangle

Who or what haunts the Broadhaven Triangle? It is mystery that has baffled scientists, military investigators and UFOlogists.

The triangle lies between Swansea, mid-Wales and Broadhaven. And it has been the subject of more UFO visitations than almost anywhere else in the world. In one year alone, more than 50 positive sightings were made.

At first it was thought the rash of reports from the triangle was connected with the intense defence activity in the area. Within a tight radius there are: the Royal Aircraft Establishment Missile Range; RAF Brawdy, an operational station; The Army's Pendine Ranges; a missile testing ground; supersonic lowflying corridors, and an American submarine tracking station. Spokesmen for the establishments are non-plussed by the flood of sightings. And very few can be explained away by defence operations.

Certainly the sight that terrified Billy and Pauline Coombs in their farm cottage has baffled the experts. They were sitting in their front room at 1 am when Pauline suddenly turned to look at the window. Blocking it was a towering, eerie figure wearing a silver suit.

Too terrified to scream, Pauline stared, transfixed, at the 7-foot figure. Sensing her fear, Billy turned in his seat and saw the monstrous outline. 'Good God! What the hell's that?' he yelled.

'It was wearing a helmet with some sort of shiny visor,' Pauline recalled. 'A pipe went from the mouth to the back of the head. I was petrified. We were rooted to the spot with terror.

'It radiated a sort of luminous light and when it touched the window, the pane started to rattle like all hell had broken loose – yet there was no wind.

'When I got my wits together, I raced upstairs to see if the children were all right. Billy put our labrador Blackie outside, but he went mad with fear. He had to be destroyed six months later.'

The Coombs telephoned for help, but by the time police arrived at their home, Ripperton Farm, near the village of Dale, Dyfed, the eerie visitor had disappeared. The couple also telephoned neighbours to report what they had seen. Billy's boss, farmer Richard Hewison, drove over as soon as he got their call. 'They were genuinely terrified,' he said. 'They were frightened out of their wits.'

The family had two souvenirs of the incident – a burned out TV set, and a rose bush near the window, which was badly scorched.

The ordeal in the early hours of April 24, 1977, was not Mrs Coombs's first brush with the unknown. Two months earlier, on February 24, she had been driving three of her five children home from nearby St Ishmael's shortly after 8 pm when one of the boys saw a light which seemed to be coming towards them at great speed.

As the children started crying with fear, Mrs Coombs, 33 and said to be a down-to-earth type by those who knew her, put her foot on the accelerator. 'I thought the thing would come through the windscreen,' she recalled. 'In the end it went just over us and did a tight U-turn.'

The craft now flew alongside them, skimming over the tops of the hedges at 80 mph as Mrs Coombs kept her foot down. For 10 minutes, the bizarre chase continued through deserted country lanes. The object was no bigger than a football, but glowed yellow and had a beam of light underneath.

Finally, the car came in sight of the farmhouse ... and the engine cut out. Hysterical, Mrs Coombs grabbed her children from the back seat and rushed into the house. As she gabbled her amazing story, her eldest son saw the object disappearing.

For a year after that, inexplicable happenings made the family's life a misery. The children frequently saw bright lights landing in the fields, and found scorch marks next morning. On a trip to the coast at nearby St Bride's Bay, they saw two silvery-suited figures, and a flying disc which seemed to disappear into rocks. Two of the children received strange burns. Five television sets and eight cars mysteriously burned out. Then, as suddenly as the incidents had begun, they stopped.

The Coombs's neighbours also reported curious happenings. Mr Hewison's wife Josephine looked out of the bedroom window one morning to see a 50-foot silver spacecraft standing beside her greenhouse.

She said: 'It was as high as a double decker bus, there were no visible windows or openings. It stood there for about 10 minutes then took off. It left no mark. Not even a broken twig.'

Teenage shop assistant Stephen Taylor, of Haverfordwest, may have come the closest to an extraterrestrial encounter – when a figure similar to the one that terrified the Coombs suddenly appeared by his side. He was walking home late one night when he saw a black shape in front of him.

He said: 'It looked about 40 or 50 feet across. I noticed a dim glow around what seemed to be the underside. Suddenly this figure popped up, right next to me. I was terrified. It was dressed in silver. It seemed to have high cheekbones.

'Its eyes were like fish eyes – completely round. I took a swing at it and ran. I don't know whether I hit it. I ran the three miles home. When I got there, my pet dog started snarling at me. It wouldn't let me near it.'

Louise Bassett, wife of a restaurant owner from Ferryside, Carmarthen, said: 'I was driving home one night when my radio went dead. At the same time, I saw flashing lights in the sky. I took a detour to avoid them but they appeared again three miles further on.' When Mrs Bassett's radio cut out, so did dozens of other people's radios and televisions in the area. Mrs Bassett's dog was in the car with her at the time. 'It has never been the same since that night,' she said.

Artist John Petts, 62, was working in his studio near Carmarthen when he saw a brilliant light in the sky. 'It was a cigar-shaped object. One minute it was there, the next it was gone,' he said.

Perhaps the best witnesses of all are the children of Broadhaven Primary School. Fifteen of them – 14 boys and a girl – were playing football when they rushed inside to tell their headmaster that they had seen a spaceship in the sky. The head, Mr Ralph Llewellyn, split them up and asked them to draw pictures of what they had seen. He compared the finished results and was astounded by their similarity. It was no prank. Mr Llewellyn said: 'I do not believe that children of this age could sustain a hoax of this nature.'

The sighting that excited investigators from the British UFO Research Association and which is regarded as the most authentic so far was by two company directors as they drove in bright daylight from Carmarthen to Newcastle Emlyn – straight through the centre of the triangle.

One of the men, Elvet Dyer, described their experience: 'A huge cigar-shaped machine, at least 20 feet long, crossed our path 100 yards ahead. It was flying so low it would have taken the top off a double-decker bus. It made no sound and we thought it was going to crash.

'We braced ourselves for an explosion as it passed out of sight into a field, but when we looked into the field there was nothing there at all.'

The two men, non-believers in UFOs, were badly shaken and unnerved.

Mr Randall Pugh, regional investigator for the UFO association, said: 'We know there is something very strange going on in this area. Many of the reports have been from intelligent, educated people who are not disposed to exaggerate or misconstrue what they have seen.'

Dozens more reports of unexplained sights have flooded into the UFO association. Randall Pugh has noticed one common link in the sightings. 'People who encounter these phenomena suffer severe headaches, trembling and sleeplessness,' he said.

It is no wonder the people living in the Broadhaven Triangle are getting jittery. They believe that they have been singled out for surveillance by interplanetary beings.

A local police inspector said: 'After what I've seen round here in the last few years, nothing would surprise me now.'

Encounters of the Sinister Kind

The Beams that Burn

The vast majority of UFO sightings are painless for the humans involved. Indeed, most people who claim to have seen the craft and met their occupants, stress that they mean us no harm, and are only here to help. But sometimes people do get hurt. One incident in America in 1968 gave a hint of what could happen if the UFOs ever decided to get nasty.

Gregory Wells was returning from his grandmother's home to his own house next door in Beaffiville, Ohio, at about 8.30 pm on March 19 when he saw a large oval-shaped object hovering over nearby trees. It was red, and brilliantly lit. Suddenly a tube came out of the bottom, and moved towards the boy. A beam of light shot from it, and Gregory was knocked to the ground, screaming with fright as the upper arm of his jacket caught fire.

He was rushed to the town hospital with second-degree burns, and his scar was still visible three months later. Sheriff F.L. Suisberger of Monroe County interviewed several other people who had seen the UFO, including Gregory's mother and grandmother, and could find no other explanation for the injuries. The jacket and the road were checked for radioactivity, but none was found.

Such unprovoked attacks are mercifully rare. Others have felt the power of UFOs, but escaped unscathed. And a few have suffered agonies that were probably not intended.

A too-close encounter with a fiery diamond-shaped UFO left two American women and a child with excruciating pains that doctors were powerless to cure. Experts who questioned the victims under hypnosis were convinced they had suffered radiation burns after coming into contact with a craft manned by intelligent aliens.

Vicky Landrum, 57, and her grandson Colby, seven, were returning home to Dayton, Texas, on December 29, 1980, after attending a bingo game in Cleveland, a small town 40 miles away, with their friend Betty Cash, 52. Some 20 miles north-west of Dayton, on a desolate stretch of tree-lined road, they all noticed a glowing object lighting up the sky. 'Suddenly it came down, over the trees, and into the road right in front of us,' said Mrs Landrum.

'It looked like the whole sky had split. We saw a massive blue diamond, hovering at treetop level, with huge red flames shooting down to the road. Colby started to scream, and I said, "Honey, if you see Jesus coming out of the sky, He's coming to carry us to a better place." I really thought it was the end of the world.

'Betty slammed on the brakes. She got out and started walking towards the object. It was as big as a water tank. I rolled down the window – it was getting so hot from the flames – and stuck my head out to have a good look at the thing. It was making a beep, beep sound.

'I felt my eyes starting to burn, and I called to Betty to come back, but she was standing there entranced. Colby went berserk and tried to get out and run to the woods, so I grabbed him and held him close and said, "Don't cry, baby, just pray."

The diamond held them trapped for 15 minutes. Every so often, they heard a rushing sound like air brakes as bigger flames scorched down to the road. Each time the UFO lifted slightly, then settled to its previous height. Finally it rose slowly, then disappeared at high speed to the west, in the direction of Houston.

As the stunned trio continued the drive home, Mrs Landrum told Colby: 'Don't tell anybody about it – they'll think we're crazy.' But it soon became clear that someone would have to be told.

Within an hour of reaching Dayton, all three were sick. Soon both Mrs Cash and Mrs Landrum found their skin turning beetroot-red. Their eyes began to burn and weep, and they felt that they were looking through a film or mist. By morning, Mrs Cash found large lumps forming all over her body. She suffered agonizing headaches, and her hair was coming out in handfuls. She was so weak she could not get out of bed to call help. When the pain did not ease, she was admitted to Houston's Parkway Hospital.

During four weeks of exhaustive tests, medical experts tried to find out what was wrong with her. 'The doctors and nurses kept asking if I was a burn victim,' she said. 'Skin was peeling off my arms and legs and face. I was blistered all over. My ears and eyes were so swollen that my own family didn't recognise me.'

Mrs Landrum also lost tufts of hair, and specialists who examined her eyes found them 'burned, swollen and extremely irritated'. She was warned that cataract-like films were forming which might make her blind. Her grandson suffered digestive problems, and for weeks was haunted by terrible nightmares which left him screaming every night.

Four months after their ordeal, the women were still living in a nightmare of pain and fear. 'I don't know what to do,' Mrs Cash told a newspaper reporter. 'I'm at my wits' end. I need help and so does Vicky. I look terrible and I'm too sick to work. There must be something that can be done to help us. We don't know where to turn.'

UFO experts who investigated their case learned that the same shape had been seen 30 minutes earlier by three people driving 20 miles farther east. But none of them left their car, and none suffered ill effects.

Mrs Landrum agreed to be hypnotised by investigators who wanted to check the authenticity of her story. During questioning, she clutched the front of her blouse, screwing the material into her fist. She was sweating profusely as she gasped: 'We can't get through, it's blocking the road ... the whole thing's burning up ... oh my God, it's coming closer, we're going to burn up ...'

After the sessions, Dr Leo Sprinkle, a professor of coun-selling services at Wyoming University, said: 'There is no doubt she had a real experience. I believe the craft was under intelligent control.'

NASA aerospace engineer John Schussler, who watched the sessions while investigating the case for an independent UFO organisation, VISIT, said: 'This is a very important case providing physical evidence of the existence of UFOs.

A radiologist who examined the women's records said they were apparently suffering from the symptoms of radiation poisoning.

Bill English, of the Arizona Aerial Phenomena Research Organisation, added: 'It's the most incredible UFO sighting reported in the US in years.'

Teenage farmworker Mark Henshall claims he was scorched by a UFO while out riding his motorbike. Mark, then 16, said he felt a prickly, heat rash sensation on his face and arms for days after the incident in June, 1976.

He was riding a bike along a lonely country road near his home in Barnard Castle, County Durham, England, when he felt he was being watched. He looked up and saw a brilliant light in the night sky behind him, to his left.

'I was riding at about 30 mph but the bike seemed to cut out,' he told researchers from the UFO Investigators Network. 'I was very frightened. I could feel heat on my face and through my jacket. The petrol tank was steaming. It seemed as if something was pulling my bike forwards.'

A Jaguar car also spluttered to a stop nearby while Mark stood fumbling to light a cigarette. 'I was trembling,' he recalled. 'I touched the car to steady myself and it was really hot.

'Next morning my hands and arms came out in a sort of rash which lasted for a couple of days. I've been ridiculed by my mates, but I'm sure what I saw was a flying saucer.'

An American truck driver was temporarily blinded on

October 3, 1973, when he stuck his head out of the vehicle's window to get a better look at a UFO in Missouri. His wife, who was with him, said a 'large ball of fire' struck him in the face, knocking his glasses off. She took over the wheel and drove him to hospital, where he was treated for burns, but it was some hours before he could see again.

A physicist who examined the man's spectacles after the incident said the frames had been subjected to intense heat, and one lens had fallen out as a result.

Two people who claimed they saw a giant glowing egg, more than 35 yards long, near Baltimore, Maryland, on October 26, 1958, also needed hospital treatment later.

The couple said their car stalled as they turned a corner and saw the shape hovering above a bridge. They got out and crouched behind the vehicle while they watched a brilliant light and a wave of heat flood from the UFO. Then it shot up with a thunderous roar, and was out of sight within 10 seconds. Doctors found what seemed like radiation burns on the faces of both witnesses.

Space-stunned

A drive through the snow-covered countryside of Massachusetts ended in terror beside a lonely cemetery for William Wallace and his wife. It was 1 am when they arrived back at their home town, Leominster, after the 90-minute outing on March 8, 1967, and ran into a thick patch of fog by St Leo's graveyard.

Mr Wallace drove slowly through the mist, then turned the car round and headed back into it to investigate a strange glow from the direction of the church. He feared the building might be on fire.

When he drew level with the cemetery, he parked the car and the couple stared in amazement. The glow was coming from a large object shaped like a flattened egg, several hundred feet above the ground. Despite warnings from his wife, Mr Wallace stepped out of the car and excitedly pointed up at the object, which was ablaze with light similar to that from an acetylene torch.

As he raised his arm, the idling engine of his car cut out, and its lights and radio went dead. Mr Wallace felt numb and immobilised. His pointing arm was dragged back by some power, and thudded against the roof of the car. 'My mind was not at all affected,' he said later. 'I could hear my wife screaming for me to get back in the car but I just could not move. I was paralysed for perhaps 30 or 40 seconds.

Then the object, which had been rocking back and forwards began to move slowly away before shooting up and out of the fog. Abruptly the car lights and radio came back on, and I could move again, slowly and sluggishly.'

The couple drove home carefully, then phoned their parents and the police. Local officers knew them as reliable people, not liable to scare easily. But they were really shaken as they told of their eerie encounter.

Temporary paralysis of witnesses has been noted in several other UFO sightings.

On June 14, 1964, 18-year-old Charles Englebrecht was watching TV alone at home in Dale, Indiana, when a bright light flashed past his window, and the electricity failed. Groping his way to the door of the house, he saw a brilliantly lit round object hovering 50 feet from him. But as he started to walk towards it, a tingling sensation swept over his body, and he found he could not move. The sensation ended when the object disappeared, leaving behind a strong smell of sulphur and burned rubber.

Local police who investigated also smelt the sulphur, and found a scorched area of earth, the size of a large dinner plate, plus three shallow indentations which could have been left by tripod legs.

A day later, William Angelos was also watching TV late at night in Lynn, Massachusetts, when a loud roaring sound interrupted him. He rushed outside his apartment block and saw a domed object with an inverted, red glowing cone underneath, slowly rising from the car-parking area. As he moved towards it, he too felt a tingling sensation sweep up his body from his feet, until he was immobilised. Only when the object was out of sight did movement return to his muscles.

Neither man accepted suggestions that fear had frozen his limbs; and Mr Wallace swore that, when his arm was pinned to the roof of his car, it felt as though some power was pulling it backwards. Could it be that UFOs have directional forces similar to the stun guns of science fiction stories? The effects were all temporary, indicating that the UFO intelligence has a good understanding of human limitations and endurance.

The air crash death of Captain Thomas Mandtell, described earlier, shows that UFOs may have the power to kill if seriously threatened. But there have also been UFO-related deaths for which no immediate motive was apparent .

Death by Appointment

It was the strangest case Inspector José Bittencourt of the Rio de Janeiro homicide squad had ever faced. In August, 1966, two small boys found the bodies of Manuel Cruz and Miguel Viana lying on the top of Vintern Hill, a 1,000-foot-high vantage point overlooking the small town of Niteroi. Beside them were: crudely-made lead masks, only inches from their faces; pieces of green and blue paper, one of which contained a formula that no one could decipher, and two notes, neither of which made much sense.

The first note read: 'Sunday, one pill after meal. Monday, one pill after breakfast. Tuesday, one pill after meal. Wednesday, one pill lying down.'

The other said: '4.30 pm be at appointed place. 6.30 swallow pill. Then protect face with metal and wait for signal to show itself'

The two men were both wearing raincoats over their ordinary clothes. A postmortem examination revealed that they had died within seconds of each other. Two doctors reported: 'All organs had been functioning normally. After detailed investigation, it is impossible to find the cause of death.'

Bittencourt at first thought the men had been murdered for money they were carrying. He checked at their home town of Campos, and found that they had taken a bus for Rio with £ 1,000 of cruzeiros in their pockets, ostensibly to buy a car. When their bodies were found, there was only money worth £30 left, and the men had gone nowhere near a car showroom.

Instead they had left the bus at Niteroi, bought two raincoats from shopkeeper Jaime Alves – even though the day was scorching hot – and set off up Vintem Hill.

Bittencourt then put two statements together, and came up with an astonishing alternative theory. On the night that the two Brazilian television engineers had climbed the hill, stockbroker's wife Gracinda Souza reported seeing a green and yellow circular object, reddish at the rim, flash across the sky and glide towards the summit.

And in Campos, Miguel's father and a friend revealed that the two men had been obsessed with space communications, and had conducted experiments, one of which had resulted in an explosion and strange lights. 'I think they somehow contacted a flying saucer,' said Miguel's father. 'They were killed because they knew too much.'

Anywhere else in the world, police would have dismissed such a theory as nonsense. But Bittencourt was familiar with UFOs. He worked in the 100,000 square miles of Brazil known, because of frequent sightings, as Flying Saucer Alley.

No alternative explanation for the deaths was ever found, and eventually the file was closed. Police were convinced that the men were not killed where they were found. Had they really been taken aboard a spacecraft, their bodies returned by the disc seen by Mrs Souza? Were they indeed killed because they had discovered a secret, because of their knowledge?

The mystery had one more baffling twist to it. The strange formula found beside the bodies was locked in a police vault. But when the safe was next opened, the paper had vanished.

On the other side of the world, the town of Martinsicuro, near Pescara, in southern Italy, was plunged into mourning when, on October 12, 1978, brothers Gianfranco and Vittorio De Fulgentiis were found dead in the Mediterranean. But police trying to find out how they died were baffled. Their fishing boat was found undamaged on the sea-bed. And nobody could explain the peculiar puncture marks on the two men's faces.

Then other fishermen reported seeing red balls of light in the sky, following their boats. And Lieutenant-Colonel Piero Gallerano, of the Pescara police department, began to think again about earlier stories he had dismissed.

'I had received reports before from sailors about strange lights in the sky, and did not believe a word,' he said. 'Now I learn that these lights often follow boats.

'A navy patrol boat saw a red light at sea level. This light shot up about 300 meters and disappeared. The boat's radar and radio jammed. The red disc was gone in about four seconds. We are sure it was not a signal rocket. It was a very fast unidentifiable flying object.'

Suddenly, UFOs were no longer a laughing matter in Martinsicuro.

Bolt from the Blue

Did a laser beam from a UFO destroy two houses in Kuala Lumpur, Malaysia, in 1980? Police investigating the blazes which gutted the homes were told by three witnesses that a red ball of light had hovered above the buildings in the Port Klang district before the fire. Suddenly, from about 100 feet up, a bolt of blue light shot towards the earth, and the houses burst into flames.

Animals at Risk

Did a flying saucer kill 15 ponies on Dartmoor? Members of the Devon Unidentified Flying Objects Centre believe it did. The dead ponies were found close together in a little valley miles from any of the roads over the moors. Their bones were crushed, their ribs were cracked, and their flesh had rotted away to leave bare skeletons in only 48 hours, far quicker than normal.

Four UFO investigators took over the case in July 1975 after animal experts declared themselves baffled. They searched the area with geiger counters and metal detectors; and though they found nothing, the group leader John Wyse, a bandsman in the army, said: 'I think the ponies were crushed by the antigravity field of a flying saucer as it took off.'

A UFO was also the prime suspect when animals died mysteriously in a zoo at Newquay, Cornwall. Three ducks, a goose, a swan and two baby wallabies were found dead on the morning after strange lights were reported over the town. One bird was decapitated. Detectives were said to have discovered that the bodies gave off positive radiation readings.

In Minnesota, top American UFO investigator Dr J. Allen

Hynek was called in after farm animals were found muti-lated. There were no human footprints near the bodies, and no signs of attack by predators. Internal organs appeared to have been removed by surgical instruments and many cows had had their blood sucked out.

Dr Hynek said that 22 cattle were killed during the late 1960s and that curious deaths recurred in 1973 around the towns of Canby, Viking, Warroad and Kimball. He appealed for farmers to contact him at his UFO Centre in Evanston, Illinois, whenever they found more bodies.

Apart from making life painful for humans – and possi-bly holding the power of life and death over both man and animals – UFOs may have the ability to control some of Earth's most sophisticated scientific achievements ...

Unexplained Power Failures

The plant manager at Consolidated Edison was satisfied that all was in order as darkness fell on New York City on November 9, 1965. The system had plenty of power in reserve to meet the peak demand at dusk. But minutes after the city lights went on, they dimmed briefly, for no reason. A quick equipment check showed everything working normally, but monitoring machines registered an immense and unusual flow of current to the north. A phone call to the next station up the line, near Syracuse, confirmed that something odd had happened even farther north. Then, at 5.27 pm, the entire city of New York was blacked out.

As the power chaos spread, the whole of the eastern seaboard of northern America and southern Canada fell into darkness. Next morning, President Johnson ordered an immediate Federal investigation. Consolidated Edison blamed transmission lines north of Niagara Falls. But the Canadian government's Electric Power Commission said a high voltage line south of the Falls was responsible. They said Ontario's Queenstown relay station had been hit by 'a surge of electricity ... flowing in the opposite direction to the normal flow at that hour'. Much of Toronto and the surrounding area had had to be blacked out at 5. 15 to prevent damage to expensive equipment.

Later a joint US-Canadian statement admitted that the investigators 'still don't know the origin of the source of power that ripped out the relay'. And in April 1966, Oscar Bakke, eastern regional director of the US Federal Power Commission's Bureau of Power told Congress that electrical workers insisted that the blackout should not – even could not – have happened.

So why did 36 million people lose power over an area of 8,000 square miles? If there was nothing wrong with generating equipment, some outside agency must have interfered with the supplies. The Aerial Phenomena Research Organisation sent investigators from its base at Tucson, Arizona. And their findings were startling.

At 5.14 pm, just 60 seconds before the Canadian blackouts, pilot Weldon Ross was flying a passenger towards Hancock Field. As they passed over the two 345,000-volt power lines carrying supplies from Niagara to the Mohawk Power Corporation's sub-station at Clay, just outside Syracuse, NY, Ross was astonished to see what he described as a bright red fireball, about 100 feet in diameter, rising from the power lines.

Ten minutes later, at the blacked-out Hancock Field airstrip, deputy aviation commissioner Robert Walsh was arranging emergency lighting for the incoming plane when he spotted a similar fireball a few miles to the south – also hovering over the power lines. New York was in darkness two minutes later.

Could UFOs have caused the most famous blackout in history? Dr James E. McDonald, senior scientist at the Institute for Atmospheric Physics at the University of

Arizona, certainly thought they could. On July 29, 1968, he gave evidence to a UFO symposium requested by the, Committee on Science and Astronautics of the US House of Representatives.

After saying that UFOs had caused the fillings in people's teeth to hurt, and had been responsible for the failure of ten cars' ignition systems at Levelland, Texas, in 1957, he astounded the politicians by declaring: 'UFOs have often been seen hovering near power facilities, and there are a small number – too many to seem pure, fortuitous chance of system outrages coincident with a UFO sighting.

'After the New York blackout, I interviewed a woman in Seacliff, NY. She saw a disc hovering and going up and down, then shooting away from New York just after the power failure. I went to the Federal Power Commission for data. They didn't take them seriously, although they had many dozens of sighting reports for that famous evening. There were reports all over New England in the midst of that blackout and five witnesses near Syracuse saw a glowing object ascending within about a minute of the blackout.

'It is rather puzzling that the pulse of current that tripped the relay at Ontario has never been identified; there is a series of puzzling and slightly disturbing coincidences here which I think warrant much more attention than they have so far received.'

If the authorities were reluctant to probe the coincidences, UFO enthusiasts were not. The National Investigations Committee on Aerial Phenomena files, showed that UFOs had been sighted over Mogi Mirim, Brazil, and Tamaroa, Illinois, during the 1957 power failures. Rome was in darkness in August, 1958, when a luminous flying object was spotted over the Italian capital, and a similar coincidence was reported 11 months later from Salta, Argentina.

News agency reports from Umberlandia, Minais Gerais,

Brazil, on August 17, 1959, told of automatic keys at a power station being turned off as a round UFO flew along the transmission lines, then being switched on again, restoring normal service, when the UFO vanished.

Observers also noted a series of curious blackouts in late 1965 and 1966, a time when worldwide UFO activity was at fever pitch. San Salvador was without power for an hour for undisclosed reasons on November 9. Toledo, Ohio, mysteriously blacked out two days later. Relays tripped at Lima, Peru (November 19), Texas and New Mexico (December 2) and Buenos Aires, Argentina (December 26). High consumption was blamed when parts of London plunged into darkness on November 15, and in East Texas the lights inexplicably went out on December 4 – just as Federal Power Commission chairman Joseph C. Swidler was explaining his New England investigations to President Johnson at the Texas White House.

At Cuernavaca, Mexico, the governor, mayor and a military zone chief all saw a glowing disc hovering at low altitude when power mysteriously cut out. And at St Paul, Minnesota, power officials, police and residents all spotted UFOs on November 26 when sudden, unexplained electricity losses hit the city. Car lights and radios also failed as the UFO, described as huge, bright blue and glowing 'like someone welding in the sky' crossed the area.

All of southern Italy lost power for up to two hours on January 8, 1966, and no reason was ever announced. Five days later, when 75 square miles of Franklin County, Maine, were blacked out, the local electricity company blamed 'an apparent equipment failure which somehow corrected itself'.

The spate of mysterious failures, coinciding with such concentrated UFO activity, convinced UFO believers that aliens were showing a growing interest in Earth's electricity production. But even they had no answer to the next question: were they interfering with it intentionally or inadvertently? And if the meddling was deliberate, what other powers do the UFOs have?

American defence chiefs fear their electro-magnetic power could play havoc with the sophisticated electrical systems controlling nuclear warheads. Pentagon experts were worried after several unidentified craft were spotted over their Minuteman Intercontinental Ballistic Missile sites in 1966 and 1967, and over sensitive nuclear silos and bomber bases in Maine, Michigan and Montana in 1975.

One incident caused special concern. On August 25, 1966, radio transmissions were interrupted in a concrete bunker 60 feet below the surface of a North Dakota missile base at the same time as UFOs were observed 100,000 feet above. Those interviewed about the incident were sworn to secrecy. And in 1975, a directive from the Air Force Secretary instructed public relations personnel to avoid linking sightings over nuclear bases unless specifically asked.

Skyway Robbery

Have UFOs snatched satellites sent up from Earth, so that they can study our space knowledge? Robert Barry thinks they have. Barry, head of Twentieth Century UFO Bureau in Yoe, Pennsylvania, put his theory forward after experts announced they were baffled by the disappearance of the $20 million communications satellite Satcom 3.

The one-ton orbiting unit, designed to relay telephone and television transmissions, simply vanished while working perfectly. Jim Kukowski, one of the NASA staff who helped launch it, said: 'We just don't know what happened to it . '

And John Williamson, a spokesman for RCA, which owned the satellite, admitted: 'We've lost it and we have no idea why.'

He added: 'If the satellite had exploded, at least one part

would have shown up on radar. The North American Air Defence Command can track an object no bigger than a basketball 23,000 miles up, so they would have certainly located something.

'If the satellite had been pushed into a different orbit by a malfunction of the engine, NORAD would again have located it when it reached its nearest point to Earth. That hasn't happened either.'

Robert Barry believes aliens grabbed the satellite to examine it for information. He said: 'I suppose they'd want the same thing from it as we would want from one of their spaceships.

'Someone out there is showing a lot of interest in our activities down here. UFOs are usually reported in heavy concentrations around Cape Canaveral before a launch.

'This isn't the first time a satellite has vanished mysteriously. The Soviet Molniya satellite disappeared the same way, and we know our Gemini missions and the Soviets' Salyut space lab were buzzed by UFOs.'

Barry added: 'Just imagine if they had plucked a manned spacecraft from orbit. The implications would be tremendous.'

The Scottish Saucers

The British UFO Research Association launched a major investigation in the Scottish border regions in May 1981 after two women reported a series of strange sightings.

Mrs Mary Watson and Mrs Joyce Byers, both of Moffat, Dumfries, said they had logged more than 100 separate UFO sightings in a diary provided by Eskdalemuir Observatory. 'We have noted everything from swirling, saucer-shaped objects to orange and red triangles,' said Mrs Byers.

The women said they believed the Moffat Hills might be

a base for UFOs, and that there could be a link with a series of mysterious plane crashes in the border country, in which 12 people had died. They also pointed out that two nuclear power stations, Chapelcross and Windscale, were within easy flying distance.

Stuart Campbell of the UFO Research Association said: 'Inquiries are being made. The two women are not the sort to make up stories.'

Encounters
of the
Aerial Kind

Air Mysteries

The best planes and pilots Earth can muster have taken off to challenge UFOs in the sky – and all have been found wanting. In the late 1940s and early 1950s, when US Air Force aces were under orders to shoot down aerial intruders, not one victim was claimed. But the interceptors suffered casualties.

On January 7, 1948, USAF Captain Thomas Mantell led three F51 Mustang fighters into action after Kentucky police were inundated with reports of a hovering 'giant air machine', in the form of a glowing disc 300 feet across. The control tower staff at Godman Field air base had seen it as well.

Mantell was an experienced pilot, a veteran of World War Two air battles. He closed in on the silvery shape over Fort Knox. 'It's a disc,' he radioed to Godman. 'It looks metallic and is tremendous in size ... it has a ring and a dome, and I can see rows of windows ... the thing is gigantic, it's flying unbelievably fast. It's going up ... I'll climb to 20,000 feet ...'

Then the voice cut out and the radio went dead. Two hours later the wreckage of the plane was found scattered over an area a mile wide. Mantell's body lay nearby. The authorities refused to let anybody see it.

Top-level inquiries were held; but the findings, announced 18 months later, were unbelievable. The Air Force announced that Captain Mantell had probably fainted from lack of oxygen as he climbed to 20,000 feet – and what he had seen was probably the planet Venus. A planet with windows? A planet chased by an experienced pilot? Later statements changed the story. The object was simply a naval research balloon.

In June 1953, an F-94C jet fighter-interceptor took off from Otis Air Force base on Cape Cod after a UFO had been reported. At 1600 feet the engine cut out and the entire electrical system failed. As the aircraft's nose dipped towards the ground, pilot Captain Suggs yelled at his radar officer, Lieutenant Robert Barkoff, to bale out.

Normal procedure was for the radar officer to pull a lever which triggered explosive bolts to jettison the canopy. He then pulled a second lever, which ejected him and his seat from the plane, and when the pilot heard the second explosion, he pulled his own ejection lever. Captain Suggs baled out before he heard the second explosion, because the jet was already down to 600 feet and only seconds away from crashing.

Suggs landed just after his parachute opened in the back yard of a house. The owner, sitting near an open window, was astonished. Suggs was equally amazed. Why had the man not heard the plane crash? And where was the radar officer?

A full-scale search was launched. Cape Cod was combed on foot and from the air, and divers scoured nearby Buzzard's Bay. When the hunt was called off three months later, not a trace of the jet or of Lieutenant Barkof had been found. They had seemingly disappeared.

On November 23, 1953, Lieutenant Felix Moncla and radar officer Lieutenant R. R. Wilson took off from Kinross Air Force Base to chase a UFO spotted over Lake Superior by Air Defence Command radar operators. The F89C jet was guided towards the object from the ground, and controllers saw the plane close in on the UFO blip. Then, 160 miles from the base, at 8,000 feet and 73 miles off Keeweenaw Point, Michigan, the two blips merged and faded from the screen. The jet and its occupants were never seen again.

At first the Air Force said the F-89C had identified the UFO as a C47 of the Royal Canadian Air Force. But the RCAF denied that any of its planes was in the area. The official line was 'that the pilot probably suffered from vertigo and crashed into the lake'.

422

Pilot and co-pilot both survived the next disastrous attempt to intercept a UFO, but four civilians were not so lucky. On July 2, 1954, an F-94C was diverted from a routine training flight by Rome Air Force base after reports of a balloonlike object over the village of Walesville, New York. Radar scanners had shown two unidentified tracks. The first turned out to be a Canadian C47, but the second could not be identified.

What happened next was contained in an official report of the incident by Air Force investigators. 'As the pilot started a descent,' it said, 'he noted that the cockpit temperature increased abruptly. The increase in temperature caused the pilot to scan the instruments. The fire warning light was on, the engine was shut down and both crew members ejected successfully.'

The plane crashed in Walesville, hitting two buildings and a car. Four people were killed, two of them children. The Air Force dismissed the second object seen by radar operators as 'probably a balloon'.

Why did the Air Force cover up what really happened in all four incidents? Documents released since 1954 reveal that, contrary to statements at the time, there was a genuine belief that the objects chased by the jet were craft manned by intelligent beings.

As early as September 23, 1947, Lieutenant General N. F. Twining of Air Material Command had sent a memorandum to Brigadier-General George Schulgen, Commanding General of the Army Air Forces, saying: 'It is the opinion (of this Command) that the so-called flying discs phenomenon is something real and not visionary or fictitious.

They reported operating characteristics, such as extreme rates of climb, manoeuverability and evasive action when sighted or contacted by friendly aircraft and radar lend belief to the possibility that some of the objects are controlled either manually, automatically or remotely.'

The immediate suspicion of the Americans was that the

discs might be some spectacular advanced technology that the Russians had captured from the Nazis during World War Two. After the Mantell crash, an urgent investigation of possible threats to national security was launched.

In August 1948, the Air Technical Intelligence Centre drew up a top-secret report concluding that UFOs were not of Russian origin, but were interplanetary craft. Air Force Chief of Staff General Hoyt S. Vandenburg ordered: 'Burn it.' And on December 27, 1948, the ATIC study on UFOs, code-named Project Sign, was wound up. The public were told: 'Reports of flying saucers are the result of misinterpretation of various conventional objects, a mild form of mass hysteria, and hoaxes. Continuance of the project is unwarranted.'

But the project was not closed down. In February 1949, it resumed inquiries under a new code-name – Project Grudge. UFO sightings continued, and in 1952 a new upsurge of reports forced the government to act again.

On July 26 of that year, three F-94 jet fighters scrambled to investigate a cluster of curious lights which appeared above the White House in Washington. The lights had also been spotted a week earlier, but this time there were more of them, nearly a dozen, zigzagging erratically at high speed.

Two of the intercepting pilots found no trace of them. But the third said he flew straight into a group of the whitish-blue lights, which travelled alongside him for 15 seconds before dispersing. All three planes returned safely, and the lights – labelled the 'Washington Invasion' by the Press – were never seen again.

That month, the UFO investigating team – now working under the diplomatically more acceptable title of Project Bluebook – were receiving between 20 and 30 sightings every day, 20 per cent of them objects that no one could identify or explain away. Embarrassingly for touchy Air Force chiefs, one of the witnesses was Dan Kimball, the Secretary of State for the American Navy. He said two disc-shaped

UFOs buzzed the plane in which he was flying to Hawaii, circling it twice before shooting off at more than 1,500 mph, then repeating the exercise round a Navy plane 50 miles away.

When Kimball later inquired what progress Bluebook was making on his report, he was told that no action had been taken and that officers were forbidden to discuss case analysis with anyone. Furthermore, no copies of reports could be returned.

By 1953, public pressure for information about UFOs forced the Central Intelligence Agency (CIA) to make some sort of gesture. It convened the Robertson Panel, under a respected Californian scientist, H. P. Robertson, and asked it to evaluate UFOs. There were three possible findings, that UFOs were explainable objects and natural phenomena, that there was insufficient data in reports to make a conclusion, or that UFOs were interplanetary spacecraft.

According to Edward Ruppelt, former chief of the Air Force UFO Project, the panel opted for the second possibility, and urged that Bluebook manpower be quadrupled, bringing in skilled scientists and observers to try to solve the problem of what UFOs really were. It also recommended that the public be told 'every detail of every phase' of UFO investigations. Privately, said Ruppelt, almost every member of the panel was convinced that UFOs were extra-terrestrial.

The CIA suppressed the report, finally releasing a censored version of it in 1966. And they ignored its recommendations, instigating instead a 'debunking' programme. A secret document released years later read: 'The debunking aim would result in reduction of public interest in flying saucers which today evokes a strong psychological reaction. This education could be accomplished by mass media such as television, motion pictures and popular articles. Basis of such education would be actual case histories which had been puzzling at first but later explained. As in the case of conjuring tricks, there is much less stimulation if the secret is known.'

While the public was told that UFOs did not exist, servicemen were ordered to shoot them down. People who reported seeing flying saucers were ridiculed. Forces personnel were threatened with jail or fines if they broadcast what they had seen. 'Only false statements and fictitious reports may be published,' read one Air Force order. 'All real reports must be treated as secret and forwarded to the appropriate authorities.'

Once the Americans realized that UFOs were not a Soviet secret weapon, the race was on to capture one before the Russians did. Insight into such advanced technology would be of incalculable value to either power. Meanwhile, public debunking of UFOs might make the Russians less interested in trying to bring one down.

The ploy did not work. Moscow had come to the same conclusions as Washington. In 1957, anti-aircraft batteries around the Soviet capital opened up on an object in the sky – until the guns electrical systems mysteriously went dead.

In 1967, American Air Force agents monitored a broadcast from one of two Cuban jet fighters sent up to intercept a curious UFO. The pilot said he had just seen his partner's plane disintegrate without smoke or flames as he tried to shoot the object down.

Stanton Friedman, who revealed the story after leaving his job as a space-related nuclear technician for the US government, claimed that tapes of the conversation were sent to the National Security Agency, which ordered the loss to be listed as equipment malfunction.

Not all UFOs proved so lethal. One spotted above the English counties of Norfolk, Suffolk and Cambridgeshire in 1956 seemed almost playful when a plane came close. The excitement started at 9.30 pm on August 13 when radar operators at USAF Bentwaters spotted an object which zoomed off the screen at what seemed to be 5,000 mph. Then a group of slow-moving shapes were tracked out to sea. They seemed to link up into one object before disappearing with a stop-go-stop motion. More sightings, were reported at 10 pm and, again at 10.55 pm when observers saw a blurred white light pass overhead. A C-47 aircraft radioed that it had passed below them at extraordinary speed.

Bentwaters alerted radar crews farther north at Lakenheath, and they too saw the object, on screen and visually. Its antics were baffling, changing direction crazily, shooting off at right angles without stopping, and scorching to enormous speeds from a standing start.

Two jet fighters diverted to intercept could find no trace of the object. Then a Venom single-seat fighter, equipped with nose radar, took of from Waterbeach, and was guided from the ground towards the UFO, at that stage motionless and clearly visible, between 15,000 and 29,000 feet, over Lakenheath.

The pilot radioed that he had radar contact and 'gunlock' – then he lost sight of his quarry. 'Where did he go?' he asked ground control. 'Roger, it appears he got behind you and he's still there,' came the reply. The UFO had zipped into position in an incredible right-angled flight too fast for most of the radar watchers to follow. Once behind the Venom, it had split into two separate units, one behind the other, and locked on to the fighter.

A bizarre game of hide and seek began. For ten minutes, the Venom pilot dived, climbed and circled, trying to shake off his pursuer. But the UFO stuck to his tail, always 100-200 yards behind. Finally the Venom headed for home, its fuel running low. The UFO followed it down, then stopped, hovered in triumph for a while, and vanished.

Cynics pointed out that the East Anglian terrain is notorious for generating false radar traces, known as 'angels', and that the incident occurred at the climax of the Perseid meteor shower, which passes Earth each year and appears as a series of luminous white blobs.

427

But the official report on the incident, filed on August 31 by Captain Edward Holt of the 81st Fighter-Bomber Wing, Bentwaters, said: 'The object . . . followed all manoeuvres of the jet fighter aircraft.'

Nearly a year later, the six-man crew of a US Air Force RB-47 jet reported another playful UFO. It chased them for more than 2 hours in a 1,000-mile flight across Mississippi, Louisiana, Texas and Oklahoma early on the morning of June 17, 1957. Curiously, they added, the object occasionally vanished from sight momentarily – and when it did so, it also disappeared from their radar screen, only to reappear within seconds in the same place.

The debunking of UFOs worked quite well for a while. Project Bluebook successfully managed to 'investigate' sightings, and to produce unsatisfactory answers.

Then, in 1964 and again in 1967, came fresh waves of UFO activity. In response to renewed public pressure, the Air Force announced that renowned physicist Dr Edward Condon would lead a University of Colorado inquiry into the sightings, to run parallel with the Bluebook investigations.

In January 1969, Condon's report said: 'Careful consideration of the record as it is available to us leads us to conclude that further extensive study of the UFOs probably cannot be justified in the expectation that science will be advanced thereby.' It admitted, however, that 30 per cent of cases it investigated remained unexplained.

The 1,000-page report was condemned as a whitewash – and worse. One UFO research group pulled out of the investigations because of Condon's negative and subjective and comments; and two of the inquiry team, Dr Norman LeVine and Dr David Saunders, were fired for leaking a memorandum that read: 'The trick would be, I think, to describe the project so that, to the public, it would appear a totally objective study but, to the scientific community, would present the image of a group of non-believers trying their best to be objective but having an almost zero expectation of finding a saucer.'

The memo was written by assistant project director Dr Robert Low, whose job was to co-ordinate the inquiry. The two doctors were not alone in having no confidence in him. Condon's administrative assistant quit, saying: 'Bob's attitude from the beginning has been one of negativism.'

Criticism of the Condon Report was loud and long. Congressman J. Edward Roush told the House of Representatives that he had 'grave doubts as to the scientific profundity and objectivity of the project'. He added: 'We are $500,000 poorer and not richer in information about UFOs ... I am not satisfied and the American public will not be satisfied.' Aviation pioneer John Northrop, the 80-year-old founder of Northrop Aircraft Company and co-founder of Lockheed, said: 'The 21st century will die laughing at the Condon Report.'

The Condon inquiry did one service for the subject of UFOs. The fact that such a distinguished scientist was prepared to study them allowed other top boffins to take UFOs seriously as well. Even after he debunked them, others felt free to continue their studies without fear of ridicule. And though the Air Force announced, on December 17, 1969, that it was closing down Project Bluebook because UFOs 'didn't exist', it too continued to monitor and analyse reports through the Aerospace Defence Command.

In the 1970s, laws concerning freedom of information, and the more enlightened attitudes of other governments, notably France and Italy, and even Russia, allowed greater access to UFO reports, and there were more frequent reports of confrontations between unidentified objects and Earth aircraft.

A squadron of F-106 jet fighters scrambled in 1975 when a fleet of mysterious shapes appeared at 15,000 feet over Montana. As they approached the shining lights, the objects simply vanished.

An even stranger encounter emerged only a few years after it happened. Captain Lawrence Coyne and three crewmen took off in a US Air Force helicopter from Columbus, Ohio, at 10.30 pm on October 18, 1973, heading for Cleveland. Forty minutes later, they were 2,500 feet up over Mansfield when one of the men noticed a red light approaching from the cast at high speed. Coyne dived to 1,700 feet but a collision seemed inevitable. He braced himself for the impact. It never came.

About 500 feet away from the helicopter, the UFO stopped abruptly. Coyne noticed a huge grey metallic hull, about 60 feet long and shaped like a streamlined fat cigar. The front edge glowed red, green lights flickered at the back, and there was a dome in the centre. A green light suddenly swivelled and flooded the helicopter cockpit.

Coyne tried to radio an SOS, but his set would not transmit or receive. Then he looked at his instrument panel and gasped. The helicopter was being lifted into the air.

'I could hardly believe it,' he said. 'The altimeter was reading 3,500 feet, climbing to 3,800. I had made no attempt to pull up. All the controls were still set for a 20-degree dive. Yet we had climbed from 1,700 feet to 3,500 with no power in a couple of seconds, with no G-forces or other noticeable strains. There was no noise or turbulence either.'

Finally the crew felt a slight bounce, and the UFO zipped away towards the north-west. Seven minutes later, the helicopter radio started working normally again, and Coyne reported the incident to incredulous ground controllers.

Double Saucer Over the Thames

The British Air Ministry was forced to take an interest in UFOs in 1955 when Flight Lieutenant James Salandin filed a report of a strange encounter over the Thames Estuary. He was flying his Meteor jet fighter at 16,000 feet in a cloudless sky when he spotted a metallic silver object approaching him. He described it as two saucers joined together, with a dome or bubble on top. He saw no visible portholes or jet pipes, and estimated that the craft, about 40 feet wide, was travelling at twice his 600 mph.

Scientists Who Knew Too Much?

Suicide verdicts were recorded on two top US scientists who died after studying UFOs, having decided that they were extraterrestrial spaceships investigating life on Earth. Atmospheric physician Professor James McDonald, of the University of Arizona, was found with a bullet in his head in 1971, and astronomer Professor Robert Jessup was discovered in his gas-filled car in 1959. A friend of Jessup claimed: 'He knew too much, they wanted him out of the way.' But fellow scientists felt both men had suffered depression after battling for years against a brick wall of governmental UFO denials and evasions, and the scorn of sceptical colleagues.

Phantoms Versus UFOs

One early morning in September 1976, an F-4 Phantom jet fighter streaked into the skies of Iran from Shahrokhi Air Force base. It had been ordered to investigate a dazzling bright light spotted by hundreds of people south of

431

Tehran. The fighter closed on the object, but when it was 30 miles away, all radio contact was lost.

As the pilot broke off and headed back to base, his radio crackled back to life, and he reported that all communications and instrumentation systems in the plane had suddenly and inexplicably cut out.

A second Phantom was already in the air and in pursuit of the UFO at a speed much greater than the speed of sound; but the craft was still accelerating away from it. The pilot radioed that it seemed about the size of a 707 passenger aircraft. Suddenly the UFO released a smaller, disc-shaped object which also glowed brilliantly. It hurtled straight for the jet.

The pilot reached for his weapon control panel and pressed a button to release an AIM-9 air-to-air missile. Nothing happened. All his electrical systems had blacked out. He swung his defenceless plane into a dive to avoid the approaching disc, which changed course to follow him for four miles. Then it zoomed back to the larger UFO.

As his instruments started working again, he again went after the mother ship, which was moving away rapidly. Then it shed another disc, which fell at great speed towards the Earth. He watched it go down, expecting an explosion, but it stopped just above some hills, casting an eerie glow over a two-mile area. The pilot looked up again, and realised that the larger UFO had used the disc to distract him while it vanished. He returned safely to base.

The Iranian government later filed reports of the incident to the Pentagon in Washington. A year later the Italian government revealed that its jets had also encountered UFOs. It listed six separate encounters during 1977 and 1978, two of which involved air force personnel, and one a civilian airliner.

On February 23, 1977, a fighter pilot spotted an intense ball of light over Milan. 'When radar gave me authorization to intercept, the object went up to 12,000 feet and kept its distance,' he said. 'It was in my sight for 23 minutes.'

On October 27 of the same year, a football-shaped UFO buzzed a helicopter during NATO exercises at Elmas Air Force base, near Cagliari, Sardinia.

The Defence Ministry quoted an air controller as saying: 'I saw a UFO that flew at the speed of a jet, around 565 mph. It was behind a helicopter that was participating in military manoeuvres.' Three other helicopter pilots and jet fighter crews also reported seeing the UFO which flew alongside some of them. Later a jet was sent up to intercept a separate cigar-shaped object, but it proved too fast.

Three other sightings in the Italian report were by air traffic controllers using binoculars. At Naples on August 4, 1977, officials watched a pulsating star-shaped object for go minutes. At Elmas on November 5, a UFO was observed for eight minutes, during which it rose from 5,000 feet to 30,000 in 30 seconds. And at Pisa on November 23, staff saw a strange glowing shape change colour from red to violet to green for two hours at 15,000 feet.

The last of the objects, all listed as 'genuine UFOs', was seen on March 9, 1978. The pilot of International Airlines flight IH-662 radioed Milan control tower to report 'a green rocket moving above and below us about a mile away'. He asked if it could be another aircraft, and was told none were in the area.

'I thought I was going mad,' the pilot later told officials who interviewed him. 'I only reported the sighting for information. When other pilots said they had seen it too, I knew I wasn't seeing things.'

Three Austrian air force jets took to the sky on May 7, 1980, after a KLM liner pilot told Vienna air controllers that a grey spherical object was flying above him over the Dachstein mountains. Two of the fighters were ordered to intercept, while the third filmed the confrontation. But both missions proved impossible. All three made visual contact with the object, but could not get close because of its unpredictable, erratic behaviour. It soon vanished completely.

Action Over the Arctic

Russian pilots have also reported seeing UFOs, and one even had a 'dogfight' with one. Professor Felix Zigel, of Moscow's Aviation Institute, said: 'His name was Arkady Apraksin. He was flying a jet fighter when he encountered a cigar-shaped UFO. Radar had also spotted it, and he was ordered to force it to land, or open fire.

'Apraksin began his approach, but the mystery craft fired a fan-shaped beam which momentarily blinded him and killed his controls and the engine. He had to glide into a landing.'

On June 14, 1980, another Soviet flier reported a UFO above Moscow that played cat-and-mouse with him. 'Its manoeuvres were too bizarre for our jet to duplicate,' said Professor Zigel. 'Suddenly it took off at incredible speed.' The pilot said the craft seemed almost 900 feet wide, and was circular.

Four months later, on October 22, Captain Vladimir Dubstov spotted a similar size saucer hovering below him as he flew his patrol bomber over the Arctic Ocean. He changed course to circle it.

'He told me it was truly immense,' said Professor Zigel. 'A cone of light protruding down from it gave it an eerie appearance, but it showed no sign of life. Then Dubstov's instruments went haywire, and he lost altitude. The UFO took off vertically and soared past him, leaving behind a greenish-blue cloud. Dubstov nursed his crippled jet home and reported the incident.

Phantoms Fight 'Cover Up'

American airline pilots were furious in 1954 when the CIA and USAF posed military-style curbs on them reporting UFOs. The clamp down followed a conference in February when Military Air Transport Service Intelligence officers met the heads of major airlines to try to speed up the process of reporting UFOs spotted during civilian flights.

Until then, pilots had reported strange objects after they landed. Now the Air Force instructed them to radio the news to MATS HQ in Washington, or the nearest air base, while in flight. And it asked them not to discuss sightings, or give information to newspapers.

A month later, regulations threatening Air Force pilots with ten years jail and a fine of $ 10,000 for 'failing to maintain absolute secrecy' were extended to cover civilian air crews. Understandably, the airline veterans reacted angrily. A protest petition was signed by 450 men, 50 of whom, all with at least 15 years service, said at a meeting that the censorship bid 'bordered on the ridiculous'. It was, they said, 'a lesson in lying, intrigue, and the Big Brother attitude carried to the ultimate extreme'.

The pilots knew that the curbs were part of a cover-up, for all had seen a UFO with their own eyes. Many had seen several. They revealed that five or ten sightings were reported every night by commercial pilots in America alone, and said that it was almost routine to warn passengers to put on seat belts when UFOs were near.

Some of the civilian sightings over the last 40 years have been every bit as spectacular as those reported by the air forces.

Early on July 23, 1948, Captain Clarence Chiles and his co-pilot John Whitted saw a craft from their Eastern Airlines DC-3 over Montgomery, Alabama. A cigar-like projectile was heading for the Dakota from the north east.

Chiles swung his plane to the left, and as the UFO passed it 200 yards away, he noted two rows of portholes emitting an uncanny light along the side of the metallic, wingless shape. 'There was a deep blue glow on the underside of the craft, and a 15-yard trail of orange-red flame,' the pilot reported.

The object stopped when it drew level with the plane, then shot upwards at great speed. The Dakota wobbled, as if caught in the blast. Chiles later found one passenger who had not been sleeping and had seen the 'great streak of light'.

Six years later, the crew and passengers of the BOAC stratocruiser *Centaurius* watched an even better in-flight show. As the plane approached Goose Bay, Labrador, on June 29, 1954, en route from New York to Shannon and London, Captain James Howard noticed a large dark object emerge from clouds four miles to his left, apparently flying parallel with him. It was surrounded by six smaller blobs.

Howard radioed ahead to Goose Bay, and two US F80 Sabre jets scrambled. What happened next was seen by the 11 crew and 19 passengers of the stratocruiser, and described later by investigator John Carnell.

He wrote: 'When 15 miles away, one of the fighter pilots radioed that he had the unknown objects and the airliner on his radar scope. At that instant the six smaller objects, which seemed like discs, moved into single file and appeared to enter the larger object, which then began to fade, disappearing as the fighter appeared overhead.'

Carnell, who described the mother ship as 'a large, shape-changing object, rather like a swarm of bees, but solid,' said the same formation was seen several times that year, over both America and Europe.

Keeping Tabs on Concorde

People living near London's Heathrow Airport claim to have seen UFOs keeping watch on the Anglo-French supersonic jet Concorde. Mrs Dee Godden, 65, of Chiswick, West London, said she first saw one in August 1979.

'A huge reddish ball of light appeared in the sky right in Concorde's flightpath,' she said. 'I thought there was going to be an almighty crash, but when Concorde reached the spot, it just flew straight through it. The shape looked as if it was keeping watch on the plane.

Her husband Ernest, 64, also saw the light. 'I was sceptical when my wife told me what she had seen,' he recalled. 'I looked out of the window of our flat, and saw a shimmering object. It stayed in the sky for about 17 minutes.'

At Heathrow, officials said: 'Nothing was picked up on radar, so we cannot explain the sighting.' But UFO researcher Barry Gooding said: 'It is quite possible that UFOs from another planet are keeping watch on technological advances such as Concorde.'

Unchartered Activities

A dream trip to a sunshine island turned into a nightmare flight for 109 German and Austrian tourists when UFOs took too close an interest in their charter jet in November 1979.

Captain Javier Lerdo-Tejeda, 34, a pilot with 15 years flying experience, was at the controls as the Caravelle took of at 9.30 am from the Mediterranean island of Majorca, bound for the Canary Islands. But soon after levelling out, he noticed two very bright red lights in the sky.

437

'I was intrigued because they seemed to be flying in formation,' said Captain Lerdo-Tejeda. 'They were moving abreast at a slight angle to me, but getting closer all the time. They were about 15 miles away when we were at 23,000 feet, but only half a mile off when we reached 28,000 feet. Soon I realised they were almost on a collision course – they were virtually on top of me.'

The pilot ordered his passengers and six crew to put their seat belts back on, and radioed ahead to Barcelona control tower. He was told there were no aircraft in his flight path, and nothing on the radar screens.

'I decided to call in help from the Spanish air force and the Madrid radar station,' said Captain Lerdo-Tejeda. 'The equipment there is more sensitive than that used for civilian traffic, and they had picked up two objects which seemed to be very close to my plane.

'I swung my aircraft away sharply from the red lights and began descending at 5,000 feet a minute to 15,000 feet – an extremely steep dive for the passengers. Madrid was still monitoring the UFOs, and said the objects suddenly dropped 12,000 feet in just 30 seconds, following me. I know of no aircraft which is capable of doing such a thing.'

He continued to take evasive action, trying in vain to shake off the two shadows. Then, 30 miles out to sea off Valencia on Spain's south-west coast, an air force Mirage fighter arrived. The pilot instantly spotted the two glowing red shapes, apparently chasing the airliner. But seconds after the fighter jet zoomed into sight, the lights suddenly vanished.

A shaken Captain Lerdo-Tejeda swung back to Valencia for an unscheduled stop, and filed a full report of his dramatic encounter. 'I have never known such danger, and I have been flying for nearly half my life,' he told stunned officials. His crew backed his account in separate interviews.

Spain's Transport and Communications Minister, Sanchez Teran, was in Valencia at the time, and spoke to Captain Lerdo-Tejeda. He said later: 'I am now prepared to believe that unidentified flying objects do exist.'

Chichester and the UFO

Probably the first air-to-air sighting of a UFO was by Francis Chichester, later to earn fame as a round-the-world yachtsman. In 1931, he was piloting a tiny plane from Australia to New Zealand when a strange airship appeared, a dull grey-white colour with brightly flashing lights. The disc followed him for some miles across the Tasman Sea, occasionally vanishing behind clouds, before accelerating out of sight.

They Never Returned

Have UFOs caused civilian planes to crash? In 1953, the pilot of a DC-6 airliner flying from Wake Island in the Pacific to Los Angeles reported UFOs approaching before his radio went dead. Searchers later found wreckage and 20 bodies. And over Michigan, as reported elsewhere in this book, witnesses saw a curious ball of light in the sky on the night a DC-4 crashed, killing 58 people.

Authors Kevin Killey and Gary Lester used the disappearance Frederick Valentich over Australia in October 1978 (also reported earlier in the book) as evidence of their claim, in 1981, that the Bass Strait was another Bermuda Triangle. They said a new four-engine plane carrying a crew of two and ten passengers had vanished there in 1932, and in 1979 a racing sloop and her crew of five disappeared without trace. They renamed the waters between Melbourne and Tasmania the Devil's Meridian.

'Flying Saucers,
the Size of Battleships ....'

Three UFOs were also spotted over the Iberian peninsula by the crew and 100 passengers of a British Airways Trident. Captain Denis Wood saw them as he flew to Faro, Portugal – and again as he made the return flight to London later the same day. It happened over the Portuguese west coast on July 30, 1976. Captain Wood, 42, from Haslemere, Surrey, was told by air traffic controllers that an unidentified flying object had been reported in the area. He scanned the skies, and saw a bright object like nothing he had seen before in 20 years of flying. 'It was not a satellite, weather balloon or a star,' he said later.

As he invited the passengers to look at the UFO, two more objects appeared in the night sky. 'They were cigar-shaped, and appeared to come from nowhere,' said First Officer Colin Thomas, 38, from Camberley, Surrey. 'They took up positions to the right and below the first object. It was just after 8 pm, and I could see them clearly for eight minutes. They did not move.' Thomas had served 12 years as an RAF fighter pilot, and had flown with British Airways for seven years, but he too had never seen anything like the UFOs.

After dropping the 100 holiday-makers at Faro, Captain Wood, Flight Officer Thomas and the third crewman, Stephen Sowerby, of Richmond on Thames, set out at once for home. As they flew through the area where they had seen the UFOs, Captain Wood switched on his radar scanner and tilted it towards the spot where the shapes had been. They were still there.

'The two cigar-shaped objects were exactly where they had been,' said Captain Wood. 'We got to within seven miles of them, then they just disappeared off the side of the screen.'

The crew described the UFOs later as 'flying saucers, the size of battleships'. But it was ten months before they told the world about them. 'We were afraid people would ridicule us,' said one of them.

Some people did just that after they announced their sighting. The Science Research Council, in London, said the main 'UFO' was probably a giant research balloon, on its way from Sicily to America. Rays of the setting sun would have caught the plastic fabric, making it appear brilliantly lit. And the secondary UFOs were probably either ballasts being thrown overboard as the gas of the balloon cooled, or clouds of fine steel shot used to measure the wind.

The Greek Cover-up

Eminent Greek scientist Paul Santorini stunned members of his country's astronautical society in February 1967 when he announced that there was 'a world blanket of secrecy' about UFO activities – because the authorities did not want to admit the existence of forces against which Earth had 'no possibility of defence'.

Professor Santorini, then over 70 and the most respected scientist in Greece, revealed that in 1947 the Greek army had called him in to lead a team of engineers to investigate what were thought to be Russian missiles flying over the country.

'We soon established that they were not missiles,' he said. 'But before we could do any more, the army, after conferring with foreign officials, ordered the investigation to be stopped. Foreign scientists flew to Greece for secret talks with me.'

Professor Santorini added that he had no doubt aliens were 'visiting Earth to collect plant and animal specimens', but he would not guess why.

Watchers on the Moon?

A top American space consultant claims that two UFOs were watching when Neil Armstrong took his 'one small step for a man, one giant leap for mankind' by walking on the Moon's surface on July 20, 1960.

The astronaut spotted them on the rim of a nearby crater as he stepped out of his Apollo 11 spacecraft, according to Maurice Chatelain, who had left the National Aeronautics and Space Administration team by the time he made the claim in September, 1979.

While Armstrong was reporting his sighting to Houston control, Chatelain said, co-pilot Buzz Aldrin filmed the alien craft from inside Apollo.

But, alleged Chatelain, NASA ordered a cover-up of the incident. Mission controllers blacked out Armstrong's radio report from worldwide broadcasts of the historic event 'for security reasons'.

NASA dismissed the story as absolutely ridiculous'. Chief spokesman John McLealsh said: 'The only breaks in transmission from Apollo 11 occurred when it was on the other side of the Moon. The only conversations we have never made public were private talks between the astronauts and doctors.'

Chatelain's story received unexpected backing – from Moscow. Physicist Dr Vladimir Azhazha said: 'We heard about this episode two years ago. I am certain it took place, but it was censored by NASA.'

Soviet space expert Professor Sergei Boshich added: 'It is my opinion that beings from another civilization picked up radio signals from Earth and spied on the Apollo landing to learn the extent of our knowledge. Then they took off without making contact.'

Other American astronauts have had close encounters with strange craft. In 1953, Gordon Cooper, later to join the NASA programme, saw a UFO while piloting a plane over Germany. He said: 'I now firmly believe in extraterrestrial craft.'

In 1965 James McDivitt and Ed White were orbiting Earth 100 miles up in Gemini 4 when they spotted a silver cylinder with protruding antennae. McDivitt started taking pictures of it, but then the two men had to prepare for evasive action as the UFO moved closer. Just when a collision seemed inevitable, the curious craft vanished. Mission control at Houston dismissed the shape as one of Gemini's booster rockets, in orbit alongside the ship. But McDivitt said: 'It was in the wrong place at the wrong time for that.'

Eight years later, astronauts Jack Lousma, Owen Garriot and Alan Bean saw a rotating red shape from Skylab 2. They spent ten minutes photographing it, 270 miles above Earth. Again NASA denied that the shining capsule was another spacecraft.

Gordon Cooper said: 'NASA and the Government know very well that intelligent beings from other planets regularly visit our world to enter into discreet contact and observe us.

'They have an enormous amount of evidence, but have kept quiet in order not to alarm people.'

'They Want to Help Us'

Psychic Greta Woodrew, of Connecticut, claimed she was contacted by aliens, and told that they are waiting to help Earth cope with future catastrophes.

She said she met beings from a planet called Ogatta, many light years away, during experiments at the Ossining, New York, laboratories of parapsychologist Dr Andrija Puharich.

The first contact came in December, 1976. Mrs Woodrew was put into a deep hypnotic trance. She claimed she found herself in a long shadowy tunnel being guarded by a man-like creature called Hshames and two bird-like 'entities'. Hshames stood just over five feet tall, and his skin was covered with minute feathers. He had large, gold-flecked, luminous, lashless eyes, and his upper lip resembled a beak. They conversed by telepathy, and the figure told her about Ogatta.

During the second experiment, Mrs Woodrew claimed that her soul left her body, and she was transported to Ogatta itself. Everything shone, and the surface was covered with dots like glistening halves of marbles. They held a precious water-like substance.

At the next session, according to Mrs Woodrew, an entity called Ogatta spoke to her. 'He said beings had set up a way-station on the minor planet Vesta in our solar system, which will be used to help Earth. An armada of spacecraft, called gattae, would come down to Earth after drastic changes occurred. Their preparations were well under way.'

Mrs Woodrew claimed she was then shown scenes of devastation which could happen to Earth in the next few decades. Floods, hurricanes, super-magnetic storms, droughts, earthquakes, volcanic eruptions, tidal waves that covered entire cities, and people dying of thirst and hunger.

'I was told by the extra-terrestrials that they were survivors of what could come,' Mrs Woodrow said. 'Then they said, "Despite what man can do to man and nature's plan, there are civilisations in the cosmos who believe planet Earth is worth helping."'

'Only When We Believe'

Aliens will not reveal their mission on Earth until enough people accept the reality of UFOs, and man can understand them on a technical and scientific basis. That is the verdict of Dr Harley Rutledge after a seven-year study of the subject.

He claims UFOs zip round the Earth constantly, reading our minds and listening to our conversations. But they usually travel so fast that we cannot see them. And they only appear when they want to attract our attention.

Dr Rutledge, chairman of the physics department of the Southwestern Missouri State University, said he was a sceptic when the study began in 1973. He and nearly 500 helpers spent 2,000 hours studying the sky over three Missouri towns, Cape Giradeau, Piedmont and Farmington, and reported 157 sightings of 178 UFOs. In 16 cases observers noted UFO reaction to the team's movements, voices, radio signals and thoughts.

'We sensed that we were dealing with an intelligence,' said Dr Rutledge. 'I felt as though something was toying with us. On one occasion we deliberately changed our viewing position and moved 110 miles to the west to get directly into the path of UFOs we had been observing. The UFOs changed direction to go round us, just as they had done before.'

He added: 'I suspect their game is gradually to create general acceptance by repeated appearances. More UFO flaps will occur from location to location, winning converts.

'When we understand them, and when most of the world's inhabitants accept the reality of UFOs, then we will meet them face to face and know why they are here.'

'They Are Already Here'

Many people in Spain believe that aliens from space are already living on Earth. For more than 30 years, a group who call themselves Ummo, have been sending papers through the post and holding long late-night telephone conversations with people all over the country, alleging that they landed from a spacecraft in France in 1950 to help mankind reach maturity. They claim to come from the planet of Ummo which, they say, orbits the star known on Earth maps of the universe as Wolf 424.

All communications from Ummo have a thumbprint seal with a curious symbol, three horizontal lines crossing one vertical one. In May 1967, members of a Spanish space flight discussion group received invitations bearing the symbol. They were to gather on June 1 at a café in Santa Monica, near Madrid, for evidence of Ummo's existence.

They were there at the appointed time, and, sure enough, an object looking like a flying saucer arrived, the Ummo symbol on its underside. It performed aerial antics over the Madrid suburb of San José de Valderas before landing briefly in view of the café. Many witnesses took photographs of the strange craft before it flew away again. No one has since been able to trace Ummo to find out whether they really are aliens – or just very clever hoaxers.

Encounters
of the
Ghostly Kind

The Bloodstone Ring

Gale force winds lashed the tiny English village of Willisham, ripping slates from the roofs and tearing limbs from trees. A huge old oak shuddered before the onslaught and then, caught by one mighty gust, toppled, its roots tearing at the earth beneath.

Villagers who rushed to the spot to see if anyone was hurt stopped in horror as they gazed between the gnarled roots. There lay some human remains.

Police Constable Klug, the only bobby in the East Anglian community, was called and he ordered that the body be taken from its strange grave. One of the dismembered hands had a ring on one finger. Acting on a hunch, the grim-faced constable carried the hand to Ellen Grey, sister of a girl who had vanished mysteriously 18 years before, in 1873. Ellen screamed and then hugged the ghastly relic to her breast.

'It's Mary's,' she sobbed. 'The bloodstone ring was my wedding gift. She was born in March, and it was her birthstone.'

Klug understood. Though the case was before his time, it was so well known in the area it had been the subject of a popular ballad.

On her 18th birthday, Mary Grey had married Basil Osborne. She had written a letter to John Bodneys, her sweetheart since childhood, asking for his forgiveness.

An hour before the groom was to take her away on the honeymoon, Mary told her sister she wanted to spend a little time alone in the upstairs room they had shared. When Osborne arrived with the carriage, she still hadn't come down. Frightened, they forced their way into the locked bedroom, but found no trace of the bride.

One window opened onto a balcony where a flight of

steps led to an enclosed garden. But the garden, too, was empty.

The abandoned bridegroom died a month later. The villagers blamed a broken heart.

Now, 18 years later, the village knew what had become of Mary – for the skeleton had a broken neck! Ellen refused to give up her murdered sister's hand. It had been brought to her for a purpose, she said. That purpose must be fulfilled.

Dying, she left a bizarre provision in her will. Her housekeeper Maggie Williams was to have her estate, but must display the hand in some public place 'where it may some day confront the murderer'.

Maggie opened what became the finest pub in Willisham and gave the hand a place of honour on one wall. Enclosed in glass against a black velvet background, the bony ringed fingers claimed the attention of everyone.

After the shock of the exhibit had worn away, the tale of Mary's murder was a frequent topic of conversation. On a dismal March night in 1895, a stranger sat listening to scraps of the talk.

'Must have been just such a night as this that the wind ripped out that old oak tree,' said the publican.

The stranger, a brooding man with a ravaged face, looked up from his glass. 'I don't understand. What oak tree?' he asked.

'Have a look at the case on the wall and then we'll tell you the story,' the barman told him.

Moments later, the stranger was screaming. He sagged against the wall, blood dripping from his fingers. An older man at the bar recognised him as Mary's missing former sweetheart John Bodneys.

When Constable Klug arrived, the bleeding man confessed to the murder of Mary Grey. In a frenzy of jealousy, he had found the bride alone in her room. Muffling her cries, he carried her from the house.

Bodneys insisted that he had not meant to kill her. But

when they reached the big oak tree, she was struggling so hard he had broken her neck.

He left her in a shallow grave under the oak and tried to put Willisham behind him forever. But there had never been a moment of peace since the crime, and inevitably he had been compelled to return.

Committed to the local jail to await trial, he died 'of no known disease' before his trial could be held. The authorities dismissed the old wives' tale that a murderer's hands sometimes drip blood when he faces the proof of his crime. But the people of the village knew what they had seen.

They buried Mary Grey's hand with the rest of her skeleton – and then ceremoniously burned the shirt smeared by John Bodneys' bloody fingers the day he came face to face with his guilt.

The Fiend in Bandages

Only four men escaped when the British square-rigged yacht *Pierrot* capsized in the Atlantic in July 1884. Huddled in a battered dinghy, they drifted for 25 days. Near death from starvation and exposure, Captain Edwin Rutt then made a last desperate suggestion.

Lots should be drawn to determine which of the four would be eaten.

Two of the sailors agreed with Rutt, but 18-year-old Dick Tomlin, the youngest crewman, protested that he would rather die than eat human flesh.

Tomlin's resistance sealed his fate. At the first opportunity Rutt crept toward the sleeping boy and drove a knife into his neck.

The mate Josh Dudley and seaman Will Hoon had no reservations about cannibalism. When they were rescued by the yacht *Gellert* four days later, it was the slain boy's flesh that had sustained them.

The horror-stricken master of the *Gellert* rejected the idea of burial at sea. Hidden away underneath a tarpaulin, the body of the victim accompanied the three survivors to the Cornish port of Falmouth.

All three were tried and condemned to death for murder on the high seas. But the Home Secretary decided that there had been horror enough and commuted the sentence to six months' imprisonment.

No one could have known that the horrors were only beginning.

When the three men were freed from jail, they found little fortune. To keep body and soul together, Josh Dudley found work as a drayman. Two weeks later his team of horses saw something that frightened them in the middle of a foggy London street. Bolting, they tossed Dudley to the cobblestones where his head shattered.

Witnesses said the thing in the fog had been a figure swathed from head to foot in bloodstained bandages. After Dudley's death, the figure mysteriously vanished.

With fear beginning to take root, Captain Rutt went to the Soho slums and sought out Will Hoon. He found the old seaman far gone in drink, a sodden derelict in desperately bad health.

Rutt told Hoon that some vengeance-crazed relative was masquerading as Dick Tomlin's ghost, and he urged Hoon to help him ferret out the plotter. But Hoon wanted only more gin, and in a last delirium, he was taken to the charity ward of a hospital where he died in a screaming fit.

Witnesses said later that another patient 'dressed all in bandages' had been holding Hoon down, apparently trying to soothe him. Then the patient vanished.

Now in a state of abject terror, Rutt went to the police. They scoffed at his tales of a 'figure in bandages'. But in view of the captain's mental condition, they offered him one night of lodging in a cell.

Rutt went gratefully to the cell, checking twice to be sure he was locked in. It was a cell block for the disturbed of London, and screams in the night were not uncommon.

But when at 3 a.m. the police heard the captain, some distinctive quality in his cries brought wardens running. They unlocked the door and went to his bunk, where Rutt lay with his knees scissored upward and his dead eyes like marbles.

Clenched in his fingers the shocked bobbies saw shreds of cotton. And bloodstained gauze.

Princess of Death

Egyptologist Douglas Murray neither liked nor trusted the dishevelled American who sought him out in Cairo in 1910. The man had a furtive manner and appeared to be in the final stages of disease. But Murray, a refined Briton, could not resist the blandishments of his disreputable visitor – for the American was offering him the most priceless find of his career.

It was the mummy-case of a high princess in the temple of Ammon-Ra, who was supposed to have lived in Thebes in 1600 BC. The outside of the case bore the image of the princess, exquisitely worked in enamel and gold. The case was in an excellent state of preservation.

An avid collector, Murray couldn't resist. He drew a cheque on the Bank of England and took immediate steps to have the mummy-case shipped to his London home. The cheque was never cashed. The American died that evening. Murray learned from another Egyptologist in Cairo why the price had been so reasonable.

The princess from Amman-Ra had held high office in the powerful Cult of the Dead, which had turned the fertile Valley of the Nile into a place only of death. Inscribed on the walls of her death chamber she had left a legacy of misfortune and terror for anybody who despoiled her resting place.

Murray scoffed at the superstition until three days later. That was when he went on a shooting expedition up the Nile and the gun he was carrying exploded mysteriously in his hand. After weeks of agony in hospital, his arm had to be amputated above the elbow.

On the return voyage to England, two of Murray's friends died 'from unknown causes'. Two Egyptian servants who had handled the mummy-case also died within a year.

Back in London, Murray found that the mummy-case had arrived. When he looked at it, the carved face of the princess 'seemed to come alive with a stare that chilled the blood'.

Although he had made up his mind to get rid of it, a woman friend convinced him that he should give it to her.

Within weeks, the woman's mother died, her lover deserted her, and she was stricken with an undiagnosed 'wasting disease'. When she instructed her lawyer to make her will, he insisted on returning the mummy-case to Douglas Murray.

By now a broken wreck of a man, Murray wanted no part of it. He presented it to the British Museum, but even in that cold and scientific institution, the mummy-case was to become notorious. A photographer who took pictures of it immediately dropped dead. An Egyptologist in charge of the exhibit was also found dead in his bed.

Disturbed by the newspaper stories, the board of the museum met in secret. There was a unanimous vote to ship the mummy-case to a New York museum, which had agreed to accept the gift provided it was handled without publicity and sent by the safest possible means.

The case must be shipped by the prestigious new vessel making her maiden voyage from Southampton to New York that month. All arrangements were successfully completed. But the mummy-case never reached New York. It was in the cargo hold of the 'unsinkable' *Titanic* when she carried 1,498 people to their doom on April 15, 1912.

The Witch's Mark

The year was 1692, and 13 people had been hanged for witchcraft in Salem, Massachusetts. It was a matter of concern to Colonel Bucks, of Bucksport, Maine, that his own village should be just as vigilant in stamping out witches. He raised the question repeatedly at town hall meetings. Shortly his one-man crusade produced a victim.

There was a public accusation of a bent and withered old lady who looked every inch a witch. Historians disagree as to her name and age, but one of them calls her Comfort Ainsworth and is sure she was more than 90 years old.

Because of her obvious frailty, the old lady went on trial without torture or pricking with needles to find 'the witch's mark'. But the crowds that surged into the courtroom knew her guilt had been predetermined by Colonel Bucks himself, who sat within whispering distance of the magistrate. When witnesses took the stand against her, without exception they looked to Bucks for approval.

One woman said she had heard the old woman muttering something that sounded gibberish. But when she reached home and her ears started bleeding, she knew it had been a curse.

A man swore he'd seen a black-garbed figure ten feet tall – obviously the devil or one of his henchmen – standing in Comfort's doorway.

The jury quickly returned a verdict of guilty. Quoting the text, 'Thou shalt not suffer a witch to live,' the judge denounced Comfort Ainsworth and sentenced her to hang next day.

No one was prepared for the scene that followed. Because she had not been permitted to testify in her own defence, people had assumed that the toothless old woman would remain mute.

Before the bailiffs could stop her, she got to her feet and pointed a bony finger at the colonel. 'In all of my life,' she screamed, 'I have cursed no other being! But I am capable of laying a curse on you sir, because you and your toadies have lied me to the gallows!

'Then mark you this, and mark it well – when you go to your grave, which will be soon, I pledge you I shall leave the print of my foot on your gravestone. And the print, Colonel Bucks, will be there forever so that the world can never forget this day!'

A bailiff clapped his palm to her mouth and carried her from the courtroom. But her words left the village uneasy, and there were few spectators when she was hanged next morning. Even Colonel Bucks failed to appear.

Three months later, he died from a 'wasting disease', and the colonel's heirs found he had written a new instruction into his will. His headstone must be of the most flawless marble, 'incapable of being stained or besmirched'.

But in a few days the relatives were secretly approached by a terrified cemetery worker. He had found a woman's footprint in the marble, and no amount of sanding could remove it.

A new stonecutter was sworn to absolute secrecy. Working in the dead of night, he cut a marker that was an absolute replica of the first. The old stone was buried secretly and the new one raised.

Ten days later, the heirs saw crowds of frightened people moving in and out of the cemetery. Joining them, they found that the trick had failed. The shape of an old woman's narrow stockinged foot was clearly visible in the new stone.

Publicly deploring the phenomenon as an act of 'grave-yard vandalism' – an explanation that convinced no one – the heirs had a still more costly headstone hauled to the cemetery. It was raised with no attempt at secrecy. Incredibly, the print of Comfort's bony little foot soon began to take shape in the stone.

This time his discouraged heirs made no effort to replace the marker. Nearly three centuries later it stands over the grave of Colonel Bucks, the footprint still scarring its surface like a wound that will never heal.

The Skulls of Calgarth

Wealthy Myles Phillipson owned huge tracts of the pictur-esque English Lake District countryside around Winder-mere, during the 16th century. But he was never satisfied with the extent of his empire, always restlessly seeking new acres to add to his estates. He eyed the small farm of Kraster and Dorothy Cook, which overlooked the lake. And he decided that their humble plot of land would be the ideal site for the new luxurious mansion he planned.

The farm was all the Cooks had in the world, and they were not prepared to sell when Phillipson made his offer. He was not a man to take no for an answer. He invited the poor couple to share Christmas dinner with him and his family. The Cooks were awed by the foods and wines, and overjoyed when the landowner said they could keep a golden bowl they admired.

Next morning, soldiers hammered at the door of their home, and arrested them. For a week they were held in separate cells, with no idea why they were being impris-oned. Only when they arrived in court did they learn of their 'crime' – stealing a golden bowl from Myles Phillipson.

The verdict was a foregone conclusion, because the

magistrate hearing the case was Phillipson himself. When he sentenced them both to death, Dorothy Cook cried out: 'Look out for yourself, you will never prosper. The time will come when you own no land. You will never be rid of us...'

Phillipson was not worried by the threats. The couple were hustled to the gallows, and strung up to die.

Within days, Phillipson had acquired their land and started work on his magnificent new home, called Calgarth Hall. When it was finished, he held a lavish Christmas feast to celebrate. Friends and neighbours joined him round the table, making merry with no expense spared.

Then a terrifying scream sent them rushing upstairs, swords at the ready.

Phillipson's wife stood halfway up the staircase, shuddering as she stared transfixed at a hideous sight on the banister – two grinning skulls. The landowner seized them, threw them into the courtyard, and swore revenge on whoever had perpetrated the tasteless joke.

But his threats failed to put the minds of his guests at rest. Several shuffled off to bed early – only to be woken in the small hours by more screams. The skulls were back on the stairs.

Over the next few days, Phillipson tried everything he knew to get rid of them. But each time they were thrown outside or buried, the skulls returned to haunt the home.

Christmas was ruined. And as the news spread, so was Phillipson. His business declined, his riches dwindled. When he died, a broken man, his beautiful home rang all night with the demonic laughter of the skulls.

The two gruesome relics continued to visit the hall, giving the landowner's heirs no rest. They appeared each Christmas Day, and on the anniversary of the Cooks' execution. Only when the family became too poor to maintain Calgarth, and were forced to sell it, did the skulls leave the building in peace.

Possessed!

Terry Palmer set out to find the resting place of the last witch in England to be burned at the stake. But his quest took him along an unnerving path – with his own body taken over by the witch he sought.

The spirit said she would be with Terry for all time, no matter where he went.

England's last witch, whose name was Elsa, was tortured and burned to death on the old village green of Dedham, Essex, in 1763. More than 200 years later Terry set out to find her burial place. But everywhere he went, inexplicable happenings took place.

He once visited a former convent where a terrier dog ran out to greet him, barking and wagging its tail. Then it ran back to its master, before bounding back – right past Terry. It was barking and leaping at an empty space.

A few days later in a shop, another dog barked twice at Terry . . . and then twice at the empty space behind him.

The extraordinary tale of Terry, a good, book publisher, began when he went to a seance on his witch-hunting trail. There, says Terry, Elsa herself took over his body and joined in the ghost hunt.

Shortly afterwards, his father became possessed by evil spirits, and a fire broke out in his factory causing thousands of pounds of damage.

Terry claimed to have found Elsa's grave, near an hotel not far from the tiny village where she was executed. He stood on the spot and felt a tingling sensation from the back of his head to the middle of his spine. But when he and a friend dug down they found nothing.

Terry's story was told in the hotel and was treated with scepticism – until one day when a barman was having lunch while keeping an eye on the empty bar through a mirror. He saw a woman standing in the room and went

459

to serve her. But the bar was empty, and all the doors were locked.

Skyway to Doom

Has the ghost of an American construction worker put a curse on the Sunshine Skyway Bridge? That was the theory put forward by a Florida fisherman after nearly 60 people died in four separate shipping disasters at the Tampa Bay bridge during the first five months of 1980.

In January, 23 coastguards were killed when their cutter collided with an oil tanker. The following month, a freighter smashed into one of the main bridge supports, and ten days later a tanker ran out of control, and slammed into the main span.

But the worst accident to hit the jinxed bridge came on May 9. Possibly blinded by the wind-lashed rain of a violent storm, the skipper of a 10,000-ton Liberian freighter, *Summit Venture*, misjudged his approach to the bridge. Instead of passing under the middle of it, the ship ploughed into one of the main supports, and a huge section of the road running over it collapsed.

Cars, trucks, and a Greyhound bus plunged 150 feet into the water; 32 people died, 23 of them on the bus. Other drivers missed death by inches, slamming on their brakes just in time as the yawning gap opened up in front of them.

The Florida House of Representatives stood for a moment's silence as news of the tragedy reached them during a meeting. One member blamed the Tampa Bay harbour pilot system, and called for an inquiry.

But a local fisherman, 27-year-old Charlie Williams, said later: 'When the bridge was being built, a construction worker fell into some wet concrete. He's still there, in the structure. The Skyway has been cursed ever since.'

The bridge, which is four miles long, was opened in September 1954. More than 40 people have committed suicide by leaping from it.

House of Evil

Actor James Brolin is certain there was an evil jinx on the film *The Amityville Horror,* in which he starred. He played surveyor George Lutz who, with his family, was driven from his home by a terrifying series of demonic happenings. The film was based on the best-selling book by Jay Anson, to whom the Lutz family told their nightmare story.

Brolin said: 'On the first day of filming I stepped into the elevator in my apartment block and pressed the button for the lobby floor. Before we'd gone three floors it shuddered to a grinding, screeching halt, the lights flickered and I was plunged into frightening darkness. I screamed for help but nobody could hear me.

'It was an eerie, frightening experience. You imagine all sorts of hair-raising things in the silent darkness. My pleas bounced back like an echo. Those 30 minutes seemed an eternity.'

The jinx hit again the next morning. 'I'd been on the set less than one minute when I tripped over a cable and severely wrenched my ankle,' said Brolin. 'I hobbled around in pain for days.'

The film recorded the horrifying events experienced by George and Kathleen Lutz and their three children after they moved to Long Island, New York, to a house which had been the scene of a multiple murder in 1974.

Ronald Defoe, 23-year-old son of a wealthy car dealer, had drugged his parents, brothers and sisters at supper and at 3.15 a.m., he stalked from room to room shooting each victim in the back with a rifle.

He claimed in court that 'voices' had ordered him to commit the crime. Defoe was sentenced to six consecutive life sentences.

For the Lutzes, the house's macabre history gave them the chance to buy a dream home at the bargain price of $80,000. Seen in the bright light of day, it was a beautiful, three-storey colonial-style residence, set on a well-kept lawn which sloped gently down to the bay, and its own boathouse. In the small, middle-class community of Amityville it was a showplace.

Soon after the family moved in they asked the local priest, Father Mancuso (played in the film by Rod Steiger) to bless the house. Author Anson wrote: 'The priest entered the house to begin his ritual. When he flicked the first holy water and uttered the words that accompany the gesture, Father Mancuso heard a masculine voice say with terrible clarity, 'Get out!'

'He looked up in shock, but he was alone in the room. Who or whatever had spoken was nowhere to be seen.'

For the first two nights in their new home, the Lutzes were awakened by strange noises at 3.15 a.m. But the real horror began on the third night.

As usual, George Lutz checked that all doors and windows were locked before going to bed. The noises roused him again at 3.15, and this time he went downstairs to investigate.

He could not believe what he saw. The heavy, solid-wood front door had been wrenched open and was hanging by one hinge. With mounting terror he realised it had been forced from inside the house. The thick steel doorknob spindle was twisted, and the surrounding metal plate had been forced outwards.

From then on, the house seemed to have an evil life of its own. Windows opened and closed at will and a bannister was wrenched from the staircase.

Two weeks after the front-door incident, George woke

462

in the night to find his wife Kathleen floating above the bed. George pulled Kathleen down by her hair and switched on the light. He was looking not at his attractive young wife, but at a hideous vision.

Kathleen caught a glimpse of her reflection in a mirror and screamed: 'That's not me. It can't be me!' Her appearance changed slowly back to normal over the next six hours.

A few nights later Kathleen was in the sitting room with George when she looked up and saw two glowing red eyes at the darkened window. She and George hurried outside and found strange tracks in the snow. Kathleen told Anson: 'The prints had been left by cloven hooves – like those of an enormous pig.'

After only 28 days the Lutzes fled the dream house that had become a nightmare.

As they hurriedly gathered a few belongings, amid a series of unearthly noises, green slime oozed from the walls and ceiling and a sticky black substance dripped from the keyholes.

Because of the curse, the film men dared not use the actual house. They found an almost identical building in New Jersey. They knew only too well of the frightening things that had happened to people connected with the story.

A photographer went to take pictures of Anson immediately after photographing the Amityville house. While he was in the author's home, his car caught fire and billowed orange smoke as it stood empty with the engine switched off.

Anson himself told of terrifying events linked with his book. He said: 'A woman to whom I loaned some early chapters took the manuscript home. She and two of her children were suffocated in a fire that night. The only item in the apartment that was not damaged by the fire was the manuscript.

'Another man put the manuscript in the trunk of his car and attempted to drive home. He drove through what he thought was a puddle. It turned out to be a 12-foot-deep hole into which his car slid. When the car was fished out the next day, the only dry object in it was the manuscript.

'And when my editor picked up the completed manuscript at my office his car caught fire and he discovered that all the bolts on his engine had been loosened.'

Anson himself suffered a heart attack, and his son and friend were nearly killed in a car smash.

The Lutzes are today alive and well in California, and planning another book about their experiences. Their Long Island house of horrors is now owned by James and Barbara Cromarty.

They say the place is not haunted.

Whatever the truth, the movie *The Amityville Horror*, will remain a chillingly realistic record of paranormal events. Director Stuart Rosenberg says that he would not have taken on the project if it was just another horror film.

He insists, 'My first reaction was that it wouldn't be my cup of tea. But I read Jay Anson's book – and it had the ring of truth about it.'

Torment of Calvados

Diaries written by a French aristocrat who lived in a gloomy mediaeval castle set among the apple orchards of Normandy tell the story of one of the most violent hauntings ever recorded. Known simply as X, he recorded in vivid detail the extraordinary events that turned his historic home into a nightmare in the year 1875.

They began without warning. Everyone in Calvados Castle had settled down for the night when they were disturbed by ghostly wailing and weeping and rapping on the walls. The noises were heard by the entire household

– X himself, his wife, his son, his son's tutor who was an abbé, Emile the coachman, and servants Auguste, Amelina and Celina.

After several nights of ever-increasing noise and disturbance, the aristocrat instructed that fine threads were to be strung across every entrance to the castle. He hoped, of course, that in the morning they would be broken, proving that someone had entered and was trying to terrorise them.

But the threads remained intact. There was no escaping the fact that the forces existed within the castle walls.

On Wednesday, October 13, 1875, X began keeping a diary. That night the abbé was alone in his room when he heard a series of sharp taps on the wall and a candlestick on the mantlepiece was lifted by an unseen hand.

The terror-stricken priest rang for X who found that not only had the candlestick been moved, but also an armchair which was normally fixed to the floor.

For the next two days, pounding on the walls, footsteps on the stairs and other unnerving phenomena continued unabated. X and the abbé armed themselves with sticks and searched the castle from top to bottom. They could find no human explanation.

By October 31 the castle was hardly ever at peace. X recorded in his diary, 'A very disturbed night. It sounded as if someone went up the stairs with superhuman speed from the ground floor, stamping his feet.

'Arriving on the landing, he gave five heavy blows so strong that objects rattled in their places. Then it seemed as if a heavy anvil or a big log had been thrown at the wall so as to shake the house.

'Nobody could say where the blows came from, but everyone got up and assembled in the hall. The house only settled down at about three in the morning . . .'

The following night everyone was awakened by what sounded like a heavy body rolling downstairs followed by blows so ferocious they seemed to rock the castle. Over the next few days, the haunting had become so violent the family felt it could not possibly get any worse. But greater ordeals were to come.

On the night of November 10, X wrote in his diary, 'Everyone heard a long shriek and then another of a woman outside calling for help. At 1.45 we suddenly heard three or four loud cries in the hall and then on the staircase.'

Cries, screams and moans which 'sounded like the cries of the damned' seemed to fill the whole castle. Heavy furniture was moved, windows flung open and – more terrifying – Bibles were torn and desecrated. The family began to wonder if the powers of darkness had taken over.

X's wife suddenly became the focus of attention. Hearing a noise in the abbé's room, she crept up the stairs and put out a hand to press down the latch on the door. Before she could touch it she saw the key turn in the lock then remove itself, hitting her left hand with a sharp blow. The abbé, who had run up the stairs after her, saw it happen and afterwards testified that madame's hand was bruised for two days. That night something hammered on her door so furiously she thought it would break down.

The New Year brought only fresh terrors to the wretched family: louder knocking, more persistent voices. The worst day of all was January 26 when the noise was thunderous. 'It sounded as if demons were driving herds of wild cattle through the rooms.' Peals of demonic laughter rang through the ancient walls. The family had had enough.

Next day a priest was called in to exorcise the evil spirit and the family saw to it that every religious medallion and relic they possessed was placed in full view. The treatment was effective and at last the hideous uproar ceased.

To the family, who believed they would be forced to abandon their home, the peace that followed came as a blessed relief.

But the ghostly tormentors of Calvados had not quite

finished. Shortly after the exorcism, all the religious relics disappeared and could not be found. Then, one morning, as the lady of the house sat writing at her desk an unseen hand dropped them one by one in front of her. There was one short burst of violent sound, then silence.

The Haunting of Willington Mill

When Joseph Procter and his family moved into the mill house in 1835 they paid little attention to rumours that the place was haunted. The house was a pleasant, comparatively new building set by a tidal stream in the Northumberland village of Willington, in England's rugged north-east. The Procters were a highly-respected, devoutly Quaker family. Mr Procter was said to be a man of high intelligence and common sense, good and kind to his family and employees.

Yet after little more than a decade, the Procters were driven to leave in distress, unable to stand any more of the weird and ghostly happenings that plagued them from the day they first arrived at Willington Mill.

Only much later were they to learn that their home had been built on the site of an old cottage, that a terrible crime had been committed there years before . . . and that a priest had refused to hear the confession of a woman who desperately wanted to unburden her conscience.

So prolific was the haunting of Willington Mill while the Procters lived there that when W T Stead, the writer and ghost hunter, first pieced together the story in the 1890s, there were still 40 people alive who had actually seen the ghosts.

The hauntings began one night in January 1835. A nursemaid was putting the children to bed in the second-floor nursery when she heard heavy footsteps coming from a room immediately above. It was an empty room, never used by the family. At first the girl took little notice, thinking it must be one of the handymen with a job to do. But they went on night after night, getting louder and louder.

Other servants and members of the family also heard them, but when they burst into the room to surprise the 'intruder', no one was there. They sprinkled meal over the floor, but there were no footprints.

One morning, as Mr Procter was conducting family prayers, the heavy steps were heard coming down the stairs, past the parlour and along the hall to the front door. The family heard the bar removed, two bolts drawn back and the lock turned.

Mr Procter rushed into the hall to find the door open. The footsteps went on down the path.

Poor Mrs Procter fainted.

It became increasingly difficult to get servants to stay in the house. Only one girl, Mary Young, whom the family had brought with them from their previous home in North Shields, loyally refused to leave.

There was a period when it seemed as though the whole house had been taken over by unseen people. There were sounds of doors opening, people entering and leaving rooms, thumps and blows and laboured breathing, the steps of a child, chairs being moved and rustling sounds as if a woman in a silk dress was hurrying by.

Until a certain Whit Monday, the haunting remained entirely by sound. On that day, Mary Young was washing dishes in the kitchen when she heard footsteps in the passage. Looking up she saw a woman in a lavender silk dress go upstairs and enter one of the rooms. That night the noises in the house were worse than anybody had heard before.

Two of Mrs Procter's sisters arrived for a visit. The first night, sleeping together in the same four-poster bed, they felt it lift up. Their first thought was that a thief had hidden there, so they rang the alarm and the men of the house came running. No one was found.

On another night their bed was violently shaken and the curtains suddenly hoisted up then let down again several times.

They had the curtains removed, but the experience that followed was even more terrifying. They lay awake half-expecting something to happen when a misty, bluish figure of a woman drifted out of the wall and leaned over them in an almost horizontal position. Both women saw the figure quite clearly and lay there, speechless with terror, as it retreated and passed back into the wall.

Neither would sleep in the room another night and one of them even left the house to take lodgings with the mill foreman Thomas Mann and his wife.

One dark, moonless night the Manns, their daughter and their visitor were walking past the mill house after paying a call on neighbours. All four saw the luminous figure of what appeared to be a priest in a surplice gliding back and forth at the height of the second floor. It seemed to go through the wall of the house and stand looking out of the window.

The focus of the hauntings seemed to be what the Procters called The Blue Room and in the summer of 1840 they agreed to allow Edward Drury, who specialised in supernatural investigation, to spend a night there. He took with him a friend who refused to get into bed, but dozed off in a chair.

Drury later wrote a letter describing what happened. 'I took out my watch to ascertain the time and found that it wanted ten minutes to one,' he said. 'In taking my eyes off the watch they became riveted upon a closet door, which I distinctly saw open, and saw also the figure of a female attired in greyish garments, with the head inclining downwards and one hand pressed upon the chest as if in pain. It advanced with an apparently cautious step across the floor towards me. Immediately, it approached my friend, who was slumbering, its right hand extended towards him. I then rushed at it . . .'

It was three hours before Drury could recollect anything more. He had been carried downstairs in a state of terror by Mr Procter. Drury had shrieked, 'There she is. Keep her off. For God's sake, keep her off!'

The grey lady was seen by others. So were unearthly animals and other startling apparitions. The Procters tried to shield their children from the worst of the haunting, but eventually they became involved. One day a daughter told Mary Young, 'There's a lady sitting on the bed in mama's room. She has eyeholes, but no eyes, and she looked hard at me.'

Then another daughter reported that in the night a lady had come out of the wall and looked into the mirror . . . 'She had eyeholes, but no eyes.'

Another child saw the figure of a man enter his room, push up the sash window, lower it again, then leave.

In 1847 Joseph Procter decided his family could endure no more. They moved away to another part of Northumberland and were never again troubled by ghosts.

The house was later divided into two dwellings and eventually deteriorated into a slum. People continued to hear and see strange things from time to time. But Willington Mill House was never again to know the terrifying days and nights that afflicted the pious Quaker family.

Knock Twice for Terror

From the outside, it looked like any house. But the many people called in to investigate the strange happenings there knew differently. Journalists, psychic investigators, even the police, came to the same conclusion: the rented house in the North London suburb of Enfield, occupied by Mrs Peggy Hodgson and her four children, was haunted.

It all began in September 1977, when daughter Janet, then 11, heard a shuffling noise in her bedroom. It sounded like someone walking in loose-fitting slippers. Four loud knocks followed and Janet was horrified to see a heavy chest of drawers sliding away from the wall.

In the days that followed, other objects, including a heavy bed, began to move unaided. A hairbrush flew through the air hitting one of the sons on the head. A policewoman, who was called, saw a chair hurled across a room.

But fright turned to terror as the thing that haunted the house extended its powers and started to influence the children's behaviour. The girls, both in their early teens, spoke in coarse language with the voices of old men. But their lips did not move – the sounds just seemed to come from within them.

It also appeared that the children's lives could be in danger. Mrs Hodgson's nine-year-old son Billy escaped narrowly when a heavy iron grate flew across his room as he lay in bed. As Janet lay asleep, she would suddenly find herself hurled into the air to wake screaming.

On one occasion the strange force nearly killed Janet. As she lay in bed, a nearby curtain wrapped itself tight around her neck. Hearing her daughter's choking scream, Mrs Hodgson rushed to the room and fought to pull the material from the girl's neck.

The family considered moving away from their home, but for a divorcee with four children such a step was not easy. And the family feared that 'the thing' might follow them, for there had been strange voices and happenings when they were on holiday in a caravan at Clacton, Essex.

Many of those who heard of the family's plight were quick to dismiss it all as childish pranks, but not so the experts. Pye Electronics specialists who visited the house were baffled to find that video recording equipment which worked perfectly well outside would not function at all inside.

A policewoman who was called in admitted: 'I saw a chair lift into the air. It moved sideways and then floated back to its original position. I have been called to the house several times, but there isn't much the police can do.'

Physical researcher Maurice Grosse tried communicating with 'the thing' using a code of one knock for No, two knocks for Yes.

'Did you die in the house?'

Two knocks.

'How many years did you live here?'

Fifty-three knocks.

This was followed by a barrage of knocks. Bewildered, Grosse asked: 'Are you having a game with me?'

In answer, a cardboard box filled with cushions leaped off the floor hitting him on the head.

It was Jane's sister Margaret who shed some light on what was happening. One night when she was asleep, she began to bounce up and down in bed and cried: 'Go away, you ten little things.'

Still asleep, she gave details about them. They included a baby, three girls, two boys and an elderly couple one of whom she identified as Frank Watson 'the man who died in the chair downstairs.'

Then frightening, throaty growls began to come from Janet's direction, but investigators were convinced she could not have made the sounds herself.

One day the voice told researcher Grosse that its name was Joe, and on another day, Bill Hobbs. He said he came from Durant's Park graveyard.

Hobbs, whose voice was being taped, told them: 'I am 72 years old and I have come here to see my family. But they are not here now.'

The hauntings lasted three years, and then ended, never to resume.

The Spinster's Grave

Spinster Hannah Beswick died more than 200 years ago. Her body was embalmed, but her restless spirit still haunts a factory built on the land where her home once stood.

Hannah was a wealthy Lancashire landowner whose house, Birchen Bower, dominated acres of fertile land at Hollinwood, on the outskirts of Manchester. She was not normally fearful, but when in 1745 Bonnie Prince Charlie crossed the border into England and advanced south, she became so obsessed with the thought of the invading Scots that she hid all her money and valuables. They remained hidden for the rest of her life.

Apart from the Scots, her only real terror was that of being buried alive – a fear quite understandable in the light of what had happened in her own family. One of her brothers had fallen ill and, while unconscious, had been pronounced dead by a local doctor. Preparations were made for his funeral and he was laid in an open coffin so that friends and relatives could pay their last respects. While lying in his shroud surrounded by flowers he began to show signs of life. The unfortunate man was hastily removed to his bed.

Hannah died in 1768 without divulging where she had hidden her fortune. Because of what had happened to her brother, she took steps to ensure that her corpse was not buried.

She left Birchen Bower to young Doctor Charles White with the stipulation that he must have her embalmed and kept in a safe and respectable place above ground.

For some strange reason, she also insisted that every 21 years her body should be taken back to the house and allowed to lie in the granary for seven days.

Old Hannah was duly mummified, her body coated with tar and wrapped in heavy linen bandages. In accordance with her wishes, the face was left uncovered.

For many years Dr White faithfully kept the body at his home, Sale Priory, but when he died it was moved to Manchester Natural History Society's museum where it became a major attraction.

A century after Hannah Beswick was embalmed, the commissioners of the society, finding the museum over-filled with relics and needing room for new acquisitions, decided it was time she was given a proper burial. She was finally interred at Harpurhey Cemetery on July 22, 1868.

Some people had already claimed to have seen her ghost wandering through the rooms of Birchen Bower, dressed in her usual black silk gown and white lace cap. After burial the ghost became more agitated. Hannah was seen hurrying between the old barn and the pond as though deeply troubled. Sometimes, it was said, the old barn glowed as if on fire.

The house remained empty for some years, inhabited only by the spirit of Hannah Beswick. Then it was bought by a developer, renovated and converted into small dwellings to be rented out to cotton workers and labourers. The new tenants were often to see her drifting by, head bent as though in deep thought.

One particular aspect of her behaviour was puzzling. Sometimes she would disappear at a particular spot – a corner flagstone in the parlour of a house occupied by a handloom weaver. Hannah seemed to hover about this room as though reluctant to leave it.

The weaver decided to pull up part of the flagstone floor to make a place where he could install a new loom. To his amazement he found hidden underneath a hoard of gold. He had found part of Hannah Beswick's fortune.

After this, Hannah was seen frequently . . . no longer thoughtful but angry and menacing. People spoke of a brilliant blue light darting from her eyes. Sometimes at night she was seen near the pond and at other times strange unearthly noises were heard in the barn. No one would venture there after dark unless they had urgent business.

It gave rise to the speculation that perhaps the rest of her valuables were hidden nearby and she was determined to protect them.

The hauntings continued until Birchen Bower was demolished. A factory was built on the spot and that was thought to be the end of the affair.

Then, people who knew nothing of her story began saying that they had caught a glimpse of a strange old world figure in a black silk gown and white lace cap . . .

A Lingering Loss

Peter Turner has had more than his fair share of encounters with the supernatural. Some people have seen the same ghost on more than one occasion; but Peter Turner's experiences were years apart and completely unconnected.

The first happened in 1945 when Peter and a group of young friends were playing in a row of derelict houses in the Camp Hill district of Leeds, Yorkshire. The houses were awaiting demolition and provided a natural if dangerous, playground for the local youngsters in austere, post-war Britain.

It was a cold November day and Peter and his friends were playing in a house which had recently had its floorboards removed, leaving only rafters and beams.

As Peter walked across one of the upper-floor rafters he glanced out of the shell of a window. There below him was a neat little garden with an old man tending rose bushes in full bloom.

There was no rubble, no bricks, no broken glass, just a well-kept garden with grass and flowers. The fact that it was impossible for such a garden to exist there at that time, still less contain flowers in full bloom in the middle of winter, didn't occur to young Peter. All he could think about was not getting caught. He and his friends knew what to expect if their parents found out they had been playing in the derelict buildings.

They rushed off, counting their luck at not being spotted by the old man. It was only later that it occurred to Peter how impossible it was for the garden to have been there. He returned to the house but where he had seen roses and flowers there was just brick and rubble.

Eleven years passed and Peter was still living in Leeds and preparing for his wedding, just six months away. Homes were not easy to come by, so Peter and his fiancée Pamela were delighted when they got the top-floor apartment in a Georgian house at 10 Woodhouse Square, previously used as a nursery. One evening Peter and two friends had been decorating the rooms. Feeling hungry, two of them nipped out to get a fish-and-chip supper. On their return they found the third friend outside the flat, too terrified to go back in after hearing strange noises and feeling eerie sensations.

With the excitement of the wedding, the incident was soon forgotten. But no sooner had the newly-married couple moved in, than more strange things began to happen. The door of a large cupboard would swing open, despite being securely fastened. Then footsteps would cross the room and the living room door would swing open.

When Mrs Turner was working in the kitchen, the cupboard door would swing ajar, there was the sound of shuffling feet and the feeling that someone was standing behind her.

One night, after the Turners had gone to bed, they heard the sound of their settee being dragged across the floor of the living room. When they went to investigate, nothing had been moved.

The sounds continued until one night there was a tremendous crash, just as if the old iron mangle they owned had been thrown on its side. They leapt out of bed and as usual found nothing amiss. But that didn't stop the neighbours in the flat below complaining about the noise the Turners had made 'moving their furniture late at night'.

The Turners found another home as soon as they could – but not before they discovered a possible cause of the mysterious noises in the night.

By chance, they met an elderly woman who had been brought up in number 10. She remembered, she said, that when she was little, the house was reputedly haunted by a Victorian lady who constantly searched the nursery for her two children who had died there.

Hands of Grisly Glory

Thieves in the 18th century tried to ensure success for their crimes by invoking the aid of the dead.

The villains believed a candle gripped in the cut-off hand of a man who died on the gallows had the supernatural power to stupefy anyone they wished to rob.

In the 1790s, a traveller in woman's clothing arrived at the Old Spital Inn, which stood at Stainmore between Barnard Castle and Brough on windswept Bowes Moor in County Durham. The new arrival asked only to be allowed to doze by the fire before continuing her journey.

The landlord ordered a maid to sit with the visitor. The maid was suspicious, for she was sure she had spotted trousers peeping below the skirts of the 'woman'. The maid pretended to be asleep. When all was quiet, the traveller pulled out a hand taken from the body of an executed felon, and wedged a candle between the dead fingers. He recited: 'Let those who sleep, sleep on, let those who are awake be awake.' The man then unlocked the door and called softly to accomplices waiting outside.

The watchful maid wasted no time. She rushed at the door, slamming it shut and bolting it, with the would-be raiders outside. She then dashed upstairs to rouse her employer, but she could not waken him. Downstairs the door was being smashed by the gang.

The candle was still burning in its macabre holder, and the girl remembered the legend that only extinguishing the flame with milk broke its spell.

She grabbed a jug of skimmed milk from the kitchen and upended it over the hand. The flame died – and as it did so, the landlord and his staff were woken by the noise downstairs. The thieves promised to leave once their candle was returned, but they were forced to flee empty-handed when the innkeeper opened fire with his shotgun.

The grisly candle-holders were called Hands of Glory, and similar stories have been heard in other parts of England, and in France, Belgium and Ireland.

One encyclopedia of superstition lists the mysterious ingredients of the candle as the fat of a hanged man, virgin wax and Lapland sesame.

The Menace of Berkeley Square

London's elegant Berkeley Square is today as calm and peaceful as any such place in the middle of a bustling capital can be.

Office workers spend the summer lunchtimes relaxing on its lawns, while after dark gamblers and revellers flock to its clubs and casinos. And traffic speeding through this fashionable area of Mayfair makes it hard for tourists to hear the nightingale immortalised in a wartime hit song.

But less than a century ago, Berkeley Square was the most feared place in Britain. For the house at No 50, today a bookshop, was the home of the deadliest spectral killer of all time. Even now, no one is sure exactly what caused the deaths that alarmed Victorian England, for few who saw the killer lived to tell the tale. Those who survived generally became incoherent with sheer terror.

The house was already the talk of the town where Sir Robert Warboys accepted a foolhardy challenge at his club, White's. The handsome adventurer scoffed when friends discussed the possible causes of the disasters at No 50, and vowed to spend a night there to prove that talk of supernatural happenings was nonsense.

The owner of the house, a man called Benson, was reluctant to allow the experiment, but Sir Robert would not be dissuaded. He agreed, under pressure, to take a gun with him. And Benson insisted that Sir Robert's friends and himself stand guard on the floor below the bedroom where he would spend the night. If anything strange happened, the young aristocrat was to pull a cord which rang a bell in a room on the first floor.

Sir Robert retired at 11.15 p.m. after a good dinner. Just 45 minutes later, the bell began to jangle. As the rescue party raced upstairs, they heard a shot. They burst into the room to find Sir Robert slumped across the bed, his head dangling over the side.

He was dead – but not from a bullet wound. His eyes bulged in terror, his lips were curled hideously above clenched teeth. He had died of fright.

Intrigued by this and other stories about the deadly house in Berkeley Square, Lord Lyttleton resolved to investigate the mystery and arranged to spend a night in the room where Sir Robert had died.

He took along two guns, one filled with shot, the other with silver sixpenny pieces – charms to ward off evil spirits. During the night, he fired the barrel of coins at a shape that leapt at him. In 1879, he published the full details of the ordeal in his book, *Notes and Queries*. He concluded that the room was 'supernaturally fatal to body and mind'.

Lyttleton's researches included the discovery that a girl who had stayed at the house as a guest had been driven mad by terror. A man who had slept one night in the haunted room had been found dead next morning. And the maid of a family renting the place had died in hospital after being found crumpled on the floor whimpering, 'Don't let it touch me'.

Not surprisingly, few people were anxious to move into the house. For years it stood empty, its paintwork peeling, its secret undisturbed. Then, in December 1887, the frigate *Penelope* docked in Portsmouth, and its crew went ashore to head for home.

Two sailors, Edward Blunden and Robert Martin, arrived in London on Christmas Eve with little money and no lodgings for the night. They wandered the streets for a while before finding themselves in Berkeley Square. They soon discovered the To Let signs outside No 50. There was no doubt that the house was vacant and the men decided to spend the night there.

They wandered through the neglected rooms, arriving at last in a second-floor bedroom which seemed in better order than the others. Martin was soon asleep but his shipmate was nervous. As he tossed and turned, Blunden heard strange footsteps scratching along the corridor outside their door. He woke Martin, and the two watched, hearts racing, as the door slowly opened and something large, dark and shapeless entered the room.

Blunden darted towards the fireplace to try to grab something he could use as a weapon. As the intruder went after the terrified sailor, Martin seized the chance to escape through the open door. He raced down the stairs and ran for help. In nearby Piccadilly, he blurted his story to a police constable and the two men hurried back to the house.

They were too late. The shattered body of Blunden, his neck broken, his face fixed in a grimace of unimaginable terror, was sprawled on the basement steps.

Beware of the Bones

People who tamper with the remains of the dead, or ignore the last wishes of the dying do so at their peril. It seems that the spirits of the departed keep a careful watch on their earthly relics, and are quick to intervene if anything untoward happens to them.

An American farmer called Walsingham was exploring his new home at Oakville, Georgia, when he came across an ancient skeleton and threw it into a lime kiln. Soon doors began to slam in the middle of the night, chairs were mysteriously overturned, and bells tolled through the rooms.

Walsingham was not a superstitious man. He put the curious sounds down to mischief by neighbours. Then a series of things happened which he could not ignore. They were described by the *San Francisco Examiner* in 1891.

First the man's dog began barking furiously at a wall. One day the animal lunged towards it and fell back yelping, its neck broken as if somebody had hurled its body backwards.

Hideous laughter, shouts and wails began coming from all over the house. One of Walsingham's daughters saw a disembodied hand grasp her shoulder. And the farmer watched aghast as the prints of a naked pair of feet formed beside him as he walked in the rain.

The last straw for the family came at a dinner party. Guests shrugged off strange groans from the room above. Then a red stain began to form on the white tablecloth, spreading as more scarlet drips fell from the ceiling.

The men in the party raced upstairs and began tearing up the floorboards. But they found only dry dust, even though the liquid – later identified by chemical tests as human blood – continued to spread on the table. The Walsinghams packed their belongings and moved out of the haunted home.

A skull called Awd Nance has caused headaches for successive occupants of Burton Agnes Hall, a beautiful English mansion in a village between Bridlington and Driffield in Yorkshire. For more than 300 years the grue-some relic has insisted on staying in the house. Those who have tried to get rid of it have been forced to think again.

The hall was built by three spinster sisters early in the 17th century, financed by a legacy from their father, Sir Henry Griffith. Soon after the building was completed, robbers ambushed the youngest sister, Anne, as she re-turned from visiting friends in nearby Harpham. Villagers gently carried her battered body home. She was well cared for, but her injuries were so severe that it was clear she would never recover.

Shortly before she died, Anne made an extraordinary request. She insisted that her sisters cut off her head before they buried her body, and preserve it in the walls of her beloved home. She threatened that if they did not do so, dire consequences would follow. The two sisters promised they would follow her wishes, but the idea horrified them. When Anne died, they buried her body intact.

Seven days later, an inexplicable crash in an upstairs room woke the whole household. After another week, the family were again roused from their sleep when doors slammed in every part of the building. Three weeks after the funeral, they lay terrified in their beds all night as the whole house shook with the clatter of crowds of invisible beings running along corridors and up and down stairs. Agonised groans echoed through the rooms.

Next morning the servants quit. The sisters called in the local vicar, and when they mentioned Anne's dying wish, he agreed to open the grave. The body was just as they had last seen it – except for the head. That had been cut off, and all the skin had shrivelled away, leaving the skull bare.

Reluctantly, the sisters carried the grim memento into the hall. The haunting ceased. The skull stayed with them for the rest of their lives.

Several subsequent owners of Burton Agnes have dared to throw out Awd Nance, but each time, mysterious shuffling and scratching sounds, slamming doors and terrifying groans have forced them to restore the relic to its rightful place.

Other ghosts have been responsible for the finding of human bones in places where they should not have been. Such spectres vanish only once the bones have been given a decent burial.

Guests at a house in Lynton, Devon, complained during the 1930s that they could never get a decent night's sleep in the room above the skullery. They told their hosts, the Ewing family, that they saw the ghost of an old woman, and heard the sobbing of a child.

Then Mrs Ewing's brother-in-law realised that the haunted bedroom was much narrower than the skullery below it. The family discovered a cupboard which had been walled up. Inside was a child's box, and a collection of ancient human bones. Only when they were removed did peace reign in the house.

Frederick Fisher was a poor Australian farmer whose profits from crops never covered the cost of his living. Eventually he was taken to court, and jailed for debt. To stop creditors seizing his farm at Cambelltown, New South Wales, he transferred it to the name of an associate called George Worrall.

Six months later, Fisher finished his sentence and returned to his home. But one night in 1826 he left a nearby inn after a heavy drinking session – and was never seen again. Police investigated the disappearance, but could find no evidence of foul play.

Months went by. Then came the dark night that neighbour James Farley got the shock of his life. A strange figure was sitting on one of his fences, pointing to a spot in Fisher's paddock. Farley was too scared to do anything until daylight. Next morning he guided a police constable to the site the phantom had indicated, and in the shallow grave, discovered the badly beaten body of farmer Fisher.

George Worrall was questioned, and found to be an ex-convict. He confessed and was hanged. His victim's ghost was never seen again.

A Rover's Return

Cecil Bathe had just driven past a wrecked German tank during World War Two when a sandstorm struck. The Royal Air Force mechanic, returning to his Libyan base after a supply trip, decided to wait it out. He sat reading a dog-eared magazine and drinking beer given to him earlier by some Australian pals.

As a gust of wind shook his three-ton truck, Cecil looked up, and saw a lone un-uniformed figure in the whirling dust. He beckoned the man to the lorry and told him to jump in. Both sheltered from the lashing sand outside, chatting and swigging beer.

The stranger, wearing a khaki uniform without any insignia, spoke in a clipped, rather stilted accent. But after two years of desert warfare, Cecil paid no heed to the rather ragged look, and put the accent down to either South African or Dutch, both fighting alongside the Allies.

As they drank, the Englishman noticed his stranded companion had a raw burn on his right hand and arm. He urged the stranger to get medical attention, but the only response was a chilling laugh and the words, 'It's a bit late for that. Anyway, it doesn't matter.'

But Cecil did discover that he and his eerie companion had something in common. Before the war both had been Rover Scouts and had attended an international jamboree in southern England.

Daylight had begun to fade as the wind dropped and the RAF man decided to try to make a run for his camp. He offered the stranger a lift, but he shook his head. 'I'm going in the other direction,' he said. 'Thank you for the beer.'

The man stepped down from the truck and they shook hands. 'God watch over you, Tommy,' said the stranger. Cecil felt that the hand he was shaking was cold and stiff.

Cecil drove away, but there was no trace of the uniformed figure in the rearview mirror. He got out to look, but all he could see was the wrecked shell of the enemy tank.

Later that week Cecil returned to the airfield by motorbike just as a recovery crew was hauling the tank wreck onto a transporter.

He stopped to look and the corporal in charge told him that the tank driver had died at the controls a month before when the tank was hit in the turret by a shell. They had laid the body out under canvas.

485

Cecil lifted the makeshift shroud.

There, with decaying skin and dead eyes, was his beer-drinking desert companion of a few days earlier.

Shaken, Cecil dropped the groundsheet. He clambered onto the tank wreck and peered into the half-light. Glinting in the sunlight he could see an Australian beer bottle.

Unexpected Guest for Tea

A tea party was taking place at the home of Vice-Admiral Sir George and Lady Tryon in London's fashionable Eaton Square. The mansion was crowded with the cream of the capital's society. Sir George, in full dress uniform, walked down the graceful curved staircase towards his guests. Lady Tryon dropped her teacup and screamed.

The guests watched aghast as the famous admiral reached the foot of the stairs, calmly and silently crossed the room, opened a door and was gone. It was June 22, 1893, and the guests knew that they had just seen a ghost. For that day Sir George was on the bridge of the flagship, *Victoria*, off Tripoli and, tragically, was guiding her into one of Britain's greatest naval disasters.

His squadron steamed along in two columns as part of a carefully planned fleet exercise. *Victoria* led one of the columns, with the *Camperdown* heading the other.

The naval squadron consisted of Britain's entire Mediterranean Fleet – eight ironclad battleships and five cruisers. Sir George's plan called for the two columns to move within six cable lengths of each other. Steaming ahead on parallel courses, they could turn inward on command and then reverse course. But were six cable lengths – about 4,500 feet – enough?

One officer meekly suggested that the two columns could come dangerously close to a collision.

486

The Vice-Admiral agreed that perhaps there should be eight cable lengths. However, within minutes he again mysteriously changed his mind, ordering the manoeuvre to go ahead as planned. There was later testimony that his eyes were strangely dull when he re-issued the order.

At a combined speed of 18 knots, the two lead ships were heading toward each other on a collision course – yet Sir George gave no signal for the turn. He stood like a statue on the charthouse deck, his eyes still vague. When other officers pleaded that they must do something, the admiral failed to answer.

At the last moment, he shook himself like a waking dreamer and whispered, 'Yes, go astern.' The order was given too late.

Even with the propellers in reverse, the *Victoria's* momentum carried her like a juggernaut toward the *Camperdown*.

As buglers sounded the call summoning 'All hands on deck,' the ironclads met in a terrifying collision. *Camperdown* pierced the flagship some 65 feet aft of the bow, on the starboard side. A dreadful shudder racked the *Victoria*.

The ship's pumps might have coped with the torrents of water, but compounding his first ghastly mistake with another, the admiral shouted at *Camperdown*, 'Go astern with both engines!'

As the great ship backed away, the fate of the *Victoria* was inevitable. The flagship was swamped by a wall of water that flooded everything in its way; men, machinery and bulkheads.

The ship's hydraulic system was submerged, and below decks hundreds of men were caught in the smothering assault of the seawater. When the order came to report topside, many were either dead or dying in flooded compartments.

Among the 600 who leaped from the ship many were

ground to fragments in the propeller blades or trapped in the suction of the foundering *Victoria*. A total of 25 officers and 259 men were picked up by the boats, with 22 officers and 336 seamen dead.

Among the dead was Sir George Tryon himself, who was still standing on the *Victoria's* bridge when she slipped beneath the waves.

Fishermen's Fear

For hundreds of years fishermen on the east and southeast coasts of England have kept a watchful eye for a phantom schooner, the *Lady Lovibond*. They wonder how many sailors who met their deaths on the notorious Goodwin Sands had first spied the ghost of the three-master.

The *Lady Lovibond* ran aground on the Goodwins and sank with all hands on February 13, 1748. Captain Simon Peel, his bride and some of their wedding guests were on board. Legend has it that the first mate, who was in love with the bride himself, killed Peel out of jealousy and steered the ship to its doom on the Goodwins.

Fifty years later to the day, a three-masted schooner identical to the *Lady Lovibond* was seen heading for the Goodwins. The crew of a fishing boat followed her and heard the sounds of a celebration and women's voices. The schooner hit the sands, broke up – and vanished.

The same apparition appeared to another ship's crew exactly 50 years later and was next seen by a group of watchers near Deal, Kent, on February 13, 1898.

Does the phantom appear every 50 years? Watchers were on the lookout on February 13, 1948, but visibility was poor and they saw nothing.

North America also has a famous phantom ship lurking off Rhode Island. The *Palatine* left Holland in 1752, packed with colonists bound for Philadelphia. A fierce winter storm blew her off course and, when the captain was lost overboard, the panicking crew mutinied.

The passengers spent Christmas Day in confusion and terror. Two days later, the *Palatine* ran aground on rocks off Block Island and began to break up.

As the storm abated, the doomed ship began to slip back off the rocks, drawn out to sea again by the tide. But before she could do so, dozens of local fishermen descended on the *Palatine*, took off the passengers and looted the ship.

When their frenzied rampage had ended and they had stripped the *Palatine* of everything in value, the fishermen set it on fire and watched it drift, ablaze from bow to stern, out to the open sea.

They watched in horror when they saw a woman appear from her hiding place on the *Palatine* and stand on the deck screaming for help until the flames swallowed her.

There have been sightings of the ghostly vessel off the New England coast ever since, blood-red flames rising from a wrecked hulk.

The Ghostly Swimmers

Seamen James Courtney and Michael Meehan were buried at sea on the morning of December 2, 1929. They had died the previous day, asphyxiated by fumes while working below decks aboard the oil tanker *Waterton*, owned by the Cities Service Corporation, and bound from California to Panama.

When the weighted bodies of the two sailors dropped into the Pacific, their fellow crewmen mourned deeply, for Courtney and Meehan were two of the most popular men on the ship.

One friend said, 'Somehow they made everyone feel good.'

But the crew of the *Waterton* were not without their dead colleagues for long. The day after their burial, the officers and crew saw two men swimming in the open sea. Captain Tracy put his binoculars on them and whispered, 'Oh, my God!'

But when the ship slowed to ten knots and drew up alongside the swimmers, they faded like morning mist – only to reappear just 40 feet from the ship. At that distance there were no longer any doubts. The men in the water were Courtney and Meehan.

For three days the swimmers kept pace with the *Waterton*. Now there was no terror aboard the ship, because everyone saw that the dead men intended no harm. At one point they swam ahead of the vessel and seemed to be trying to divert her from the path of an approaching squall.

Reporting later in the New Orleans office of the shipowners, Captain Tracy told his employers about the death and reappearance of Courtney and Meehan. Tracy was provided with a camera and film and asked to substantiate his tale on the return voyage.

It was a voyage without incident until they were in the Pacific again and the deckhands saw two pale figures bobbing in the wake of the *Waterton*. By dawn they were once more alongside the vessel, and with full light the captain snapped eight pictures at close range. Within a few hours the swimmers had vanished.

They were not seen again in the days that followed, and back in port the captain took the film to company headquarters.

Still wet from processing in a photographic laboratory, the roll of film was closely examined in company offices. One by one the negatives were rejected.

The one of the executives lifted the eighth frame to the light. 'There they are!' he said.

When prints were made from the negative, the two pale

faces emerging from the waves were positively identified by friends and relatives as those of James Courtney and Michael Meehan.

Voyage of the Frozen Dead

The Yankee whaling ship *Herald* was cruising off the west coast of Greenland, inside the Arctic Circle. From the bridge, Captain Warren peered ahead at a three-masted schooner drifting through the ice floes like a ghost ship. Warren took eight men in a longboat and rowed to the silent vessel. Through the encrusted ice, they could make out the schooner's name: *Octavius*.

Warren and four of the sailors boarded the schooner. They crossed the silent, moss-covered decks, opened a hatch and descended to the crew's quarters. There they found the bodies of 28 men, all lying on their bunks and wrapped in heavy blankets.

They fumbled their way aft to the captain's cabin, where the nightmare continued. The master of the *Octavius* slumped over the ship's log, a pen close to his right hand as if he had dozed at work. On a bed against one wall of the cabin, a blonde woman lay frozen to death under piles of blankets. And in a corner there were a sailor and a small boy whose bodies told a tragic story.

The sailor sat with his flint and steel clutched in frozen hands. In front of him was a tiny heap of shavings, silent evidence of a fire that had failed to ignite. The little boy crouched close to him, his face buried in the seaman's jacket as if he had huddled there in pathetic search for warmth.

The men from the *Herald* clambered back onto the deck, taking with them the schooner's log book as proof of what they had found. Back aboard the whaler, they could only watch helplessly while the derelict schooner drifted away from them among the icebergs, never to be seen again.

It was well they had taken the log book. The world would not be ready to accept their story, which remains one of the strangest tales of the sea.

The last log entry was dated November 11, 1762. The dying captain wrote that the *Octavius* had been frozen for 17 days. The fire had gone out and they could not restart it. The location of the ship at this time, said the captain was Longitude 160W, Latitude 75N.

Captain Warren looked at the charts in disbelief. In those last hours of human life, the ship had been locked in the Arctic Ocean north of Point Barrow, Alaska – thousands of miles from where the whaler had found her. Guided by some unknown force, year after year the battered schooner had crept steadily eastward through the vast ice fields until she entered the North Atlantic. In doing so she had then achieved the dream of all mariners.

For centuries men had sought the legendary Northwest Passage – a navigable route around the Arctic Ocean between the Atlantic and the Pacific. On that historic 13-year voyage, the ghost ship *Octavius* with her crew of frozen dead had been the first to find it.

The Lonely Lighthouse

On the night the Eilean Mor Lighthouse went dark, two sailors on the brigantine *Fairwind* saw a strange sight. Cutting diagonally across their bow a longboat with a huddle of men aboard was bearing toward the lighthouse on the rocky Flannan Islands off the west coast of Scotland.

The sailors called, but there was no answer. The boatmen wore foul weather gear and when moonlight slashed through a rift in the clouds, their faces shone like bone. One of the would-be rescuers testified later, 'Our

492

first thought was that they were floating dead from some shipwreck. But then we heard the oarlocks and saw the movement of their arms.'

Later on that night of December 15, 1900, a squall broke. Without the guardian light ships were in dire peril. There was one angry question from the skippers: Why was the lighthouse dark?

On the day after Christmas, the supply vessel *Hesperus* hove to off the islands to investigate. When there was no answer to repeated signals, crewmen set out in a small boat to the landing dock. Tying up, they were chilled by the strange silence.

The lighthouse had been staffed by three men, but no one was there to welcome the *Hesperus*. There were no signs of violence and the larders were well stocked. Lamps were all trimmed and ready, the beds made, the dishes and kitchen utensils shining.

As the searchers climbed through the empty lighthouse, they found only two things that struck them as unusual. On the stairway and in a cubbyhole office where the log was kept, there were shreds of seaweed unknown to them.

There were no oilskins or seaboots in the building, which seemed to imply that all three men had left the lighthouse together.

No lighthouse keeper had ever been known to abandon his post, even in the worst weather, and this was a point repeatedly made during the inquiry which followed – an inquiry which was hushed to silence by the reading of the log book kept by keeper Thomas Marshall:

'December 12: Gale north by northwest. Sea lashed to fury. Never seen such a storm. Waves very high. Tearing at lighthouse. Everything shipshape. James Ducat irritable.' and later that day: 'Storm still raging, wind steady. Stormbound. Cannot go out. Ship passing sounding foghorn. Could see light of cabins. Ducat quiet. Donal McArthur crying.

'December 13: Storm continued through night. Wind shifted west by north. Ducat quiet. McArthur praying.' And later: 'Noon, grey daylight. Me, Ducat and McArthur prayed.'

Inexplicably, there was no entry for December 14. The final line in the log read: 'December 15, 1 p.m. Storm ended, sea calm. God is over all.'

No explanation could be offered but that the men had been seeing visions. While the log entries had reported gales lashing in Flannan Isles, there had been none at all 20 miles away on the island of Lewis.

Locals pointed to an even more mysterious cause of the disappearance of the lighthouse men. For centuries, the Flannan Isles had been haunted. Hebridean farmers might sail there during daylight to check on their sheep – but few except the 'foolish sassenachs' at the lighthouse dared stay overnight.

Final 'proof', if the locals needed any, was the evidence of the sailors of the *Fairwind* – of the longboat crowded with ghosts.

Ghost Fliers of the Florida Sky

'You may unfasten your seatbelts.' The indicator lights in the bulkheads above the passengers went out. At the rear of Eastern Airlines Tri-Star 318, stewardess Fay Merryweather left her seat near the emergency exit and went to the galley. There was less than two hours to serve the 180 sunseekers heading from New York to the holiday beaches of Florida.

In the galley, Fay reached for the handle of the oven door, and as she did so, she fell back, stunned, against the galley wall as though hit by an electric shock. Staring at her from the glass door of the oven was the face of a man. The image wavered. Then the lips moved, but Fay heard nothing. Her hands were clasped, her mouth fell open, but somehow she stopped herself screaming. It wasn't true, it wasn't true. It couldn't be!

Fay shut her eyes, then opened them. There was nothing there. She had imagined it. She took a deep breath. The in-flight meals must be served. She leaned forward again towards the oven door, then reached for the handle.

There it was again, blurred at first, then forming solid lines. It was the same face. The mouth moved, the eyes blinked. A frown furrowed the brow, a look of urgency crossed the phantom face.

Fay staggered from the galley, her brain swirling. You mustn't panic, she told herself. Was she sure of what she had seen? Do not alarm the passengers. The lessons of her training flooded through her head. She smoothed her skirt, swallowed a lungful of air, and walked as steadily as she could to the flight deck.

Fay shook the flight engineer's shoulder. 'Quick,' she said, 'there's a problem.'

She gave him no time to think, but strode back down the aisle towards the galley. The engineer followed, puzzled. At the door of the galley, Fay gripped his arm. She said it as matter-of-factly as she could.

'There's a ghost, a man's face. It's in the oven door.'

The engineer stepped in front of the oven, leaned forward and stared into a face he knew, the face of his Eastern Airlines former colleague, flight engineer Don Repo who had been dead for a year.

Then, beneath the whine of the Tri-Star engines, and just audible in the metallic cell that was the galley, a voice whispered, 'Beware, beware. Fire in the jet.' The words faded with the face.

Fay made the report in the unemotive words of unsuperstitious crew members who knew the dangers of the job and were too down-to-earth to believe in fantasies.

Neither she nor the flight engineer travelled in Tri-Star 318 again. Months later, the plane developed engine trouble on a flight to Mexico. It returned to New York for repairs and trials. As it took off for a routine test flight, an engine burst into flames.

At Washington's Flight Safety Foundation, records of the Federal Aviation Agency's report on the incident shows that only luck and the skill of the flight maintenance crew got the plane down without loss of life. Packed with passengers, it could have been a disaster.

The startled aircrew of Tri-Star 318 were not the only Eastern Airlines Tri-Star staff to come face to face with one of the Ghost Fliers of the Florida skies – spirits who haunted the airline's great planes to prevent horrors like the one that hurled them into limbo.

The story began on a warm autumn day in 1972 as the sun glittered on the patchy swamps of the Florida Everglades, and a soft, southern breeze gently bent the swamp grass.

In the sky above, Captain Bob Loft and Flight Engineer Don Repo were bringing their Lockheed L-1011 Tri-Star, Flight 401 from New York to Miami, to the end of a routine journey. The 176 passengers were ready to fasten their seatbelts.

The order to activate the undercarriage was carried out. Bob Loft studied the panel of instruments. Strange, the indicator light showed that nothing had happened. The nosewheel should have lowered and locked. But the small light which should have confirmed this had not come on.

Bob Loft put the aircraft on to a circling course and locked in the automatic pilot while Don Repo scrambled into an observation bay, from where he could see if the wheel was locked into position. Meanwhile, Loft decided to check the indicator light bulb for a fault. He swivelled around in his seat to reach the light cover. He took his eyes off the panels and the flight path in front of him as he tried to unscrew the bulb. As Loft twisted he had without knowing it, knocked off the automatic pilot switch.

He was still fixing the light bulb, when something made him look up. Through the cockpit windscreen there was the flash of glinting water speeding past. One glance told him the frightening message from the instrument panel, but there was not time to do or say something. Flight 401 smashed into the swamps in a stream of flying water, mud and vegetation. Bob Loft, Don Repo and 97 other people died.

The plan was not entirely smashed. Some parts were hardly damaged. Seats, and the galley in the rear of the plane were in sufficiently good condition to be salvaged. Accident inspectors sent them back to Lockheed, where they were reconstructed. Down on the factory assembly line, the parts were built into new Tri-Stars.

The galley was fitted into No 318.

Early in 1973, 318 was airborne and an Eastern Airlines vice-president flew in it, along with airline staff being ferried back as passengers from destinations where their duties had ended. Cabin staff had checked in an off-duty captain, who was sitting in the first class compartment. The vice-president joined him. They chatted amiably for a while.

The captain suddenly turned and looked full face at the airline chief, who gasped. It was Bob Loft.

The vice-president dashed from his seat to seek help from the cabin staff. But when a stewardess returned with him, the seat was empty. The ghost fliers were riding the skies.

The next mysterious visitation occurred when a startled flight engineer stepped onto the flight deck to check the instruments before a routine Florida trip.

A uniformed officer was already in his seat, and he turned to face the duty engineer. The face was unmistakably that of Don Repo. His voice said, 'You don't have to check the instruments. I've already done that.'

Weeks later an Eastern Airlines captain, aware of the flight engineer's report on the eerie cockpit incident, decided to check the instruments himself before taking off from Miami for Atlanta, Georgia. He ran through the checks, but staring at him from the face of the panel was the ghostly, wavering outline of Don Repo's face.

Then came words like a distant echo. 'There will never be another crash on an L-1011. We will not let it happen.'

One other captain saw the ghostly travellers of Eastern Airlines, and a stewardess, sent to check smoke coming from a Tri-Star bulkhead during a flight, came face to face with the misty figure of dead pilot Bob Loft.

The Flight Safety Foundation has studied detailed reports of the ghost sightings. Liaison executive Doris Ahnstrom said, 'The reports were given by experienced and trustworthy pilots and aircrew. We consider them significant.

'The appearance of the dead flight engineer in the Tri-Star galley door was confirmed by the flight engineer. Later records at the Federal Aviation Agency record the fire which broke out on that same Tri-Star. We published reports of the ghost sightings in our safety bulletin issued to airlines in 1974.'

The ghost appearances of the dead fliers ended after 18 months, following an amazing ceremony in the galley of the haunted jet, Tri-Star 318. A religious devotee, who was also a technical second officer with Eastern Airlines, was granted permission to hold an exorcism ceremony.

Aircrew, distressed by the increasing frequency of the apparitions, recited prayers. The officer sprinkled water in the galley and, as he did so, the anguished face of Don Repo stared despairingly at him.

The ghost fliers of the Florida skies were never seen again.

The Unlucky Car

Archduke Franz Ferdinand wanted a car that would impress the public when he and his wife, the lovely Duchess of Hohenburgh, toured the tiny Bosnian capital of Sarajevo. There were reasons for putting on a brave show. Europe seethed with political unrest, and the Archduke's good-will trip could be hazardous.

The royal couple arrived at Sarajevo on June 28, 1914, in a blood-red, six-seat open tourer. It made a splendid target. A young fanatic armed with a pistol leaped onto the running board of the car. Laughing in the faces of the Archduke and Duchess, he fired shot after shot into their bodies.

The double assassination was the spark that touched off the Great War, with its casualty list of 20 million.

After the Armistice, the newly appointed Governor of Yugoslavia had the car restored to first-class condition, but after four accidents and the loss of his right arm, he felt the vehicle should be destroyed.

His friend Dr Srikis disagreed. Scoffing at the notion that a car could be cursed, he drove it happily for six months – till the overturned vehicle was found on the highway with the doctor's crushed body beneath it.

Another doctor became the next owner, but when his superstitious patients began to desert him, he hastily sold it to a Swiss racing driver. In a road race in the Dolomites, the car threw him over a stone wall and he died of a broken neck.

A well-to-do farmer then acquired the car, which stalled one day. While another farmer was towing it in for repairs, the vehicle suddenly growled into full power and knocked the tow-car aside, killing both farmers.

Tiber Hirshfield, the car's last private owner, decided that all the old vehicle needed was a less sinister colour

499

scheme. He had it repainted in a cheerful blue and invited five friends to accompany him to a wedding. Hirshfield and four of his guests died in a head-on smash as they drove to the festivities.

Finally the rebuilt car was shipped to a Vienna museum where it was lovingly cared for by attendant Karl Brunner, who revelled in the stories about the car's 'curse' and forbade any visitor to sit in it.

During World War Two, bombs reduced the museum to rubble. Nothing was found of the car – or Karl Brunner.

Check-in for Terror

Ghosts are normally associated with lonely moors, eerie castles or old mansions, but London's Heathrow Airport is haunted by not one ghost but three.

Airline girls have been terrified by an invisible ghost that pants like a dog. They say it creeps up behind them and breathes down their necks – at a spot said to have been haunted by the spirit of highwayman Dick Turpin.

One Pan American employee said, 'I had just left my car in the staff park when the panting started. It sounded like an animal. I turned round but there was nothing there. The panting got close – right up to my neck. When I turned round again two other people, a girl and a man, had moved away. They had exactly the same experience.'

An airline engineer said, 'A lot of people have heard the weird noises and there seems no logical explanation.'

The second ghost that haunts the airport is not invisible. It doesn't pant, but is just as frightening. It is known as 'the ghost in the light grey suit.'

One man who saw the ghost was a distinguished diplomat from an African high commission in London. He fled from one of the airport's VIP lounges.

The ghost appears in the VIP suite at the airport's European Terminal 1, which is used by the Queen, foreign heads of state and ambassadors.

A catering supervisor said, 'The African diplomat was petrified. All we could get out of him was that he had seen the bottom half of a man in grey trousers standing in front of him. Ever since, he has refused to go anywhere near the area.'

The supervisor believes she saw the ghost in grey on another occasion. 'When I looked again he had vanished,' she said. 'There was no way he could have got out of the lounge without coming past me. I'm not afraid. I think he's friendly.'

An airport policewoman is another who has experienced a strange 'presence' in the lounge. She said, 'I'm a sceptic about the supernatural, but I can't explain this.'

A bowler-hatted ghost has been seen on Runway 1 at Heathrow many times over the last 20 years. A spiritualist once described the figure as that of a Guardsman, about six feet tall and in his late forties, who seemed to be looking for something. On March 2, 1948, a DC3 airliner burst into flames and crashed on the runway, killing 22 passengers, mainly businessmen.

As rescue workers searched the wreckage, they were accosted by a bowler-hatted man concerned as to the whereabouts of his briefcase.

Horseman of Bottlebush Down

Ghosts and spirits have come back across hundreds of years. But none have come further than the mystery horseman who rides the north Dorset countryside.

The horseman of Bottlebush Down is thought to date back to the Bronze Age, which makes him and his mount around 2,500 years old.

His appearances are nearly always close to the A3081 road which runs between Cranborne and Sixpenny Handley in Britain's beautiful West Country. Although the area is today mainly quiet and devoted to agriculture, it was thought to have been a hive of activity thousands of years ago.

The fields are dotted with low, round burial mounds. A strange earthwork, known locally as the Cursus, runs for some six miles across the fields. It consists of two parallel ditches about 80 yards apart and it is here that the horseman is usually seen.

The horseman's ghostly rides have become legend in this part of the West Country. Farmworkers and shepherds have reported seeing him galloping across the fields towards the Cursus and in one incident in the 1920s two young girls were terrified as they cycled one night from Handley to Cranborne.

The wide-eyed girls told police that they were cycling along the road when suddenly a horseman appeared from nowhere and rode alongside them for some distance before disappearing.

But it was a few years earlier that the most documented sighting was made, and one that put the time stamp on the horseman.

Archaeologist R C Clay had such a close encounter with the rider that he was able to date the apparition by his clothes. Clay, leader of a team excavating a Bronze Age settlement near Christchurch, in Dorset, came face to face with the horseman in 1924 as he was driving home from the excavations to Salisbury, in Wiltshire.

As he passed the spot where an old Roman road crossed the modern highway, he saw a horseman galloping across the fields ahead of him. The horseman was riding full tilt towards the road, but as Clay slowed down to let him cross, the rider swung his horse round and galloped parallel with the vehicle some 40 yards away.

For some 100 yards or more, the horseman kept pace with the car while Clay watched fascinated. In spite of his surprise, the archaeologist managed to take in a great deal of detail about horse and rider.

He said later, 'The horse was smallish with a long tail and mane. It had neither bridle nor stirrups. The rider had bare legs, a long flowing cloak and was holding some sort of weapon over his head.'

As suddenly as he had appeared, the horseman disappeared. Clay stopped his car and tried to gather his thoughts. His first instinct was to get out and look round, but as it was getting dark he decided to press on home.

The next morning he was back to see if he could discover anything which would help explain his ghostly encounter of the previous evening. He searched the road and the surrounding area for several hundred yards on either side of the spot but found nothing – except a low burial mound almost exactly where the horseman disappeared.

For weeks afterwards he tried to find an explanation for what he saw. He drove along the section of road at the same time, evening after evening to see if in the gathering dusk he might have mistaken an overhanging tree or bank or anything for his galloping horseman. There was nothing.

His one encounter with the horseman, coupled with his expert knowledge, enabled him to date the figure as from the late Bronze Age – somewhere between 700 and 600 BC.

Dead Man's Quest for Justice

A row in Britain's House of Commons during the First World War may have been responsible for the mysterious reappearance of a dead pilot. The flier returned to haunt an airbase – three years after the crash that killed him.

Member of Parliament Pemberton Billing, head of the Southampton company which later developed the Spitfire aircraft, accused the Government in 1916 of doing nothing while certain Royal Flying Corps men were 'murdered rather than killed by the carelessness, incompetence or ignorance of their senior officers or of the technical side of the service'.

One of the examples he cited was that of Desmond Arthur, an Irish Lieutenant with No 2 Squadron, who died in a crash over the Scottish airbase of Montrose on May 27, 1913. Arthur was gliding down from 4,000 feet, preparing to land, when the starboard wing of the BE2 biplane folded in mid-air. As the tiny aircraft plunged, the pilot's seatbelt snapped, and he was thrown out of the cockpit. Ground staff watched him fall to his death, arms and legs flailing helplessly. There were no parachutes in 1913.

The Royal Aero Club's accidents investigation committee began a probe immediately, and concluded that an unauthorised repair job on the plane's right wing had been botched, then covered up. Someone had broken a wing span near the tip, repaired it with a crude splice, then concealed the work by stretching new fabric over the affected area. To Arthur's friends in the Royal Flying Corps, it added up to murder, but the offender could not be pinpointed.

Billing used the Aero Club's findings as a basis for his 1916 onslaught in Parliament. He was as astonished as anyone when the Government, anxious to avoid any scandal that might undermine public faith in its war effort, issued its own report on the crash. It said the wing repair explanation was based on the evidence of only one of 23 witnesses, and was completely without foundation. In other words, Arthur had only himself to blame for his death.

The interim report was issued on August 3, 1916, and a detailed version was promised before Christmas. In September, airmen based at Montrose began to notice curious things. Twice one officer followed a figure in full flying kit towards the mess, only to see him vanish before reaching the door. A flying instructor woke one night to find a strange man sitting in a chair beside the fire in his bedroom. When he challenged the intruder, the chair was suddenly empty. Two other men woke simultaneously one night, convinced a third person was in their room.

Was Desmond Arthur trying to rally his old friends in the Royal Flying Corps to clear his name? As the story of the hauntings spread around Britain's other airbases, two of the Government's committee of inquiry revealed that they had not even seen the Royal Aero Club findings that their interim report had denigrated.

After studying the results of the earlier investigations, Sir Charles Bright, an engineer, and a lawyer called Butcher added an amendment to the final report when it was issued at Christmas.

They declared, 'It appears probable that the machine had been damaged accidentally, and that the man (or men) responsible for the damage had repaired it as best he (or they) could to evade detection and punishment.'

The ghost of the dead pilot appeared to settle for that as vindication of his innocence in causing the crash. After one last appearance in January 1917, phantom flier Desmond Arthur was never seen again.

The Headless Cyclist

George Dobbs was determined that the bitter weather and shortages of wartime Britain were not going to get him down. It was 1940 and the country was in the grip of one of its worst winters for years. Snow covered the countryside and, to make matters worse, the war news was gloomy.

George wrapped himself up against the hostile night and set out from his home near Northampton to walk to the Fox and Hounds pub for a few beers to cheer himself up.

With his hands deep in his pockets, he struggled up the slippery slope past the cemetery when he noticed the dim headlights of a car slowly approaching, its wheels running in and out of the icy ruts. Framed against the lights of the car, George saw a cyclist pedalling towards him. He too was having difficulty in steering his machine because of the snow and ice.

At first George thought that the cyclist had no head, but he quickly dismissed the idea as being a trick of the light or the fact that the rider had muffled himself up against the cold.

The next time George glanced up, the cyclist was still fighting for control of his machine, completely unaware of the approaching car. But before George had a chance to cry out, the car drew level with him and chugged past towards Market Harborough.

George could not believe it. The car must have hit the cyclist, he thought. He ran through the snow as fast as he could to the spot where he had last seen the cyclist – expecting to find the result of a terrible accident.

There was nothing. No cyclist, no cycle and no accident. George searched both sides of the road in vain.

He fled as quick as his legs could take him to the Fox and Hounds pub at nearby Kingsthorpe. As he thawed out in the pub, he told his story.

When George finished there was silence – until Lid Green, who was for many years the local gravedigger, leaned across the bar.

He said, 'That sounds just like the chap I buried 25 years ago. He was knocked off his bike in deep snow outside the cemetery gates.

'His head was torn off in the crash.'

The Royal Haunted House Guests

The British Royal Family are well aware of the truth of the old saying, 'Uneasy lies the head that wears the crown'. For the troubled spirits of their predecessors still haunt the historical royal homes. The Queen, Princess Margaret and the Queen Mother have come face-to-face with ghosts.

Windsor Castle, the royal retreat in Berkshire, is said to have at least 25 different spectral skeletons in its cupboards, four of them former monarchs. It was there that Princess Margaret saw the figure of Queen Elizabeth I, the last Tudor monarch, who has wandered the 12th century building since her death in 1603. She is spotted most frequently in the castle library. An officer of the guard once followed her into the room, but when he reached the door, Good Queen Bess had vanished.

King Charles I, who lost his head in 1649 during the Civil War, has been reported many times standing by a table in the library, while George III, who died on January 29, 1820, and was confined to the castle during the last years of his lunacy, has been seen and heard in several rooms, often muttering one of his most-used phrases, 'What, what?'

The bulky figure of Henry VIII is another nocturnal visitor. Two guards saw him fade into a wall on the battlements as recently as 1977. They later learned that there had been a door at that very spot during Henry's reign.

Soldiers on sentry duty at Windsor have often seen the ghost of a young guardsman who killed himself in 1927. Many who have spotted him in the Long Walk believe at first that he has come to relieve them.

A Coldstream Guardsman found unconscious in the Great Park in 1976 had experienced a very different kind of ghost. He told those who found him that he had seen Herne the Hunter, a man clad in deer skins and a helmet with antlers jutting from the forehead.

Hundreds of other people claim to have seen the same apparition over the last 250 years, silently speeding through the castle grounds with his spooky pack of hounds. When the tree from which he allegedly hanged himself was cut down in 1863, Queen Victoria reserved the oak logs for her own fire, 'to help kill the ghost'. But the sightings of Richard II's forester have continued.

In the 17th century, a terrified servant called Parker approached one of the castle guests, Sir George Villiers, with an extraordinary story. He said he had had three visits from the armour-clad ghost of Sir George's father, the Duke of Buckingham, and had been told that unless Sir George mended his callous ways, he had not long to live. Sir George laughed off the warning. Six months later he was assassinated.

Hampton Court, the palace by the Thames presented to Henry VIII by his disgraced Chancellor, Cardinal Thomas Wolsey, is still haunted by the spirit of the King's fifth wife, Catherine Howard, who was beheaded in 1542. She had been seen so frequently, running screaming to the chapel door in search of sanctuary, that she is now mentioned in the official guide issued to the thousands of tourists who visit the palace.

The third of Henry's six wives, Jane Seymour, has also been seen at Hampton long after her death. She emerges from the Queen's apartments carrying a lighted taper, and walks around the Silver Stick Gallery on the anniversary of the birth of her son, later Edward VI, on October 12, 1537. Jane died one week later, and the weakling boy, crowned for a short reign when he was only ten years old, was fostered by a nurse, Mistress Sibell Penn, who also appears at Hamilton.

Mistress Penn was buried at St Mary's Church after she died of smallpox in 1568, but the church had to be rebuilt in 1829 after being struck by lightning. The nurse's remains were disturbed as her tomb was moved, and soon strange whirring sounds and mutterings were heard coming from behind a wall at the palace where there was no known room.

When the wall was knocked down, a spinning wheel was uncovered along with other relics which indicated that the nurse had once lived on the site. Many witnesses have also seen Mistress Penn wandering the corridors of the palace's south-west wing, where her old room was. She is a tall, thin hooded figure in a grey robe, her arms outstretched as if in appeal. In 1881, a sentry watched her walk through a wall.

Two male figures once haunted Hampton's Fountain Court, making loud noises in the middle of the night. The ghosts were never seen or heard again after workmen un-covered two skeletons in Cavalier dress buried beneath the courtyard. The skeletons were given a Christian funeral.

Perhaps the most bizarre Hampton phantoms were those encountered by a police constable on duty at the palace one cold February night in 1917. The officer, identified only as PC 2657, who had 20 years service in the force, opened a gate in the grounds for two men and seven women wearing strange old-fashioned costumes. He swore that they then walked on for 30 yards, turned to one side of the path – and simply faded away.

Today the Queen and her immediate family have aban-doned Hampton Court to the tourists, and divide their time in Britain, between Buckingham Palace, Windsor, Sandringham in Norfolk, and Balmoral in Scotland. All have their own curious ghosts.

It was at Balmoral that the Queen is said to have seen the phantom figure of John Brown, confidant and some say lover of the widowed Queen Victoria. He has often been reported stalking the castle's corridors and entrance hall, a magnificent sight in his kilt.

Sandringham has for years played host to a mischievous yuletide poltergeist. It livens up Christmas Eve in the second floor servants' quarters by flinging greetings cards about, ripping sheets from newly-made beds, and breathing heavily in the ears of unsuspecting maids. Prince Philip's uncle, Prince Christopher of Greece, once saw a mysterious masked woman while staying in one of the Sandringham guest rooms. He glanced up from his book and saw her head and shoulders framed in the dressing table mirror. She had soft, curly brown hair, a dimpled chin, and a mask over the top of her face.

Next day, while visiting Lord Cholmondely at nearby Houghton Hall, the Prince found out who she was. He saw a portrait of the same woman, carrying a mask, and wearing the dress he had noted in his mirror. It was Dorothy Walpole, unhappily married in the 18th century, whose ghost was also seen by King George IV in 1786.

The ghost of Buckingham Palace is that of Major John Gwynne, a private secretary in the household of King Edward VII early this century. Fearing that mention of his name in a divorce case had brought dishonour on the Royal Family, he shot himself at the desk of his first-floor office. And it is there that his dim shape has been seen several times since.

Gatcombe, the Cotswolds mansion home of Princess Anne and Commander Tim Lawrence, is said by locals to be haunted by a huge black dog. They call it the Hound of Odin, named after the God of the Vikings who pillaged Gloucestershire 1,000 years ago, and was always accompanied by a fierce four-legged fiend.

Kensington Palace has three ghosts. A man in white buckskin breeches strolls the arcaded courtyards, Queen Victoria's Aunt Sophia sits working a spinning wheel, and, on the roof, King George II has been seen staring at the weather vane, and asking, in his thick German accent, 'Why don't they come?' The King died at the palace on October 25, 1760, waiting for messengers with news from his native Hanover.

No ghostly goings-on have yet been reported from the Gloucestershire home of Prince Charles, but St Paul's Cathedral in London, where he married Princess Diana in July, 1981, was once the scene of a strange incident. Workmen preparing foundations for a building beside it unearthed a wooden box containing jewels. The gems were taken to the British Museum, London, and an expert took them home to clean, polish and value.

As he and his daughter worked, the room suddenly grew cold. A psychic friend who called a few hours later found out why. He saw a tall thin man in Elizabethan dress standing behind the couple, clearly angry that his hidden treasures had been disturbed.

There was another surprise when the jewels were put on display at the museum. A woman looking at them suddenly fainted, and explained after attendants revived her, that she had seen blood on one of the necklaces. Staff could find no trace of stains on the gems, but the woman remained convinced that the person who last wore the necklace had been murdered.

Of all the hauntings connected with royal buildings, the most intriguing are those of Glamis Castle. The towered and turreted fortress beside Dean Water, near Forfar, Angus, is the Scottish family home of the Queen Mother's family, the Bowes Lyons. Princess Margaret was born there. But its 16-feet thick walls have been cloaked with mystery ever since Macbeth usurped the Scottish throne by murdering King Duncan in one of the rooms in the year 1040.

The Queen Mother delights in telling the younger members of her daughters' families the spine-chilling tales that have sprung up about Glamis. How Lord Beardie Crawford and fellow revellers diced with the devil in a tower room, and were condemned to stay there, drinking eternally until the Day of Judgement. Of how the Ogilvies, fleeing from the Lindsays during a clan war, were locked in a room and forgotten, starving to death.

The Queen Mother herself is one of the many who have seen the sad Grey Lady of Glamis, who haunts the Clock Tower. She is believed to be the ghost of Janet Douglas, wife of the sixth Lord Glamis, who was burned to death on Castle Hill, Edinburgh, in 1537 after being falsely accused of witchcraft and of plotting to poison King James V.

There is one Glamis phantom that few in the family have ever been prepared to talk about. Victorian high society was alive with rumours that a hideously mis-shapen beast of a man had been born into the Strathmore clan, an immensely strong and hairy egg-shaped creature whose head ran straight into a huge body which was supported by toy-like legs and arms.

Unable to reveal the monster's existence, yet unable to kill it, the family were said to have locked their odd off-spring in one of the secret rooms built at Glamis in the last years of the 17th century. There it lived for years, known only to the Earl of Strathmore, his lawyer and land agent, and, when he reached 21, the Earl's heir.

Guests returned from Glamis with strange stories that fuelled the gossip. Many said they had been woken in the night by the howls and snarls of an animal. One woman claimed she saw a pale face with huge, mournful eyes staring at her from a window across a courtyard. When it disappeared, she heard appalling screams, and watched an old woman scurry across the yard carrying a large bundle.

In 1869, a Mrs Monro woke in her bedroom at the castle to feel a beard brush her face. As she fumbled for a light, the shape that had been standing over her shambled into the next room, where her son was sleeping. The boy's screams of terror brought Mrs Monro and her husband racing to his bedside. As he explained that he had just seen a giant, they all heard a crash.

At breakfast next morning, other guests said they too had heard the crash, and one said she had been woken by the mournful whines of her small dog. But their hosts could offer no explanation.

In 1865, a workman who found a secret passage, and claimed to have seen something alive in a room off it, was 'subsidised and induced to emigrate'.

In 1877, essayist Augustus Hare watched the Bishop of Brechin offer to share the burden that was making the then Earl of Strathmore morose at at a house party. Hare reported, 'Lord Strathmore said that in his most unfortunate position, no one could ever help him'.

Andrew Ralston, land agent to the Strathmores from 1960 to 1912, was once asked for the full story by the Countess, grandmother of Queen Elizabeth II. He replied, 'Lady Strathmore, it is fortunate you do not know it, for if you did you would never be happy'.

Dowager Lady Granville, the Queen Mother's sister once admitted, 'We were never allowed to talk about it when we were children. Our parents forbade us ever to discuss the matter or ask any questions about it. My father and grandfather refused absolutely to discuss it.'

And the 12th Earl, the Queen's great-grandfather, was quoted as saying, 'If you could only guess the nature of the secret, you would go down on your knees and thank God it was not yours'.

His warning failed to deter historians and ghost-hunters. For years they tried to unravel the nature of the secret. Once, towels were flown from the window of every known room, to try to locate the possible hideaways. Experts combed the family tree for clues, and in the 1920s journalist Paul Bloomfield came up with what seemed a plausible explanation.

According to Burke's Peerage the 'bible' of British nobility, Lord Glamis, heir to Thomas, the 11th Earl of Strathmore, married Charlotte Grimstead on December 21, 1820, and was presented with an heir, Thomas George, later the 12th Earl, on September 22, 1822. But when Bloomfield checked Douglas's Scots Peerage, he surmised that Lord Glamis and his wife also had a son 'born and died on October 21, 1821'. Another reference book gave the date as October 18.

Bloomfield guessed that the first-born son did not die, but was badly deformed. He could never inherit the title and estates. Expected to live only days, he was kept alive and well cared for, but he survived both his father and his younger brother. A third son, Claude, became the 13th Earl, and was succeeded in 1904 by his own boy, Claude George, born in 1855 and who became father of the Queen Mother.

It is believed that he was the last heir to be initiated into the grim family secret, when he reached the age of 21 on March 14, 1876. His son and successor, Timothy, was never told the story, although he once said, 'I feel sure there is a corpse or a coffin bricked up in the castle walls somewhere, but they are so thick that you could search for a week without finding anything.'

Today, the legend of the monster lives on only in the name of the rooftop lead path where he may have been exercised at night – the Mad Earl's Walk.

The Grey Lady still prays silently in the chapel. Spirits still haunt the room where the Strathmore's personal hangman used to sleep. And Earl Beardie is still seen, a huge old man with a flowing beard, sitting by a fire in one of the castle bedrooms.

Yet for all their ghosts, none of the royal palaces and castles rank as the most haunted place in Britain. That title rests with an ancient fortification that kings and queens of earlier ages used for less civilised purposes . . .

The Bloody Tower

The Tower of London's ancient battlements, colourful Beefeaters and legendary ravens attract millions of tourists each year from all over the world. They stroll the picturesque buildings and courtyards, and listen enthralled as guides bring to life the spectacular and violent happenings that shaped Britain's past. For the Tower was once the most blood-drenched spot in England, and for more than 700 years it has had the ghosts to prove it.

The first, in 1241, was that of Saint Thomas Becket, who had been murdered at Canterbury Cathedral 71 years earlier. Becket was a Londoner, and had been Constable of the Tower before becoming Archbishop of Canterbury in 1162. His spirit was seen by 'a certain priest, a wise and holy man'. It was said to have returned to demolish extension walls which were upsetting people who lived near the Tower. The priest saw the apparition strike the walls with a cross, whereupon they fell as if hit by an earthquake.

Later Tower ghosts had more personal reasons for returning to the London landmark. Anne Boleyn, the second of Henry VIII's six wives, is the most frequently seen spirit. Several sentries have spotted her over the years, and one even faced a court-martial because of her.

He was found unconscious outside the King's House on a winter's night in 1864, and accused of falling asleep on watch. At the hearing, he told how a strange white figure had emerged from the dawn mist. It wore a curious bonnet which appeared to be empty. The private, who served with the King's Royal Rifle Corps, challenged the figure three times, but it continued to move towards him. When he ran the bayonet of his rifle through the body, a flash of fire ran up the barrel and he passed out.

Two other soldiers and an officer told the court-martial

they had seen the apparition from a window of the nearby Bloody Tower. After hearing that the incident had taken place just below the room where Anne Boleyn spent the night before her execution for adultery on May 19, 1536, the court-martial cleared the unfortunate sentry.

Anne, Queen for 1,000 days, had a horror of English steel in English hands, and her husband agreed to import a French executioner with a French sword for the beheading. But after her death, there were no more niceties. Her headless body was bundled into an old arrow chest and buried in unseemly haste in the Tower chapel of St Peter ad Vincular.

Sentries have often seen her ghost pacing up and down outside the tiny church, and one night, one of the guards noted an eerie light shining from inside the chapel. He climbed a ladder to peer through a window, and saw a ghostly procession of knights and ladies in Tudor dress file slowly up the aisle, led by a woman who looked like Anne. When she reached the altar, they all vanished, leaving the chapel in darkness again.

Anne's restless spirit does not confine itself to the scene of her death. Her headless body is said to arrive in a phantom coach at her childhood home, Blickling Hall, Norfolk, on the anniversary of her execution, and she has also been reported wandering through the grounds and attics of Rochford Hall, Essex, during the 12 days after Christmas.

But it is at the Tower where most of her rambles have taken place, often near the time of other executions. At 2 a.m. on February, 1915, Sergeant William Nicholls and his watch saw a woman in a brown dress with a neck ruff. She walked quickly towards the Thames, which runs past one side of the Tower, then disappeared into a stone wall. Five hours later, a German spy was shot in the moat, one of 11 executed there during World War One. Anne was last seen in February, 1933, when a guard reported a headless apparition floating towards him close to the Bloody Tower.

The ghosts of three other 16th century ladies who lost their heads have also been spotted in the Tower. Catherine Howard, the fifth of Henry VIII's wives, was beheaded there in 1542, and has been seen walking the walls at night. Margaret, Countess of Salisbury, re-enacts the horrors of her 1541 execution on its anniversary. She was dragged screaming to the block, and the axeman chased her round the scaffold, missing his target three times before finally severing her head. Lady Jane Grey, who reigned as Queen for only nine days, has been reported several times.

Two sentries recognised her when they saw the figure of a woman running along the battlements of the inner wall, near the Salt Tower, at 3 a.m. on February 12, 1954 – the 400th anniversary of her beheading on Tower Green, less than 150 yards away. Exactly three years later, two Welsh Guardsmen spotted a white shapeless form on the Salt Tower, where Lady Jane was imprisoned before her execution, at the age of 17.

In 1970, a man from Grays, in Essex, wrote to a London evening newspaper, saying his girlfriend had seen what seemed to be a ghost during a visit to the Tower. He said the figure of a long-haired woman, wearing a long black velvet dress and white cap, was standing by an open window in the Bloody Tower. A large gold medallion hung round her neck. The girl was the only one in her party to see the figure, and she said that, when she went up to the window, both on the visit and on a later occasion, she found it difficult to breathe. Experts believe she is one of the very few people to have seen Lady Jane Grey in daylight.

Phantom men and children also roam the Tower. In 1890, a sentry described an encounter so vivid that he 'Nearly died of fright'. He was on duty in the Beauchamp Tower when he heard someone call his name.

He said, 'I turned and there, floating in mid-air, was a

face, red and bloated, with a loose, dribbling mouth and heavy-lidded pale eyes. I had often seen it in the history books – it was Henry VIII, with all the devil showing in him.

'I was so scared I did not stop running until I came upon two of my comrades. They were beginning to clamour, when they suddenly broke off – the face had followed me.

'The affair was hushed up, and were were all told not to breathe a word that the Tower was haunted.'

During World War One, another sentry reported seeing a ghostly procession pass him near Spur Tower. A party of men were carrying a stretcher bearing the headless corpse of a man, his head tucked in beside his arm. Historians said this was the practice in earlier centuries, when bodies were returned to the Tower for burial after executions on Tower Hill.

The ghost of a former Duke of Northumberland has been seen so often on the battlements between the Martin and Constable towers that sentries have nicknamed the pathway Northumberland's Walk. Sir Walter Raleigh, the favourite explorer of Elizabeth I, who was imprisoned in the Tower by her successor, James I, and executed in 1618, has also been reported by guards; and two little children seen walking the Bloody Tower hand-in-hand are believed to be Edward and Richard, the two princes allegedly murdered on the orders of their uncle in 1483, so he could claim the throne as King Richard III.

Some of the Tower's many ghosts are not recognisable as personalities. The Keeper of the Jewel House, Major-General H D W Sitwell, woke one morning in 1952 in his quarters in St Thomas's Tower, on the outer wall, to see a monk in a brown habit through the open door of his bedroom.

Some of the ghosts are not even human. In October 1817, Edmund Swifte, then Keeper of the Crown Jewels, was dining with his wife, son and sister-in-law in his parlour in Martin Tower. As he offered his wife a glass of wine, she exclaimed, 'Good God, what's that?'

He followed her startled gaze, and saw a cylindrical glass tube, about as thick as a man's arm, and filled with white and blue liquid which seemed to be constantly churning. It hovered between the top of the table and the ceiling, moving slowly from one person to another until it passed behind Swifte's wife, pausing over her right shoulder. She crouched and clutched her shoulder, shouting, 'Oh Christ, it has seized me.' Swifte lashed out at the cylinder with a chair, then rushed upstairs to check that the couple's other children were all right. When he returned, the apparition had vanished.

Only days later, a sentry outside Martin Tower watched vapour pour through the narrow gap between the closed door and sill, and take the shape of a giant bear. The guard lunged at it with his bayonet, but the cold steel passed through the figure and stuck in the wooden door. The man collapsed from shock, and never recovered. When Swifte visited him the following day, he declared him 'changed beyond recognition'. Within days, the sentry was dead.

The Tower also has its share of mischievous spirits. Several Yeoman Wardens have found themselves bundled out of bed in one particular small bedroom in the Well Tower. In the autumn of 1972, a photographer was sent tumbling from a ladder as he set up his camera to take pictures of a mural in the Beauchamp Tower room where Lord Lovat, the last Tower prisoner to lose his head, had been held.

Perhaps the most curious fact about the Tower of London hauntings is that none have ever been reported in the White Tower, the largest and oldest of all the buildings, and the heart of the entire complex. Guy Fawkes, the man who tried to blow up the Houses of Parliament in 1605, was just one of the celebrities incarcerated there in cruel conditions before his gruesome death – but none have ever returned to the scene of their ordeals.

Masons restoring one of the walls in the late 1850s may have uncovered a clue to the reason for that. In the 11th century, when work on the White Tower began, it was believed that buildings could be protected against malevolent spirits by sacrificing an animal in them. Eight centuries later, the repair men broke into one of the thickest stone walls . . . and found the skeleton of an ancient cat.

The Tragic Queens

The ghosts of two tragic Queen Marys still haunt the houses where both found temporary respite from their unhappy lives in the 16th century. Both were Catholics caught on the wrong side of the power struggle in England following the death of Henry VIII.

Mary Tudor, Henry's elder daughter, was staying at Sawston Hall, near Cambridge, when her weakling brother, Edward VI, died. Before the news reached her, the powerful Duke of Northumberland tried to stage a coup. He seized the Tower of London, proclaimed his daughter-in-law, Lady Jane Grey, queen, and sent his men north to capture Mary.

She was asleep in the Tapestry Room at Sawston when her host John Huddleston learned of the approaching danger in the early hours of July 8, 1553. He woke her, and smuggled her out of the hall's back door disguised as a milkmaid. From a nearby hilltop, she reined in her horse to watch the Duke's frustrated soldiers set fire to Sawston.

After she became Queen on July 19, Mary showed her gratitude to the Huddleston family by helping them to build an even finer hall, which was completed in 1584. Since her death in 1558, Mary has often been seen there in the Tapestry Room, which survived the fire, or walking majestically in the grounds. Sometimes the haunting strains of the virginal she used to play there for her father have been heard.

Mary Stuart, Queen of Scots, spent years as a captive in English mansions and fortresses before Queen Elizabeth I had her executed at Fotheringhay Castle on February 8, 1587. She was moved from home to home, the unwitting focus of attention for every Catholic plotter with a grudge against the Protestant Good Queen Bess, and her ghost has been reported in many of the places that were her prisons.

She has also been seen in places with happier memories. When she was held at Fotheringhay, she used to attend mass undetected by visiting Southwick Hall, three miles away, via an underground tunnel. Her spirit has been sighted there since her death.

In 1878, a guest at Nappa Hall, Wensleydale, Yorkshire, claimed he had met a 'very lovely ghost' whom he recognised from portraits as Mary. She was tall, slim and wearing a black velvet gown. Mary visited the hall while under arrest at nearby Castle Bolton.

Beaulieu, Hampshire, was the scene of one of Mary's escapes from her long years in captivity. Even today, ghostly footsteps are heard rushing down the Palace House staircase she used for her freedom dash.

The most curious story concerning Mary Stuart came after her son King James VI of Scotland and later James I of England, ordered that Fotheringhay, where his mother had died, be destroyed.

Many of the fittings from inside its stone walls were bought by a local innkeeper, William Whitwell – including the oak staircase by which Mary reached her room. Whitwell later installed them in his inn, now The Talbot at Oundle, Northamptonshire, and found to his dismay that a phantom was part of his purchase.

The ghost of Mary, Queen of Scots.

Curse of Civil War

The greatest crimes against nature, according to Church and government teaching in bygone centuries, were killing a king and waging civil war. Both happened in Britain more than 300 years ago, when Oliver Cromwell's Round-heads defeated the Royalist Cavalier armies, and beheaded King Charles I. Ghostly echoes of that catastrophic conflict have lingered around the country ever since.

Edge Hill, on the border between the counties of Warwickshire and Northamptonshire, was the scene of one of the bloodiest battles. More than 40,000 fighting men clashed there on Sunday, October 23, 1642, as Prince Rupert led the King's troops into action against Cromwell's Parliamentarians. At the end of the day, the fields were littered with dead and dying, and both sides withdrew to continue the war elsewhere.

The following Christmas Eve, a group of shepherds were hurrying home at around midnight when they passed the battlefield. The sound of approaching drums, the clatter of arms, and the awful groans and screams of dying men stopped them in their tracks. Before they could take to their heels, the rival armies materialised all around them, eerily-lit colours blowing in the wind as they blazed away at each other with muskets and cannon. The bizarre action continued for more than three hours, finally fading at just after 3 a.m. on Christmas morning.

When it was all over, the bemused shepherds ran to the nearby village of Keinton, and woke the local justice of the peace, a man called Wood, and the local minister, a Mr Marshall. Both swore an oath that the men were not drunk, and agreed to accompany them to Edge Hill the following night. News spread quickly during the day, and when darkness fell, the crowd included 'all the substantial inhabitants of that and neighbouring parishes'.

They were not disappointed. The two armies 'appeared in the same tumultuous warlike manner, fighting with as much spite and spleen as formerly'. Terrified spectators arrived home in the early hours to pray for deliverance from what they believed to be a hellish visitation. For a week, it seemed their prayers had been answered, but the following Saturday night, the horrifying scenes of bloodshed and cruelty were re-enacted, 'with far greater tumult', for four hours.

Eventually, rumours of the phantom battles reached King Charles in Oxford. He sent three officers and three other 'gentlemen of credit' to investigate the stories. Mr Wood and Mr Marshall led them to Edge Hill, and they saw for themselves the gory action replay. The officers had seen the fighting on the actual day of the battle, and recognised many of the spectral clashes – even the faces of some of the combatants. The King was convinced. He declared that the nightmare tableau was a sign of God's wrath against those who waged civil war.

Over the years, the sounds and sights of war have been reported many times by people passing Edge Hill, although the fighting has never been as vivid as it was that first Christmas.

Three years after Edge Hill, Cromwell's forces routed a Royalist army at Naseby, Northamptonshire, and for nearly 100 years, generations of villagers from miles around gathered at the site on the anniversary of the battle to watch it re-enacted in the skies, and listen to the din of the guns, and the groans of the victims. With time, the phenomenon faded, but the echoes of a third civil war battle continue to this day.

A Royalist army of 4,000 was slain at Marston Moor, Yorkshire, during another Roundhead victory on July 2, 1644, and drivers travelling through the area have seen groups of dazed, bewildered men in Cavalier clothing staggering along by the roadside as if trying to escape pursuers. Two motorists who saw them in 1932 described vividly the long cloaks, high boots, long hair and wide-brimmed hats with cockades, typical of Royalist dress of the period.

The ghosts of the two leaders of the rival armies have also been seen since their deaths. King Charles was beheaded in 1649 at Westminster, and his body taken to Windsor for burial. It has been seen in the castle library there. He is also said to be the headless phantom of Marple Hall, in Cheshire.

Cromwell's spectre has been reported both at the Golden Lion at St Ives, Huntingdon, his regional headquarters, and walking in Red Lion Square, London, with two ghostly aides, John Bradshaw and General Ireton. The bodies of the three men were said to have been exhumed and carted there, on their way to Tyburn jail, after the Restoration of King Charles II in 1660.

Cromwell's most dramatic reappearance came in the winter of 1832. England was again seething with revolt because of a controversial Reform Bill, and an angry mob was besieging Apsley House, the London home of the Duke of Wellington. As the Duke paced his room, deeply troubled over what line he should take when the Bill was debated, he met an armour-clad figure he recognised from portraits as that of Cromwell.

The phantom did not speak, but pointed meaningfully at the crowds outside the house. Long after the Bill was passed, Wellington revealed that he had seen the ghost – and that it had changed his attitude to the reforms.

A ghost also changed the mind of an army leader during the civil war. The Duke of Newcastle had occupied Bolling Hall at Bradford, Yorkshire, on behalf of the King, and ordered that everyone in the Parliamentary stronghold be executed at dawn next day. His soldiers were puzzled when he withdrew the order shortly before they were due to carry it out. They learned that a female figure in white had appeared by their leader's bed three times in the night, wringing her hands and pleading, 'Pity poor Bradford'.

Forty years after the war England was again in the grip of internecine fighting. The Duke of Monmouth led a rebellion against James II, but his West Country army was crushed in the last battle fought on English soil, at Sedgemoor, Somerset, on July, 1685. The ghosts of some of the 1,000 men slaughtered have since been seen at the site of their deaths, and a phantom Cavalier horseman is said to be the Duke himself. He escaped capture at the battle, but was beheaded nine days later.

A cruel sequence to the bloodshed was witnessed by a group of schoolchildren walking up Marlpit's Hill, near Honiton, Devon, in 1904. They saw a wild-looking man in a black wide-brimmed hat and a long, brown coat. His dazed look troubled the children, though their teacher saw nothing.

Research showed that the bedraggled ghost may have been that of a man who had escaped the carnage of Sedgemoor, and made his way back to his wife and children, who lived in a cottage on the Hill. As he neared his front door, a troop of soldiers rode up, and cut him down with their swords.

Monmouth supporters who survived the fighting were tried by Judge George Jeffreys in a series of cases which became known as the Bloody Assizes. Slavery, transportation, flogging and execution were the sentences he meted out in a legal reign of terror. Since then, his ghost, complete with black cap used to deliver death sentences, has been seen in rooms where he stayed during his West Country tour of duty – at Clough's Hotel, Chard, Taunton Castle, the Great House, Lyme Regis, and Lydford Castle.

The American Civil War, which lasted from 1861 to 1865, has also left phantom reminders for future generations. One of the most horrific battles was at Shiloh, where

20,000 men died. Next day, locals reported that a nearby river ran red with blood. And the sights and sounds of the battle have been re-enacted in the skies over the battlefield.

Wars between nations have also left their mark on the supernatural world. The eerie footsteps of marching knights in armour have been heard at the historic English West Country site of Glastonbury, and headless war horses have been seen galloping through a Wiltshire valley near Woodmanton, the scene of an ancient battle between the Britons and the invading Romans.

In 1745, more than 30 Cumbrians watched a phantom army march through the sky above Souter Fell at the time of the Jacobite rebellion, and ghostly soldiers have been spotted at the site of the 1746 Battle of Culloden in Scotland.

During World War One, soldiers from both the German and Allied armies told tales of supernatural intervention in the fighting of August 26, 1914.

The British Expeditionary Force had taken a battering and looked like being over-run by the Kaiser's troops. Then the so-called Angels of Mons appeared, causing consternation in the German trenches. The British had time to retreat and regroup.

Author Arthur Machen, who wrote a story for the London *Evening News*, described the angels as phantom bowmen from the 1415 Battle of Agincourt. But he later claimed that he had made the whole thing up. The paper was deluged with letters from officers and men saying they had seen the spectres.

An officer from Bristol told his story in a local church parish magazine. He said his company had been cut off by German cavalry, and he expected certain death. Then the angels appeared between the two forces, and the German horses were terrified into flight. A brigadier general and two of his officers told the same story to their chaplain. And a lieutenant-colonel claimed that, during the retreat, phantom horsemen guarded his cavalry battalion for 20 minutes, escorting their flanks in fields by the road.

After the war, it was learned that both German and French troops involved in the Mons bloodshed had seen unearthly allies helping the British. Cynics argued that the three armies were exhausted by heavy fighting, and could have been hallucinating. By then the Angels of Mons had served their purpose. Morale in the British trenches after the battle was sky-high.

On August 4, 1951, two English women on holiday in the French town of Dieppe awoke to the sound of gunfire. For three hours they made a note of every sound, and experts who examined their record found it a carbon copy of what had happened on August 19, 1942, when more than half a 6,000 strong Anglo-Canadian force was wiped out trying to storm the German-held Normandy port in a dawn raid.

The women asked fellow guests at their hotel about the sounds but no one else had heard a thing.

Hail to the Dead Chief

Abraham Lincoln is still in the White House in Washington, but he is far from alive and well. For the President who led the United States out of the bitter Civil War, only to be assassinated by a fanatic in April, 1865, now haunts the corridors of power as a ghost. American leaders and other celebrated visitors all claim to have seen him or felt his presence over the last 100 years.

Sir Winston Churchill, Britain's wartime Prime Minister, did not enjoy sleeping in Lincoln's old bedroom, and frequently moved to another room across the hall during the night. Queen Wilhemina of the Netherlands is said to have fainted after answering a knock on the Rose Room door to find Lincoln standing outside. And President Theodore Roosevelt once said, 'I see Lincoln – shambling, homely, with his sad, strong, deeply-furrowed face – in different rooms and halls.'

It was in 1934, during the presidency of Franklin Roosevelt, that Lincoln made his most dramatic appearance. Mary Eben, one of the White House staff, entered a bedroom on the second floor to find a figure in an old-fashioned black coat sitting on the bed and pulling on a pair of boots. She stared, stunned, at the man for several seconds before he vanished.

More than 40 years earlier, another White House aide made a public appeal to Lincoln's ghost to leave him alone. John Kenney was personal bodyguard to President Benjamin Harrison between 1889 and 1893, and his nerves were frayed by footsteps in corridors and rapping on doors which seemed to have no natural explanation.

On a visit to Baltimore, he attended a seance, at which Lincoln's spirit was present. Kenney is said to have said, 'Please don't do it again, Mr Lincoln. I am guarding the life of President Harrison now, and you've got me so scared I can't do my duty.'

Kenney never heard the ghost again.

Lincoln is said to step up his visits to his old offices in times of crisis. The chief White House usher saw him several times during World War Two, and one of Theodore Roosevelt's valets fled shrieking from the building. President Eisenhower said he sensed Lincoln's presence many times.

Even while Lincoln was alive in the White House, as the 16th President of the United States, there were ghosts there. His wife, a confirmed spiritualist, saw her brother Alexander after he was killed while fighting on the Confederate side in the Civil War.

Lincoln himself had a vision of his own death. He told

an aide shortly before his assassination that he had been woken by quiet sobbing.

He said, 'I wandered downstairs until I came to the East Room. Before me was a catafalque with a corpse whose face was covered. "Who is dead?" I demanded of the mourners. "The President," was the reply. "He was killed by an assassin." '

Franklin Roosevelt's death in 1945 came in chilling circumstances, and many lay it at the door of an ancient Shawnee curse.

What is known as the Indian's Revenge started nearly 180 years ago when Shawnee chief Tecumseh died in a pitched battle with William Harrison, then Governor of Indiana.

In revenge, the Shawnee placed a curse on Harrison. Medicine men told how the Governor would become president in a year ending in zero – but would die in office. From then on, any President elected in a year divisible by 20 would also die before his term ended.

Harrison – grandfather of Benjamin – was duly elected president in 1840 – and died a month after taking office.

Abe Lincoln was elected 20 years later – and was assassinated.

The deaths of five other presidents have also been attributed to the Shawnee curse . . .

James Garfield was elected to office in 1880 and was assassinated in 1881.

William McKinley was re-elected in 1900 and was assassinated in 1901.

Warren Harding was elected in 1920 and died of a stroke in 1923.

Franklin Roosevelt was elected in 1940 and died in 1945.

John Kennedy was elected in 1960 and was assassinated in 1963.

White Lady of Bohemia

Tiny Petr Vok was a very special baby. He was born in 1539, the sole heir and last in the line of the aristocratic Rozmberk family of South Bohemia. His father, Josta, was desperate that nothing should stop him continuing a lineage that stretched back for centuries.

He hired a team of nannies to maintain a 24-hour watch on the boy at his home, Krumlov Castle, on the banks of the river Vltava. They were with him constantly, caring for him by day, sleeping in his room at night.

One night, one of the nannies woke with a start. The room was strangely bright, glowing with moonlight. As she stared round it, she saw a curious, misty figure beside Petr's cradle. Speechless and shaking with fright, she watched the intruder, a woman in white, gaze down at the sleeping child. As the babe started to cry, the woman gently picked him up, cuddled and stroked him, kissed him, and tenderly placed him back in the crib. Then she disappeared.

Still shaking, the startled nanny woke her colleague, and told her what she had seen. Gingerly, they crept to the cradle. Little Petr was sleeping peacefully, a smile between his pink cheeks.

The two nannies had heard the legend of the White Lady, a ghost said to haunt the castles of the Rozmberk family. They had dismissed it as folklore. Next night, they both stayed awake to see if the woman returned. The room was locked, the windows shut, but just after midnight they saw a pale light, and the phantom nanny appeared again. She rocked the cradle, caressed the child, then, seemingly satisfied that all was well, dissolved into a wall.

Each successive night, the two girls waited for the White Lady to appear. She never let them down. Content that she meant no harm, they took her for granted, and did not bother to stay awake. Then one of the regular nannies fell ill, and a temporary replacement moved in. Nobody told her about the White Lady, and as she lay tossing and turning, unable to sleep, she saw the ghost arrive at her usual time. Next morning, she told one of the other girls what she had seen, and was told, 'Don't worry, the White Lady takes care of Petr at night'.

But the girl was worried. What would the master say if something happened to the boy? How could she explain that she had left him in the care of a ghost?

Next night, she again lay sleepless as the phantom appeared, walked up to the cradle, and rocked it. When Petr started to cry, she picked him up. Then the anxious nanny leapt from her bed, and walked bravely to the figure.

She grabbed the child from her arms. The White Lady put up no resistance. She stood motionless, then turned to the girl and said sternly, 'Do you know what you are doing, bold one? I am a relative of this newborn child, and it is my right to be with him. You will not see me here any more.' The ghost made a cross sign on the wall, then disappeared into it.

Petr grew up to inherit the castle when his father died. He was told the story of his mystery guardian, and often discussed her with friends and relatives. One day he decided to check the wall where she had last been seen. Workmen began knocking a hole – and discovered a cache of coins and gems.

Who was the caring White Lady? There are two theories. An historian believed she was lady Perchta, a Rozmberk who married an aristocrat called Jan Lichtenstein. But he proved cruel and merciless, and eventually she left him and his selfish family, and fled to Krumlov. Later she moved to Vienna, to live with her daughter, and died there in 1476. A portrait of her and her husband still hangs in the castle at Jindrichuv Hradce in South Bohemia, today part of Czechoslovakia.

Another theory is that the White Lady is Marketa, daughter of the Archduke Maidburce. She married Jindrich, from the Hradce castle family, but when he died in 1362, became a recluse at a convent in Krumlov. From time to time she visited her children and friends, wearing an all-white nun's habit.

Whoever she was, there are many documented sightings of her in the homes of the Bohemian nobles. Apart from Krumlov, she was seen in castles at Telci, Bechyn, Trebon and Jindrichuv Hradce. The accounts of her are always the same: a woman of breeding, all in white, with a hood over her head. She was seen just before anything happy or sad was to happen – if the news was bad, she wore black gloves instead of white.

Workmen renovating part of the Jindrichuv Hradce castle once spotted her at midday, at a window in a tower. They were startled, because no one had been in the tower for years and the staircase had been destroyed by fire. They watched her for some time before she slowly faded from sight, as if moving across the room.

Servants at the castle often saw her in the corridors at night, sometimes with a bunch of keys hanging at her waist. Her face was always serious, but never frightening. As she flowed along landings, opening doors, some servants spoke to her. She would answer with a movement of her head or hand, or even with a few words.

The servants believed she was taking care of the families to whom she was related, looking after the young, warning the adults of danger, and preparing the dying for their fate.

In 1604, one of the Hradce family, a man called Jachym, fell suddenly ill, but nobody thought his condition could be fatal. Then, on a cold snowy January night, a priest at the castle woke with a start, and thought he heard someone calling him. Dressing hurriedly, he opened his door and found a woman in white.

'Do not waste any time,' she said in an urgent voice. 'Follow me.' The priest turned back to look for a light, but the woman took his lantern, breathed on the glass, and a flame flared up.

She led the way to the castle chapel, where the astonished priest found candles burning everywhere, as if in readiness for a Mass. The woman told him to collect everything he needed to perform the last rites. Still puzzled, he did as she directed, then followed her to Jachym's bedroom door.

Here both she and the light vanished, but by now the priest realised who she was. He went into the bedroom, and found the servants asleep. On the sickbed, Jachym was clearly gravely ill and fighting for his life. The priest performed the last rites, and the master, the last of the Hradce line, died in peace.

The caring nature of the ghost points to her being Marketa, as Sedlacek suggested. She was once supervising renovations to part of Jindrichuv Hradce castle, and to encourage the workmen, she promised them all sweet pudding when they had finished – and every year after that.

When the work was complete, by the late autumn, Marketa kept her promise, and prepared a great feast. But as the men and their families sat down at long tables in the castle courtyard, snow began to fall on to the sweet pudding.

The following year, the hostess switched the date of the feast to early spring, and the tradition of feeding the poor on 'Green Thursday' continued throughout her lifetime, and the lives of her successors, long into the 16th century.

The Spooky Chess Player

Maurice Tillet was grotesquely deformed. A gentle giant, he was a professional wrestler with the soul of a poet. Highly intelligent, he could speak 14 languages.

Tillet died in 1955. Yet 25 years later American businessman Patrick Kelly claimed to have regularly played chess with him – from beyond the grave.

Years earlier, he and Kelly had often played chess in the businessman's home near Braintree, Massachusetts. During the game Tillet often raised his terrible head, looked sadly at Kelly and groaned, 'How awful it is to be imprisoned in this body'.

Kelly said that once his friend Tillet's spirit was free of its heavy burden, he often returned to the chess board 'and we play as before'.

The ghostly games began after Kelly bought an electronic, computerised chess set. The businessman had always played on his library desk, where a plaster-cast of Tillet's death mask had stood for almost a quarter century.

Late one evening the computer deviated from its programmed plays and used an 18th century opening invented by a French master – a play Tillet had used constantly. Kelly recalled, 'I played out the game, and next morning noticed that the computer was not plugged in.

'I thought nothing of it at the time, but a few weeks later the computer suddenly used a similar opening – and again it was not connected to any power supply.'

Kelly had electronic engineers check the system. They found the computer would operate without electricity so long as Tillet's death mask was near. Puzzled, the businessman had the mask X-rayed for concealed electronic devices – but it was solid plaster.

According to Kelly the unplugged set would not operate for days at a time, indicating that Tillet's spirit was absent.

Kelly says, 'When I want a game, I set up the pieces without plugging the set in. If there is no response I know Maurice is not present. But often in mid-game the computer will play above its normal level, and I know he has stopped by. I prove this by pulling out the plug, but the game still goes on.'

Tillet was born in France but in his 20s developed 'acromegaly', a horrific disease which causes an uncontrollable growth of the bones.

After coming to the United States, he became a wrestler. He died when he was 45, and Kelly firmly believed that his old friend's poetic spirit was finally allowed to roam free of its grotesque body.

Who Haunts Ya, Baby?
Film stars and the friendly ghosts

Many stars of stage and screen claim to have seen ghosts, but few can equal the uncanny experience that has haunted Telly Savalas.

The actor, best-known as the TV cop Kojak, was driving home from a friend's house in a remote part of Long Island, New York, in the early hours of the morning. Glancing at his dashboard, he noticed that the clock registered 3 a.m. He also noticed that the fuel gauge registered empty.

He was in luck. The yellow lights of an all-night café filtered through the darkness. He pulled in, ordered coffee, and asked the way to the nearest gas station. He was told to take the path that ran through the woods at the back of the café and walk until he reached a freeway.

Savalas recalled, 'I was just about to set out when I heard someone ask in a high pitched voice if I wanted a lift. I turned and saw a guy in a black Cadillac. I thanked him, climbed into the passenger seat and we drove to the freeway.

'To my embarrassment, I had no wallet – it must have fallen out of my pocket. But the man loaned me a dollar. I insisted I must pay him back and got him to write his name and address on a scrap of paper. His name was Harry Agannis.'

The next day Telly Savalas looked up his good samaritan in the phone book. A woman answered the call. Yes, Harry Agannis was her husband. But, no, it was not possible to speak with him, for he had been dead for three years.

The actor's first reaction was shock. He came to the conclusion that there must have been some mistake, but he could not put the incident out of his mind. Eventually, he visited the woman, taking with him the piece of paper on which the stranger had written his name and address.

Savalas said, 'When I showed her the paper she was obviously deeply affected and told me that without doubt it was her husband's handwriting. I described the clothes the man had worn. She said those were the same clothes Harry Agannis had been buried in.'

The famous actor and the widow sat and looked at each other for a long time, hardly daring to admit to themselves the implications of what had happened. Was Savalas helped by the ghost of Harry Agannis?

He said, 'That was a case Kojak could not solve. I doubt if I'll ever be able to explain it.

Roger Moore, famous as screen master-spy James Bond, admits to being 'absolutely petrified' by the experience he had of the supernatural.

Moore had gone to bed early one night after a hard day's filming. He had been sleeping soundly, but suddenly woke up with the feeling that someone was in the room.

He said, 'Lying there with the light of the moon coming through the window I saw a misty substance – that's as near as I can describe it – floating across the bed. I was rigid with fright. When it disappeared I looked at my watch and saw that it was precisely 2 a.m.'

The following night he woke at exactly the same time and again saw the strange, drifting mist move across his bed. 'I was not keen on going to sleep in that room again, I can tell you. But our home help, a devout Jehovah's Witness, told me to leave a Bible, opened at the Twenty-third Psalm, on my bedside table. Rather reluctantly I agreed to do that and try one more night. It worked and I never had the experience again.'

It is surprising how many famous stars live in haunted houses. Some have come to terms with their domestic ghosts, others have fled in search of new property. Peter Sellers' ghost travelled with him. He was always sure that the great comic, Dan Leno, was his guiding spirit through life.

Glenn Ford and his wife Cynthia have experienced strange goings-on in their Hollywood home. Sometimes at night there have been echoes of some long-gone party in their garden. Cynthia Ford says, 'There is laughter, as though someone has just told a joke, and the clink of glasses.

'Once when we went downstairs in the morning all the garden furniture had been arranged in a circle. At least its a happy manifestation. Once I even smelled cologne of a type that isn't made any more – the sort that Valentino wore.'

Actress Elke Sommer and her husband Joe Hyams had no idea when they bought a Beverly Hills house that it was haunted. Soon after they moved in they began to hear strange noises coming from the dining room after they had gone to bed.

Any possibility of burglars or intruders was soon dismissed. The thumping, banging and strange noises went on night after night, always in the same room. Then, in the early hours of one morning when they were both asleep there was a frantic pounding on their bedroom door.

Joe Hyams opened it to find no one there – but thick, black smoke was billowing up the staircase. He rushed downstairs to find the dining room on fire. Elke Sommer sought the advice of several mediums and each one gave the same answer. Whoever the ghost was, it had probably set fire to the dining room as an act of mischief, then re-pented and decided to warn them.

British actress Adrienne Posta has managed to live with her ghost for years. 'He's very friendly,' she insists. 'The house has such a nice atmosphere I would hate to leave it.' Adrienne almost accepts the ghost as part of the family. 'Sometimes he gets a big agitated, then he crashes about and opens and closes doors. That's a nuisance, but not really frightening.

'The only time I got really annoyed was when he persis-tently threw open my bedroom door in the middle of the night. I was really beginning to suffer from lack of sleep and felt I had to do something about it. So one night I retaliated. I threw a book towards the door when it opened and gave him a piece of my mind. The pestering stopped.'

Another star who decided to tackle a household spirit was Chad Everett, famous for his role in the TV saga *Centennial*. His home on Hollywood Boulevard was haunted by a poltergeist. When he and his wife, Shelby, arrived home after an evening out they would be greeted by lights flicking on and off all over the house.

They were not particularly worried by this, but when the ghost became more and more boisterous, throwing things across the kitchen, upsetting furniture and generally making itself a nuisance, Chad decided to take things in hand.

He decided that the only way to tackle the spirit was by treating it like a naughty child.

He said, 'One day I just stood in the middle of the kitchen while things were being hurled around and said firmly, "I think you're a fool to waste so much energy. It is difficult enough to communicate yet here you are wasting your precious energy just frightening us. You're just making yourself miserable. Settle down and don't do it again". And it didn't.'

From that day on, the Everetts lived in peace, though they were always aware of that other presence. The actor says, 'I like to think I gave the poor ghost some good advice'.

Two stars – Vincent Price and Kim Novak – had startling experiences that lasted only a few seconds but which were never to be forgotten.

Price, master of the macabre, had his extraordinary glimpse of the unknown while flying from Hollywood to New York on November 15, 1958. He was immersed in reading a classic French novel for most of the journey, but at one point glanced idly out of the window. To his horror he saw huge, brilliant letters emblazoned across a cloud bank, spelling out the message TYRONE POWER IS DEAD.

Price admits, 'It was a terrific shock. I began to doubt my senses when I realised that nobody else on the plane appeared to have seen them, but for a few seconds they were definitely there, like huge teletype, lit up with blinding light from within the clouds.'

When Price landed in New York he was told that his close friend, actor Tyrone Power, had died suddenly a couple of hours earlier.

Kim Novak's encounter with the supernatural was just as startling, but alarmingly physical. The lovely blonde actress was working on location in England, making the film *The Amorous Adventures of Moll Flanders* at Chilham Castle in Kent, a 17th century house built round the keep of a Norman castle.

One evening after filming she was relaxing in her room, playing records.

She said, 'I put on one of my favourite tunes and couldn't resist dancing to it. As I whirled round the room I suddenly felt rather cold.

'A powerful force seemed to grab me round the waist. I

was lifted off my feet and slammed against the wall.'

That was all, but it was enough to make Kim turn off the record player and hurry in search of company.

British singing star Cilla Black feels protective towards her ghost. Four times within a period of seven years, she woke in the night to find the figure of a young girl, no more than 16, standing by the side of her bad.

She said, 'When it first happened, in my drowsy state I thought one of the children had come into the bedroom, perhaps wanting a drink of water or upset by a nightmare. The second time, I knew it was a ghost.

'She was so sweet I didn't feel frightened. In fact, believe it or not, I quite liked her being there. She appears to be wearing a long dress, like a nightie and just stands there. She never looks at me.

'I have actually spoken to her, asked her why she has come back and why she can't find peace. But she never answers. She just floats away through the door.'

Some famous people have had ghostly experiences in their childhood that have haunted them ever since, staying fresh and vivid in the memory. Actress Stephanie Lawrence, star of the hit London musical *Evita* is one. She says, 'When I was a child, our home was an old house on Hayling Island that had once been an army training school. It was full of long corridors and dark corners. One evening I was walking along one of the upstairs corridors when I heard footsteps behind me and what sounded like the patter of dog's paws.

'I spun round quickly and there was a soldier in bright red and gold uniform with a golden labrador by his side. I was absolutely terrified and rushed downstairs to find my brother. I've never forgotten it. I can still see him standing there.'

Stephanie discovered that the house had been called 'Stonehenge' because a great pile of stones stood in the garden, apparently marking the spot where a soldier's horse had been buried. Her parents found many army relics under the floorboards, including an old scarlet and gold braid uniform from the time of the Crimean War. She often wonders if it had once been worn by the ghost she saw.

Tiny blonde actress Charlene Tilton, who plays the part of Lucy Ewing in the TV series *Dallas*, believes she was haunted by the ghost of her grandfather when she was a child. Even when she became a teenager and left home, the feeling that she was being haunted stayed with her.

As a little girl, she lived with her mother and grand-father in a small modest flat in Hollywood. The old man died and six-year-old Charlene had to be left along while her mother went out to work. One day, Charlene remem-bers, her mother returned from work to find her crouched on the front doorstep, shivering with fright. The little girl explained that there was 'somebody scary' in the flat.

Charlene said, 'Without hesitation my mother told me that my grandfather's spirit was still there.'

Strange things happened as the years went by. No one actually saw the old man's ghost after that but his presence was felt. Charlene said, 'I remember very clearly some-thing incredibly scary when I was a teenager. One day a neighbour called in and asked if I would turn down the radio, which was blaring out pop music. In a teenage tantrum I refused.

'Then things began to happen. The plug jerked out of the wall without anyone being near it. When I pushed it back into the socket it was jerked out again so violently that sparks flew. The neighbour stood there with her mouth open, struck dumb by what she had seen, but we knew it was the ghost again.'

When Charlene left home to live in another district she thought she had left the disturbed spirit behind. She was determined to think no more about it. She and her boy-friend found a pleasant apartment and settled down there together. But soon the boyfriend began to complain that the place was 'spooky'.

Charlene agreed. 'You can sense something or someone invisible when you walked into an empty room. There were pockets of cold air and nearly every night doors would open silently and close again.'

Eventually Charlene could stand it no longer. She moved out, hoping that whatever it was that was haunting her would grow tired of following in her footsteps. She bought a beautiful home of her own in California overlooking a canyon. It seemed to be one move too many for the ghost. If it *was* her grandfather, she believes he has now found peace.

Matthew's Paranormal Pal

One of the most intriguing supernatural mysteries of modern times involves a young Englishman named Matthew Manning.

Matthew was 16 when in 1971 he first came face-to-face with a 300-year-old ghost. It was a meeting which had a profound effect on his life – a life affected by the supernatural, paranormal and bizarre.

Since then he has developed strange psychic powers – such as the ability to bend forks, stop watches and prevent electricity flowing.

But all this was yet to come on the day Matthew saw a shadowy figure on the staircase of his 17th century house in the village of Linton, in Cambridgeshire.

At first, he thought it was a burglar; then he realised it was a ghost. He spoke to the ghost – and the spirit replied. It even remained long enough for Matthew to sketch it. This was the start of a long and fascinating association.

Matthew learned that the ghost's name was Robert Webbe and that he was born in the house in 1678. He added the front portion of the house, where he was usually seen, in 1730, and died there about three years later.

Webbe's clothes and wig were of the 1720s style, and he walked with the aid of two sticks, complaining of his 'troublesome legs'.

Matthew said, 'When I first saw him, I thought he was completely solid. He was wearing a green frock coat with frilled cuffs and a cream cravat. He said, 'I must offer you my most humble apology for giving you so much fright, but I must walk for my blessed legs.'

'I grabbed an old envelope and pencil, and sketched him where he stood. A few moments later he turned, walked up the stairs and disappeared.'

Some time later, Matthew discovered he had the power of automatic writing: when the writer lets his hand and pen be guided by another mind. In this way, he exchanged many messages with Robert Webbe, and was able to check the historical accuracy of some of the things he was told.

Then Webbe began writing on the walls of Matthew's bedroom, although he was never seen doing it. Over a six-day period in July, 1971, more than 500 pencilled names and dates appeared on the walls. They were in a variety of styles of handwriting and were the names of Webbe, his family and other families who had lived in the area. The majority were from his own lifetime, but there were some from periods ranging from 1355 to 1959.

Matthew was not the only member of his family to be caught up in this supernatural world. His father, architect Derek Manning, told of vividly real sensations he had experienced while lying in bed. He was often awakened by the sensation that someone had climbed into his bed and superimposed another body on his. He even heard the scratching of a man's unshaven chin on the sheets beside him.

Sometimes he felt as though he was 'standing in a cage, looking out on to a purple sky, with rocks all round the entrance.'

Then it seemed as though he was 'in the centre of some-one's mouth, looking out through their teeth'.

At the same time, he experienced a prickling and tingling in the lower right leg, which eventually spread through both legs – similar to the symptoms of gout. This is almost certainly what pained Robert Webbe and may eventually have caused his death.

The ghost has sometimes played mischievous games on Matthew and his family. Strange antique objects have been found on the stairs. Some of the family's possessions have been spirited away. Things like old prints from the walls, a scarf, and a 50p piece from a money box which was later found abandoned on the stairs.

The bed in Matthew's parents' bedroom has often been found with the covers thrown back, and sometimes the pillows have been dented as though a person has been resting his head on them. Pyjamas left unbuttoned and neatly folded under a pillow have been found buttoned up.

Other strange happenings have included the sudden smell of strong pipe tobacco – although no one in the family smokes – the sound of footsteps ringing out from empty rooms, the aroma of old, musty books, and the stink of rotten fish. Sometimes the family would hear the ringing of a handbell from the hallway – although there was no such bell in the house. On other occasions, a candle was found on the cloakroom floor.

When Matthew asked Webbe about these strange happenings, he admitted he was responsible. It was his house, he told Matthew, and he could do whatever he wanted in it.

The eeriest experience of all was when Matthew came face-to-face with Webbe in his parents' bedroom and attempted to shake hands with him. His outstretched hand went right through the hand of the ghost.

Nevertheless, Matthew managed to give him a present

– a doll's wooden clog which belonged to his sister.

He held it out in friendship – and experienced an eerie feeling of timelessness. The ghost grabbed the clog and thrust it into the large pocket of his coat.

The next occurrence startled even Matthew, by now well used to the peculiar. The ghost of Webbe gradually faded away to grey and then nothing, and the doll's clog vanished with him.

Who is the ghostly Robert Webbe, and why has he haunted this house so consistently?

Matthew said, 'He was a grain trader who was very proud of the house he enlarged so grandly but did not live long enough to enjoy. He wanted to take the house with him.

'I think that is why he is going round and round in a strange sort of time loop, trapped by his own will in infinity. From time to time, someone in the house provides him with enough psychic energy to allow him to make contact.'

Ironically, when Matthew once got in touch with the ghost to ask him questions raised by his research into local history, Robert Webbe said he did not believe in ghosts, and that there were none in the house!

Children From Beyond The Grave

When the noise becomes too loud in Edna Rugless's home, she shouts out, 'Children! Please play more quietly.' It always works. The two little girls who kick up such a din obey her.

Mrs Rugless has a fine understanding with the two children. Which is unusual, because the youngsters have been dead for many years.

The children love to run around upstairs in the 300-year-old farmhouse in Devon, in the west of England. Sometimes there is a creaking noise, as though someone were riding an old-fashioned rocking horse. But Mrs Rugless said, 'There's nothing frightening or creepy about them. I'm very fond of children, having had four of my own. They're more than welcome.

'Their noise doesn't usually bother me. But one afternoon I heard them run out of the bedroom and start jumping around on the landing. I went into the hall and shouted up to them to stop, and they did. Of course, in their day children were taught to be obedient.'

Mrs Rugless and her husband Bill, a retired engineer, discovered the ghosts when they moved to the village of Farway. The noises from the bedroom began immediately. Their pet cat and dog refused to go anywhere near the room.

Then a friend with an interest in psychic happenings visited the couple on holiday. Mrs Rugless said, 'I deliberately didn't mention the noises of the children to her, but merely asked if she could feel anything unusual in the atmosphere. She told me that the spirits of two girls, aged about four or five, were active in the house and that their names began with an E and A.

The psychic visitor said she thought the girls were friends rather than sisters. One wore fine clothes, the other not so good clothes.

Mrs Rugless called in the local vicar, the Rev Frederick Gilbert, who checked parish records and found that two four-year-old girls belonging to the same family had died in the house. One was Elizabeth, who was buried in 1844; the other was Ann, whose death was 1902.

Mr Gilbert said, 'It is quite possible that these children were so happy in this house that they have been reluctant to leave. Because of the 58-year gap between their deaths it may seem odd that they should be playmates. But we have our own limited concept of time and it may mean nothing to them.'

Mr Rugless said, 'I was most disturbed at first, but my attitude now is that we're lucky to have such pleasant little spirits about the place.'

Marine Ghost Demands Justice

The restless ghost of a young American haunted a house in Portland, Oregon, for two years trying to convey to his mother the truth about his violent and mysterious death.

Lieutenant James Sutton had been one of a large middle-class family living in Portland at the beginning of this century. His parents, upright, respectable people, were proud when he was accepted for training as an officer in the Marine Corps at Annapolis Military Academy.

They looked forward to his letters, which were cheerful and affectionate. When the postman arrived on the morning of October 11, 1907, his mother rushed to the letter box as usual and tore open the envelope with happy anticipation. The neatly written pages were full of good humour – but as Rosa Sutton, held, the letter, her hand began to tremble and she had the feeling that something was terribly wrong. While sitting with her family the next evening she suddenly had a sharp attack of pain, a feeling of shock. She went upstairs to read her son's letter again to make sure she had not missed anything.

The following day, still feeling troubled, she went to Mass at her local Catholic Church. Afterwards she tried to busy herself with household chores but could not shake off a strong premonition that James would come home unexpectedly. She was so sure that her daughter, Louise, was sent to prepare his bedroom.

They were taken completely by surprise when Mr Sutton suddenly arrived home from work looking distressed and very pale. 'I have some bad news,' he said – and told his wife gently that he had received a telegram from Annapolis informing them that their son had shot himself and was dead.

Mrs Sutton, a devout Catholic, refused to believe that her son had committed suicide. 'At that instant,' she wrote later, 'Jimmie stood right before me and said "Mamma, I never killed myself . . . my hands are as free from blood as when I was five years old".' No one else in the room saw or heard anything and when Mrs Sutton kept insisting he was there they merely thought the news of his death had been too much for her.

What they could not dismiss, however, were the facts that came tumbling from her lips as she listened to the unseen presence. She said her son was trying to tell them something vitally important . . . he had been hit on the head with the butt of a gun . . . three other men jumped on him, beat him and tried to rub his face into the ground . . . they had kicked him and broken his watch. 'Oh, if you could see my forehead and put your hand on my forehead, you would know what they did to me,' moaned the ghost. 'But I did not know I was shot until my soul went to eternity.' Before disappearing, he pleaded with Mrs Sutton to clear his name and said he would never rest until it was cleared.

The phantom was persistent. On October 16 he appeared again. According to Mrs Sutton's testimony, he gave further details about his death, describing how his attackers had tried to bandage his head to hide what they had done. 'My face was all beaten up and discoloured, my forehead broken and there was a lump under my left jaw.'

As though to give final proof, the next time Mrs Sutton saw the ghost of her son it was a face hideously disfigured and discoloured. Still wrapped in a great coat, he seemed to be looking for something. 'It's my shoulder knot I can't find,' he complained piteously.

By now the whole house was alive with the ghost's presence. The young Marine's brother, Dan, swore he had seen him on one occasion and his sister, Louise, was keenly aware of his presence. Another sister, Daisy 'dreamed' one night that she had been shown a photograph of a crowd of young marines and could not take her eyes off the face of one of James's fellow officers, a man called Utley. Soon after, Mrs Sutton said her son had tried to tell her that his body had been hidden in a basement and that a lieutenant called Utley had been responsible.

First confirmation of Mrs Sutton's extraordinary experiences came three week's after James's death when Louise returned from the funeral at Annapolis bringing her brother's belongings with her. Among them was a shattered wrist watch.

The Suttons listened with mounting disbelief to the authority's official story of their son's death. According to official records Lieutenant Sutton and some friends had been to a naval dance. After it had finished they set off back to camp but were very drunk and a fight broke out. During the fracas James was thrown to the ground and heard to utter threats that he would kill the other two before morning. He returned to his tent to fetch some pistols. This led to his arrest and, during the attempt to seize him, he suddenly turned a gun upon himself.

Lieutenant Sutton had been buried at Arlington cemetery, but his ghost continued to haunt his old home in Portland. At first the Suttons felt there was little they could do. The naval doctors at the inquest had sworn that James's face was not disfigured. A verdict of suicide seemed in order. But after nearly two years the agonised parents made a dramatic decision. They asked for their son's body to be exhumed.

An independent inquiry revealed some staggering facts. The remains showed that he had indeed been seriously disfigured, as though he had been beaten up. His forehead was broken in and a lump formed by an injury was visible under the left side of the jaw. After the autopsy it was admitted that the angle of the bullet's entry into the body was not consistent with a self-inflicted wound. It was also found that the shoulder knot of his uniform was missing.

Soon after, the Suttons received an anonymous letter confirming the fact that Lieutenant Sutton had been murdered. The handwriting was traced and identified as that of a young serviceman who had been in the party after the naval dance. All attempts to track him down failed.

But it seemed the ghost was satisfied. As a good Catholic, he needed to remove the stigma of suicide from his name. Mrs Sutton still caught a glimpse of him from time to time but the image grew fainter and then disappeared for good. Justice had been done.

The Bank of England Ghost

For almost 200 years Britain's famous Guardsmen patrolled the corridors of the Bank of England in the City of London.

Their nightly vigil was to protect the gold in the Bank's vaults and started back in the 1780s at the time of the Gordon Riots in the capital.

The patrol ended in 1973 – but the Bank's other nightly visitor still makes her rounds of the Threadneedle Street buildings and gardens, searching for a lost loved one.

She's the Black Nun – so called by those who have seen her because of her thick black clothes and the dark veil which hides her face.

The apparition is supposed to be the ghost of Sarah Whitehead who roams the building looking for her brother Philip. At one time Philip worked as a clerk in the Bank but in 1811 he was arrested and charged with forgery. He was convicted and hanged – the punishment in those days for such a crime.

When Sarah didn't hear from her brother for some time she went to the Bank to look for him. The news that Philip had ended up on the gallows sent Sarah out of her mind.

The next morning she was back at the bank, dressed in

mourning clothes and a thick black veil, asking if anyone had seen her brother.

For the next 25 years she would walk up and down Threadneedle Street looking for Philip, stopping passers by and going into the Bank itself.

But her lonely search didn't end with her death. Shortly after she was buried in the graveyard of the City church of St Christopher-le-Stocks, which later became part of the Bank's gardens, the lady in black was seen again.

The Black Nun, as she was then christened became a legend among the clerks who worked in the Bank at the turn of the century. Many of them claimed to have seen the desperate woman as she searched for her long-lost brother. One man claimed to have seen her in the old graveyard gardens, sobbing and pounding a stone slab with her hands.

The Guards and the gold may be long gone from the Bank, but Sarah Whitehead still keeps up her search for the brother who never came home.

What Are They & Why Are They Here?

Do UFOs, aliens and ghosts exist, or are they just a figment of man's imagination? If they are real, where do they come from?

Paranormal encounters have been reported by too many serious, sensible people to be merely dismissed by sceptics as hallucination, mass hysteria, a mystical yearning of the human psyche, or a rebellion against impersonal science. Millions all over the world report having seen them – 15 million in the United States alone in 1973. At Dr J. Alien Hynek's Illinois Center For UFO Studies, more than 50,000 sightings are contained in a computerised data bank, all of them sightings that defy explanation.

Researchers admit that up to 90 per cent of reported paranormal events turn out to be natural phenomena or freak conditions. The planet Venus, advertising planes, military and civilian aircraft, comets, meteors and falling stars, giant balloons, saucer-shaped lenticular cloud formation, ball lightning, even army flares and flights of migrating geese, have all been mistaken for spacecraft. But there always remain 10 to 20 per cent of sightings for which nobody can find a rational cause.

Cover-ups by governments have possibly prompted UFOlogists to make exaggerated claims on occasions. Anxious to prove that UFOs exist, they have often embroidered what was really seen, or ignored evidence conflicting with their version of the facts.

But today more and more governments admit the existence of objects in the sky which come from somewhere beyond human control. Despite America's policy of denying the possibility of UFOs, its armed forces have drawn up procedures to deal with them. In 1957, a CIA source admitted: 'One thing is for sure, we're being observed from outer space.' Russia, Italy, Brazil and Argentina, having issued official reports of sightings, unequivocally accept the existence of UFOs. And in 1974,

French Defence Minister Robert Galley said: 'There is a steady accumulation of sightings of luminous phenomena that are sometimes spherical, sometimes ovoid, and which are characterised by extraordinarily rapid movement.

'Reports from the gendarmerie, from pilots, from people who are heads of air establishments, and a lot of other material, are absolutely impressive ... and disturbing. It is certain that there are things we do not understand and that are at present relatively inexplicable.'

Even Britain does not deny the possibility of manned flights by non humans, though it has always stuck closely to the cynical American public line. An RAF spokesman said: 'The Ministry of Defence does not discount the possibility of intelligent life existing in other parts of the galaxy. However, we have yet to have irrefutable proof that such life exists. So far, no one has provided 100 per cent cast-iron evidence.'

In 1977, when the flood of UFO reports from the Broadhaven Triangle in Wales was at its peak, a Defence Ministry spokesman said: 'We accept that reports are made by sane, rational people, and that a hundred people do not imagine they saw something. But no physical evidence was found that anything had happened.

'We only investigate UFO reports to find whether there is a threat to our defences. If there is no threat, that is the end of the matter. We do not investigate whether UFOs exist, or what causes them.' When asked who decided there was no danger to the nation's defences, the spokesman said: 'We are not prepared to discuss how we investigate.'

Private UFO researchers are more willing to discuss how they operate. Dr Hynek, Stanton Friedman and Raymond E. Fowler in America, Norman Oliver, Jenny Randles and Stewart Campbell in Britain, all take detailed statements from witnesses, and check their background carefully with friends, relatives and employers, to make sure they are reliable people not given to hallucination or hoaxes. They then search painstakingly for possible alternative explanations, and often find them.

. Over the years, patterns have emerged from those sightings that have survived rigorous scrutiny. UFOs are usually saucer-, cigar- or egg-shaped, often with illuminated domes and almost always with navigational or warning lights in patterns different from those used by Earth aircraft.

They seem to arrive over Earth in waves – 1947, 1952, 1954, 1966-7, 1973 and 1975 were peak years for sightings in America; 1962 and 1977-8 in Russia; 1954, 1968, 1973 and 1977-9 in Britain; 1952-4, 1968 and 1973 in Western Europe; 1957, 1962 and 1965 in South America, particularly Brazil; 1959, 1965 and 1978-9 in Australasia and the Far East; and 1946 in Scandinavia, when thousands of mystery rockets were seen over Norway and Sweden.

UFOs seem able to defy all the laws of nature as we understand them – possibly the reason why many scientists deny their existence, preferring a world where all is rational and explicable. They move at a pace that would tear human beings apart, and fly at supersonic speeds without sonic bangs. They change direction and height in ways that make gravity a joke, and produce high-tension electrical charges that not only turn them luminous, but disrupt Earth's power sources.

Many of them are manned by beings who seem to fall into roughly three categories – small creatures less than 4 feet tall with outsize heads and one-piece silver or green uniforms; man-size aliens with wide eyes and thin lips; and giants of about 7 feet. There is also a rare group of fur-covered or hairy beings, approximately 4 feet tall.

Where do they come from? The most popular theory is that they are visitors from another planet. UFOlogists noted that the sighting peak years of 1967 and 1973 coincided with the time when the orbit of Mars brought it closest to Earth; they wondered whether Martians had to wait for suitable conditions to travel, just as Russia and America had to select exactly the right moments to launch their Venus probes.

The Dogons of Mali, incredibly, knew of the star Sirius centuries before human astronomers located it. Other UFO witnesses, too, have spoken of meeting beings who said they came from planets as yet undiscovered – planets from galaxies other than our Milky Way. Earthly science says this is impossible. It knows of nothing that travels faster than the speed of light, and scientists argue that it would take too long for extraterrestrials to reach us, even if they thought the trip worth making.

Yet recent discoveries about telepathy open the door to new thinking about the possibility of teleportation. And it is not beyond human imagination to believe that, if beings of superior intelligence have developed flying saucers far more manoeuverable than anything we possess, they may also have come up with ways of journeying in suspended animation.

Three other schools of thought say UFO aliens come from Earth itself. Einstein first developed the theory that two worlds can coexist in different dimensions, that they intertwine, each invisible to the other for most of the time. Many UFOlogists believe that is what UFOs do, crossing over into our consciousness only when they want to, or when they can.

Others argue that interplanetary travellers settled on Earth long ago, adopting the language and customs of the countries where they landed. Ralph Blum, three times winner of top American Science Foundation awards, says experts have never satisfactorily explained why some people are more intelligent than others, or born leaders, and he believes that superbeings could be conducting experiments with human life. 'Seriously, the person you are married to could be descended from beings from beyond Earth,' he says.

Kenneth Huer, former astronomy lecturer at New York's Hayden Planetarium, says: 'It is possible that aeons ago our ancestors came from outer space as whole beings in spaceships. Or they could he here in great numbers, but we are unconscious of their presence. They may be here in extraordinary unrecognisable forms.

Such a theory would explain the puzzling 'men in black' reported by some UFO witnesses. The woman whose claim of being raped by an alien in Somerset, England, has been reported in this book later told UFO investigator Barry King she received letters and phone calls warning her not to talk about it; and two mysterious men visited her husband and herself several times to stress the desirability of secrecy. The third idea is that UFOs originate from the centre of the Earth. Throughout the ages, some scientists have argued that the Earth is not solid, but hollow. Plato spoke of 'tunnels both broad and narrow into the interior'; and Buddhist doctrine teaches of a subterranean world called Agharta, where millions live in a subtropical paradise ruled over by the King of the World, who relays messages to surface humans via monks who travel in secret passages that possibly emerge in the Himalayas. Other academics have seriously suggested that survivors of Atlantis, even the fairies and goblins of world folklore, live below the Earth, far more advanced technologically than ourselves.

Late in the 19th century, Norwegian sailor Olaf Jansen claimed that he and his father had sailed into this wonderful underworld, and lived with the giants there for two years. He said the inhabitants lived for 500 years, had the power to propel machines by drawing energy from the air, and were well aware of what was happening to humans on the surface of Earth. Jansen's story was so ridiculed that he stopped telling it – but on his death bed, he repeated the details to an American journalist.

In the 20th century, Adolf Hitler launched massive searches for tunnels to an inner Earth. But the belief in a subterranean wonder-world really took off when Rear Admiral Richard Byrd flew 1,700 miles beyond the North Pole in 1947 and 2,300 miles beyond the South Pole in 1956. On both flights, he claimed to have come across iceless lands of mountains, lakes and green vegetation recorded on no maps.

Then, on November 23, 1968, pictures from the American satellite ESSA-7 showed the North Pole without its normal cloud cover – and revealed a perfectly round, dark circle. Advocates of the hollow Earth idea instantly claimed this to be the entry to the underworld. They said the world was not round, but indented top and bottom, so that the true Poles were in mid-air. This, they said, is why compasses go awry 150 miles from both the North and South Poles; and such holes, both in the Arctic and Antarctic, are where UFOs are supposed to emerge.

Wherever they come from, what are UFOs up to? Why are they interested in Earth? Are they, as claimed by the late Douglas Adams, author of the *Hitch-hiker's Guide to the Galaxy*, simply rich young galactic playboys in interstellar sports cars who enjoy tantalising Earth? Or have they a serious purpose?

Many people believe they may be beings from a planet that has become uninhabitable, who are looking for a new place to live. Others argue that they are worried about man destroying himself with toys such as nuclear power, which he can neither understand nor control responsibly.

Others see humanity as being in some sort of zoo, given regular medical checkups to see how it is developing, and even abducted occasionally for interbreeding to ensure that it improves over the ages.

The problem is, we can only judge the behaviour of UFOs by our own standards. We may assume that the UFOnauts want to land, but are waiting until we are ready, or less aggressive towards them; perhaps they are holding back for fear of causing panic here, or triggering a breakdown of society once people realise that Earth has no defence against powerful forces seemingly able to invade its skies at will.

Lord Clancarty, who persuaded the British House of Lords to hold a debate on UFOs, says: 'I believe that with our nuclear and pollution problems, there is concern for us coming from outer space. I think we are on the verge of an official landing on Earth.' Others say the recent UFO sightings are merely reconnaissance flights heralding the Second Coming.

But perhaps the most realistic assessment comes from Dr Stanton Friedman, who was an American government physicist before becoming a full-time UFOlogist. He says: 'They are not interested in settling here, they are just worried about what we will do when we get out there.'

Dr Friedman, who says he has spoken to more than 90 former senior military officers about messages from UFOs picked up by tracking stations, adds: 'They know it is only a matter of time – say about 100 years, nothing in galactic terms – before we send out starships and attempt to become part of the Galactic Federation. Before that happens they want to make sure they know everything about us.

'They see a primitive society which is mostly engaged in tribal warfare, so of course they want to know a lot more about us.'

Jim Lorenzen, director of the Aerial Phenomena Research Organisation, says: 'For them to attempt to land here in any numbers would be just like us intruding on an ancient civilisation deep in the jungle and imposing our civilisation on them. The result would be destructive.

'In the end it is up to us. We now hold the key to a universe that we never thought existed. It is up to us whether we use that key – or perish in the insanity of war.'

561